Investigations Into

PHYSICAL GEOLOGY

A LABORATORY MANUAL

Investigations Into

PHYSICAL
GEOLOGY

A LABORATORY MANUAL

Jim Mazzullo

Department of Geology and Geophysics
Texas A&M University

SAUNDERS COLLEGE PUBLISHING
Harcourt Brace College Publishers

Fort Worth Philadelphia San Diego New York Orlando Austin
San Antonio Toronto Montreal London Sydney Tokyo

Text Typeface: Goudy
Composition: Progressive Information Technologies
Publisher: John Vondeling
Developmental Editors: Christine Rickoff/Anne Gibby
Managing Editor: Carol Field
Senior Project Editor: Anne Gibby
Copy Editor: Pat Daly
Manager of Art and Design: Carol Bleistine
Art Director: Caroline McGowan
Art and Design Coordinator: Sue Kinney
Text Designers: Robin Milicevic/Caroline McGowan
Cover Designer: Rebecca Lemna
Text Artwork: George V. Kelvin/Science Graphics
             Rolin Graphics
Director of EDP: Tim Frelick
Production Manager: Charlene Squibb
Marketing Managers: Angus McDonald/Sue Westmoreland

Cover Credit: © Tom Van Sant, Inc./The Geosphere Project

Photo credits: page 1, sandstone slot canyon near Lake Powell (Westlight © Charles Campbell); page 75, Grand Canyon, Arizona, view of Mount Hayden (© Tony Stone Images/Gary Yeowell); page 233, aerial view of the Green and Colorado rivers confluence, Utah (Tony Stone Worldwide/David Heiser).

Printed in the United States of America

Investigations Into Physical Geology: A Laboratory Manual

ISBN: 0-03-020294-9

Library of Congress Catalog Card Number: 96-70423

8901234   032   10987654

PREFACE

This is the first edition of *Investigations Into Physical Geology: A Laboratory Manual.* It is intended to be used in the laboratory classes of college level introductory courses on physical geology and to complement any of the currently available textbooks on this subject. The central focus of this laboratory manual is its many diverse exercises, which are designed to develop the observational and analytical skills of students as they study the structure, composition, and origin of the Earth's crust and the tectonic and surficial processes that shape the continents and the sea floor. These exercises are accompanied by reviews of the basic principles of physical geology, as well as by a rich collection of maps, photographs, block diagrams, tables, and other resources and references.

GENERAL DESIGN

This laboratory manual has been designed to provide maximum **choice** and **flexibility** to instructors, so that they can choose materials which are most relevant to their students, most appropriate for their academic level, and most suited to the specific objectives of their courses. This design goal is achieved in three general ways.

First, this laboratory manual is written to be comprehensible to students who are not science majors or who have a limited scientific background and vocabulary. Each chapter is divided into two or more sections, each of which consists of a block of text followed by exercises. The text is intended to supplement students' lecture notes and textbooks with brief reviews of basic concepts and terminology that they will encounter in the exercises. The exercises require students to apply these concepts and terms to their descriptions and interpretations of rock and mineral samples; geologic, topographic, and bathymetric maps; seismograms and seismic profiles; satellite and aerial images; hydrologic data; and other types of geologic data. This laboratory manual does not shy away from the use of simple algebraic equations and graphs to represent and analyze these data, but such methods are always carefully described and often demonstrated for the students. In addition, this laboratory manual contains ample materials to interest and challenge science majors, particularly geology and geophysics students.

Second, this laboratory manual consists of fourteen large chapters, with more material than can be covered in a typical semester. The Table of Contents provides detailed outlines of the contents of each chapter, so that instructors can review and select specific materials to present to their classes. Extra materials can be used by students to reinforce the lessons of the classroom and by the instructors for homework, extra credit assignments, and laboratory examinations. All the exercise pages have spaces for students' names and class sections so that exercises can be handed in for grading. They are also positioned in the laboratory manual so that their removal will not destroy its integrity and subsequent use to students for study.

Third, the exercises in this laboratory manual are varied in their intent and graduated in their difficulty. Some are fairly straightforward queries about specimens, maps, and aerial photographs; some require mathematical and graphical manipulation of geologic data from the literature; and some require careful thought and integration of facts and observations. In addition, there are some questions in these exercises that require students to conduct experiments and engage in classroom discussions and group activities. The exercises also present topographic and geologic maps, aerial photographs, and other geologic data from every corner of the United States, so that students will encounter material and geologic problems that are familiar to them.

Format

This laboratory manual contains fourteen chapters organized into three parts:

- *Part I: Minerals and Rocks* examines the origins, properties, and classifications of minerals and rocks. Its four chapters teach students to describe and identify minerals and rocks, and to understand the origins and geologic significance of their physical properties. The chapters contain reference tables that summarize the physical and diagnostic properties of the common and important minerals and rocks, and color photographs that illustrate their physical properties and appearances. These photographs are intended to supplement (not replace) hand specimens that the instructor will provide for the exercises in this section.

- *Part II: Structure and Tectonics* examines the structure and tectonic origins of the Earth's crust. Chapter 5 instructs students how to read and analyze the topographic maps and satellite and aerial images that are used extensively throughout Parts II and III of this manual. Chapters 6, 7, and 8 teach students how to interpret the structure, composition, relative ages, and geologic evolution of the rocks of the continents and ocean floor from geologic and topographic maps, satellite and aerial images, and seismic profiles. Chapter 9 recounts the development of the concepts of continental drift and sea floor spreading and an examination of the evidence for the theory of plate tectonics. The exercises in these chapters make extensive use of maps and photographs from major geologic provinces and several national parks in the United States, and draw on the most current bathymetric maps and seismic data from the ocean floors.

- *Part III: Surficial Processes and Landforms* examines the interaction between water, wind, and ice and the rocks of the crust. Its chapters use topographic maps and satellite and aerial images of the United States to teach students to recognize, describe, and interpret the origins of fluvial, eolian, glacial, and coastal landforms. They also examine the occurrence and movement of ground water and the effects of mankind's use and abuse of this valuable resource.

Required Resources

The beginning of each chapter contains a short list of supplies that the **students** are asked to bring to their laboratory in order to complete the exercises. This list includes colored pencils or marker pens (at least 4 different colors); lead pencils (with erasers); a hand lens or magnifying glass ($10\times$ or $25\times$ power); a ruler with English and metric units; a protractor/contact goniometer; a calculator; and a drafting compass (for drawing circles and arcs). You may also want your students to bring extra graph paper to the laboratory for the topographic profiling exercises in Parts II and III.

The following sections describe the supplies and equipment that the **instructor** must provide for these exercises.

Part I: Minerals and Rocks

Several exercises in Part I ask the instructor to provide representative samples of the common and important rocks and rock-forming minerals, as well as rock and mineral samples that offer good examples of specific properties such as crystal habit, grain size, and foliation. Students are asked to bring a hand lens or a magnifying glass to the laboratory to examine the samples, but the instructor may also want to provide binocular microscopes for this purpose.

The following exercises in Part I also ask the instructor to supply these additional resources:

Exercise 1–1 requires a saline solution and other supplies for an experiment on crystallization. The solution yields crystals visible with a hand lens in 10 to 15 minutes, and it is a safe, satisfactory alternative to thymol. For coarse crystals, dry the solution at the *lowest* possible temperature setting on the hot plate, and replenish the solution 3 to 5 times until a thick crust develops. You can also repeat this experiment under slightly higher temperatures to produce finer crystals.

Exercise 1–2 requires several samples each of crystalline quartz and fluorite (or other minerals with good habits). The crystals should be varied in size to test for the constancy of interfacial angles. Wooden or ball-and-stick models of crystal structures and habits can also be used in the optional part, which is recommended for advanced students.

Exercise 1–3 is a group participation exercise in which students construct models of silica tetrahedra and silicate structures. This requires styrofoam balls to represent the Si and O atoms (preferably two sizes, with a smaller one for Si and a larger one for O) and sticks to represent the chemical bonds between them. The author uses shish-kabob sticks for the latter purpose, for they are cheap and readily available.

Exercise 1–5 requires glass, copper pennies, and steel nails for the hardness tests, as well as streak plates, dilute hydrochloric acid, and magnets. It is also recommended that the instructor issue a warning about the use of HCl to the students.

Exercise 3–3 requires several sheets of sturdy white cardboard or posterboard and a razor blade to construct the bases of the grain-size comparators. (Precut pieces of plexiglass can also be used, if available.) It also requires pre-sieved sediments to mount on the comparators. The author recommends using sediments in the following textural classes: granules, very coarse sand, coarse sand, medium sand, fine sand, very fine sand, silt, and clay.

Exercise 3–5 requires the same supplies and techniques as Exercise 1–1.

Exercise 3–6 requires a sample of poorly sorted sediment and a tall water-filled Mason jar (or other sealable glass jar) in order to conduct an experiment on the formation of grading. The sediment should contain subequal amounts of sand, silt, and clay. Generally, it requires a minimum of 30 minutes for some of the finer sediments to settle and the grading to become apparent, so it is suggested that this experiment be run early during the class and set aside for later study.

Part II: Structure and Tectonics

Several exercises in Part II ask the instructor to provide a Heezen-Tharp Map of the World Ocean Floor, a world atlas, and stereoscopes.

The 1994 wall-sized (1:23,230,300 scale) edition of the Heezen-Tharp map, which shows the structure of the sea floor and continents in great detail, is highly recommended, but the smaller (1:46,460,600 scale) version can also be used. Maps can be ordered directly from Marie Tharp Oceanographic Cartography, 1 Washington Avenue, South Nyack, New York 10960 U.S.A.

The world atlas should contain good physiographic maps of the continents. *Goodes' World Atlas* and the *Rand-McNally Contemporary World Atlas* are two inexpensive atlases that were used during the writing of this laboratory manual and hence would be suitable for this purpose, as would be the more costly atlases published by the National Geography Society or the New York Times Company.

The following exercises in Part II also ask the instructor to supply these additional resources:

Exercise 5–1 requires a topographic map of any part of the United States. Generally, the topographic maps in this laboratory manual are severely cropped and edited in order to fit on the pages, so this map will be the only unedited one that the student shall see in the laboratory. The Bright Angel quadrangle is recommended because its topography is familiar to most students.

Exercise 8–3 can be supplemented with a geologic map of the San Francisco Bay area, so that the students can also gauge the effects of distance on earthquake intensity.

Please note also that Map 6–1 Stable Interior of the United States, and Map 7–4 Grand Canyon, are extra-large foldouts that are enclosed within the back cover of the laboratory manual.

Part III: Surficial Processes and Landforms

Once again, several exercises in Part III ask the instructor to provide a Heezen-Tharp Map of the World Ocean Floor, a World Atlas, and pocket stereoscopes. See the notes on Part II for more information about these resources.

In addition, several exercises refer students to geologic and geomorphic maps of the United States. The wall-sized map *Fold and Thrust Belts of the United States* (#GB–2345) and the desk-sized map *Relief of the United States* (National Atlas of the United States, sheet 56) are both recommended. They can be purchased from any USGS Earth Science Information Center (1-800-USA MAPS).

Exercise 11–4 asks the instructor to provide samples of unconsolidated sediments, coffee filters, funnels and graduated cylinders for a classroom experiment on permeability. The samples should vary markedly in grain size and sorting. For example, you can run the experiment with a variety of well-sorted sediments (such as granules, coarse sand, fine sand, silt, and clay) and poorly-sorted sediments (such as mixtures of granules and coarse sand; granules and fine sand; coarse and fine sand; sand and silt; and sand, silt, and clay). The graduated cylinders should be small (100 ml) and the funnels appropriately sized to fit them. The coffee filter paper can be cut to fit the funnels. You can either prepare one experimental set-up and repeat the experiment with different sediment samples, or construct multiple set-ups and run several samples simultaneously.

Acknowledgments

The author expresses his gratitude first and foremost to his friend and former colleague, Dr. Vicki Harder, for her help in launching and completing this laboratory manual. For several years, Vicki taught physical geology and coordinated the geology laboratory classes in the Department of Geology at Texas A&M University. She generously shared these teaching experiences with me at the beginning of this laboratory manual project, reviewed its contents from cover to cover, and provided sound advice on the academic level of its contents and the practicality of its exercises. Whatever success this laboratory manual achieves will be due in no small part to her.

I also wish to express thanks to my publisher, John Vondeling, who had sufficient faith to publish this laboratory manual, and his assistant, Margaret Crocker, for her help in acquiring maps and other graphics; to the developmental editor, Christine Rickoff; and most of all, to the senior project editor, Anne Gibby. Despite the complicated nature of this book and the many stumbles and errors of a first-time author, Anne managed to produce this finished book with the grace, efficiency, and smoothness that mark a true professional. I thank her for all the aggravation she may have suffered in my hands. Thanks are also extended to the graphic artist, Mr. George Kelvin, for the beautiful works of art that grace these pages, and to the art director, Caroline McGowan, for her excellent managing of the art program.

There are several people at Texas A&M University to whom I owe thanks for their contributions to this laboratory manual. They include Drs. Robert Berg, Tom Hilde, and Philip Rabinowitz, Department of Geology and Geophysics, and Ms. Shawn Webb, Ocean Drilling Program, for the seismic profiles and sea floor maps in the second part of the book; Dr. Patrick Domenico, Department of Geology and Geophysics, for his assistance with the chapter on Ground Water and Karst; Dr. Steve Harder, Department of Geology and Geophysics, for his assistance with the chapter on

Seismology; Drs. Andy Hajash, Ray Guillemette, Will Lamb, Mel Friedman, Karl Koenig, and Rick Carlson, Department of Geology and Geophysics, for their critique of the first edition; and Mr. Mark Thomas and the staff in the Sterling Evans Library for their assistance with locating topographic maps. I would also like to thank Dr. Jim Coleman of the Department of Geology at Louisiana State University for the false-color image of the Belize delta.

In addition, I would like to express my appreciation to the following people for their reviews of the contents of this laboratory manual.

John Ciciarelli, *Pennsylvania State University*
W. B. Clapham, *Cleveland State University*
Thomas Donnelley, *State University of New York at Binghamton*
Cydney Faul-Hauser, *Wilkes University*
Terry Engelder, *Pennsylvania State University*
Pamela Gore, *DeKalb College*
Sidney Halsor, *Wilkes University*
Lance Kearns, *James Madison University*
Peter L. Kresan, *University of Arizona*
Albert M. Kudo, *University of New Mexico*
Michael Lyle, *Tidewater Community College*

Glenn M. Mason, *Indiana University, Southeast*
Eric Mysona, *Wright State University*
Philip Reeder, *Valdosta State University*
Richard Stenstrom, *Beloit College*
J. Robert Thompson, *Glendale Community College*
Charles Waag, *Boise State University*
Lorraine W. Wolf, *Auburn University*

Finally, I must acknowledge my spouse, D. S. Linthicum, who provided me with support and space, both physical and mental, during the often grueling writing of this laboratory manual.

Despite all this help, however, I must accept full responsibility for the contents of this laboratory manual and for any errors that may have worked their way into its pages. I would appreciate feedback from users of this laboratory manual on its contents and accuracy, as well as their evaluations of the usefulness of the exercises and recommendations for future revisions of the text and especially the exercises.

Jim Mazzullo
College Station, Texas
July 1996

CONTENTS

PART II STRUCTURE AND TECTONICS

PART III   SURFICIAL PROCESSES AND LANDFORMS

MAPS

Investigations Into

PHYSICAL
GEOLOGY

A LABORATORY MANUAL

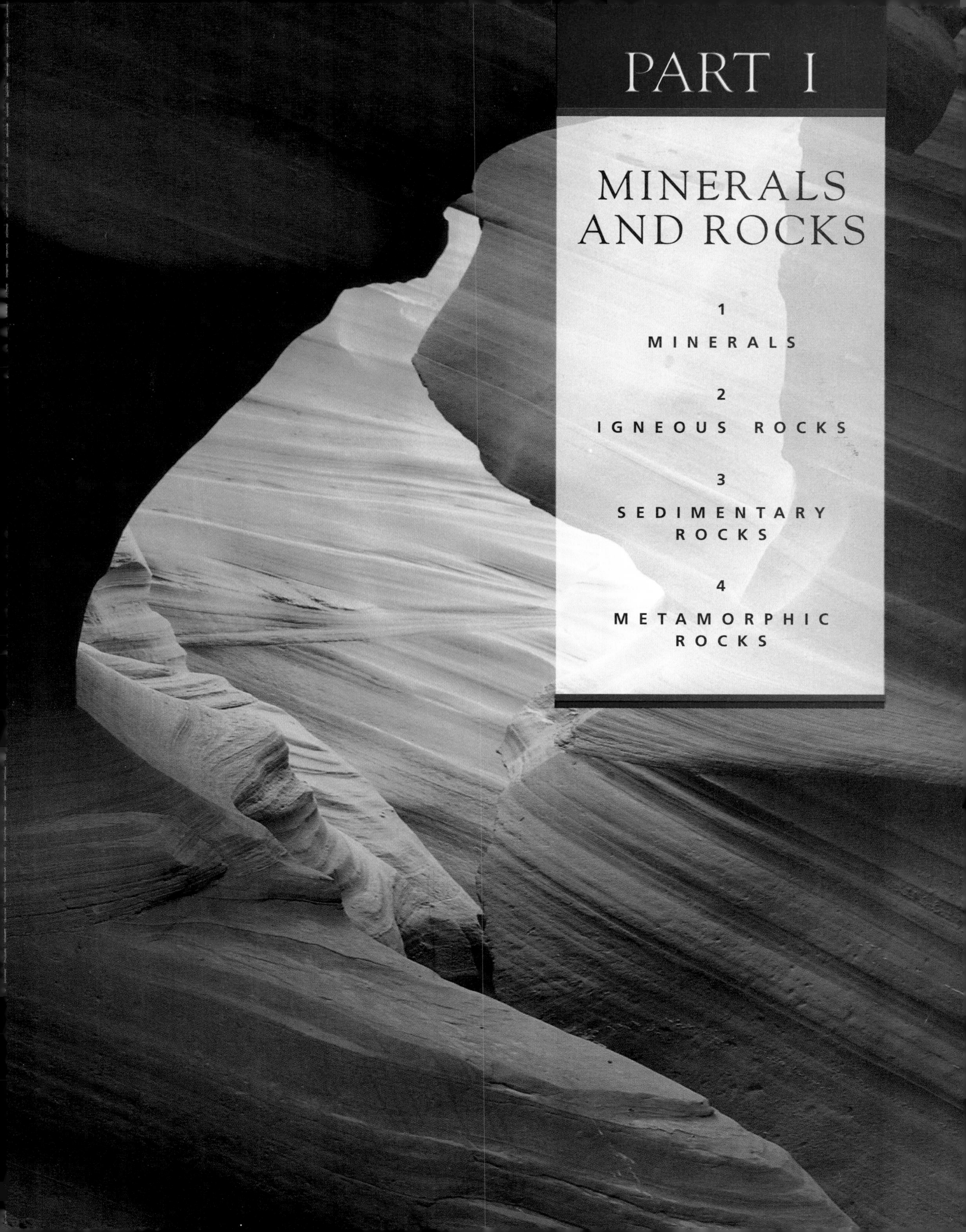

PART I

MINERALS AND ROCKS

1

MINERALS

KEY WORDS

Minerals . . . crystals, crystal structures, and unit cells . . . silicate structures and silica tetrahedra . . . crystal habit, twinning, luster, hardness, cleavage, fracture, color, streak, specific gravity, effervescence, fluorescence, and magnetism

MICA

Materials Needed *Colored pens or pencils, lead pencils with erasers, hand lens or magnifying glass, ruler, protractor*

The crust of the Earth is composed of solid rock. If you examine these rocks closely, you will find that they are composed of discrete grains of different sizes, shapes, and colors. These grains are **minerals**, which are the building blocks of all rocks. Minerals are naturally occurring inorganic solids each of which possesses a specific chemical composition and a precise internal crystalline structure. They can be formed by many different types of processes and in many different types of environments: they can crystallize from molten magma deep within the crust, from lava which spews from

volcanoes, and from salty lake waters as they evaporate under a hot desert sun.

Geologists study the mineral compositions of rocks for important information about their origins and the conditions of the Earth's crust and surface during their formation. For example, we study the mineralogy of igneous and metamorphic rocks for evidence about the compositions of their parent magmas and rocks (respectively), the temperatures and pressures within the Earth's crust during their formation, and the tectonic processes which led to their formation. Similarly, we study the mineral compositions of sedimentary rocks for evidence of the tectonic settings and climatic conditions in which their parent rocks weathered and the compositions of the sea waters in which they crystallized. Minerals are also important for their economic value, because they are a major source of metals, chemicals, and other materials which are necessary for industry and manufacturing. Lastly, minerals are an essential part of our culture: We value them for their beauty; we use the rarest minerals for currency; and we decorate our bodies with them to indicate our material wealth and marital status.

This chapter describes the fundamental properties of minerals and the common types of minerals which are found in the Earth's crust. It begins with an examination of the two most fundamental properties of minerals: their crystallinity and compositions. It then concludes with a discussion of the diagnostic properties which geologists use to identify minerals.

MINERAL COMPOSITION

Every mineral has its own specific chemical composition. This composition may be fixed and invariant. For example, the mineral **quartz** is composed of one silicon (Si) atom for every two oxygen (O) atoms, and rarely do other cations substitute for the silicon atom in its crystal structure. However, there are also many minerals whose compositions vary within a well-defined range, due to the ability of cations of the same size and electric properties to substitute for one another in their crystal structures (a phenomenon known as **ionic substitution**). For example, there is a group of minerals called the **olivines,** which share the same general formula of $(Mg,Fe)_2SiO_4$—that is, they contain two similar-sized and equally charged metal cations (Mg^{2+} and/or Fe^{2+}) for every one Si atom. However, there are some olivines which contain two Mg atoms (but no Fe) for every Si atom (the mineral **forsterite**) and some which contain two Fe atoms (but no Mg) for every Si atom (**fayalite**). Other common mineral groups with variable compositions of this sort include the **pyroxenes, amphiboles, plagioclase feldspars,** and **micas**.

Mineral Classes

There are 88 naturally occurring elements in the universe, but only eight of them—oxygen (O), silicon (Si), aluminum (Al), calcium (Ca), magnesium (Mg), potassium (K), sodium (Na), and iron (Fe)—comprise the bulk (greater than 96 percent) of the Earth's crust. The most abundant minerals in the crust are compounds of these elements. Nine common **mineral classes** can be defined on the basis of chemical composition:

- The **silicates**, which are compounds of silicate (SiO_4^{4-}) and other elements and the most common minerals in the crust;
- The **oxides**, which are compounds of oxygen (O^{2-}) and other elements;
- The **sulfides**, which are compounds of sulfur (S^{2-}) and other elements and the most common economic ore minerals;
- The **carbonates**, which are compounds of carbonate (CO_3^{2-}) and other elements;
- The **hydroxides**, which are compounds of hydroxide (OH^-) and other elements;
- The **sulfates**, which are compounds of sulfate (SO_4^{2-}) and other elements;
- The **phosphates**, which are compounds of phosphate (PO_4^{3-}) and other elements;
- The **halides**, which are compounds of oppositely charged ions such as Na^+ and Cl^-; and
- The **native elements**, each composed of only one element, such as gold (Au) and silver (Ag).

THE INTERNAL STRUCTURES OF MINERALS

The basic differences between gases, liquids, and solids are the arrangement and motion of their constituent atoms. In gases, the atoms are widely separated and move independently from one another; in liquids, they are more closely packed together and slower moving; and in solids, they are essentially motionless and arranged in a definite order. This property of solids is called **crystallinity**, and it is the second of the fundamental properties of all minerals.

The atoms within a mineral assume a specific geometric arrangement which is called a **unit cell** (Fig. 1–1). These unit cells are stacked in orderly and repetitive three-dimensional patterns to form **crystals**. This stacking can occur in many different patterns to form a variety of different **crystal structures** or **crystallographic systems** (Fig. 1–2).

Silicate Structures

The silicate minerals provide a good illustration of the complexity and diversity of crystal structures. All of the silicate minerals contain the same fundamental building block, the **silica tetrahedron**, which consists of a single silicon atom surrounded by four oxygen atoms (Figs. 1–3A and 1–3B). However, silica tetrahedra can be linked in different ways to

Diamond is the hardest mineral. (Courtesy of Smithsonian Institution)

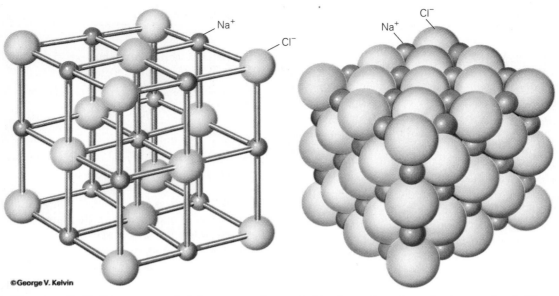

Figure 1–1 The unit cell of halite is composed of alternating sodium and chloride ions bound together by ionic bonds. The spatial arrangement of the Na and Cl ions forms a cubic structure with six faces which intersect one another at right angles. This structure is modeled in (A) and shown in a more realistic form in (B). The repetitious stacking of such cubic unit cells produces the distinctive cubic shapes of large halite crystals.

form several types of **silicate structures**. They can be linked by simple ionic bonds to form **isolated** structures; they can share two oxygen atoms with adjacent tetrahedra to form **single-chain** structures and be linked with other single chains to form **double-chain** structures; they can share three oxygen atoms with adjacent tetrahedra to form **sheet** structures; and they can share four oxygens with surrounding tetrahedra to form **framework** structures.

Silica tetrahedra have negative charges (except in framework silicates) which must be balanced by bonding with positively charged cations. However, this charge varies with the silicate structure: it is highest in isolated structures and decreases with an increase in the number of shared oxygen atoms (Fig. 1–3B). This variation in charge has a great effect on the compositions of silicate minerals, because it controls the types and numbers of cations that they can incorporate into their structures.

For example, a silica tetrahedron in an isolated structure contains 4 oxygens (each with a charge of -2) for each silicon (with a charge of $+4$), which results in a net charge of -4. (Calculate this value for yourself.) Thus, it can potentially bond with either (a) four singly charged cations such as K^{2+} an Na^+), (b) two doubly charged cations such as Ca^{2+} and Mg^{2+}, or (c) two singly charged and one doubly charged cations. On the other hand, a tetrahedron in a ring structure contains one silicon ($+4$), two nonshared oxygens (with a total charge of -4), and two shared oxygens (with a total charge of -2), which results in a net charge of -2. Thus, it can potentially bond only with two singly charged cations or one doubly charged one.

Figure 1–2 There are six major classes of crystal structures, or crystallographic systems. Each crystallographic system can be defined on the basis of the shapes of the crystals and specifically the relative lengths and angles between crystal axes. ▶

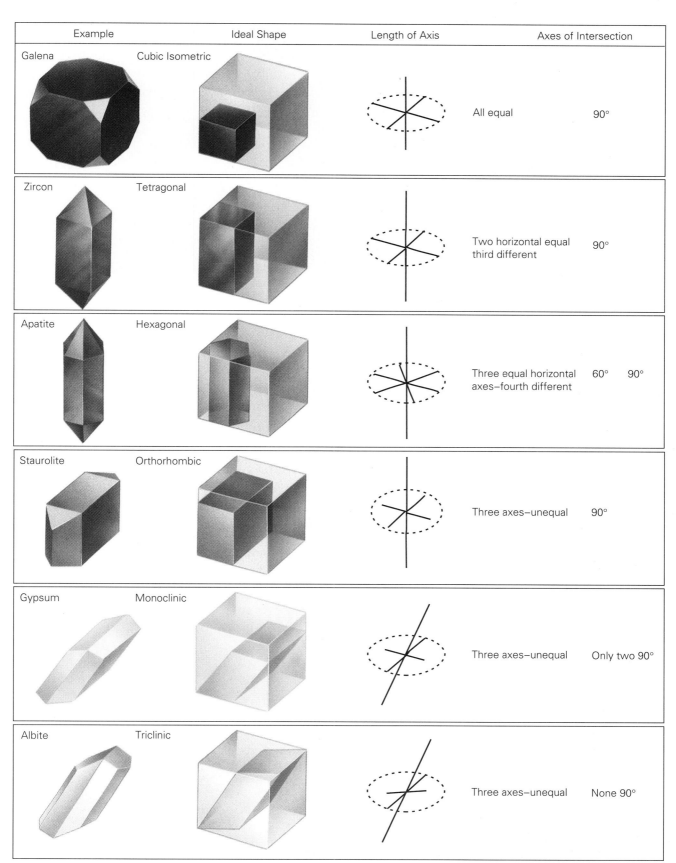

	Example	Ideal Shape	Length of Axis	Axes of Intersection
Galena	Cubic Isometric		All equal	90°
Zircon	Tetragonal		Two horizontal equal third different	90°
Apatite	Hexagonal		Three equal horizontal axes–fourth different	60° 90°
Staurolite	Orthorhombic		Three axes–unequal	90°
Gypsum	Monoclinic		Three axes–unequal	Only two 90°
Albite	Triclinic		Three axes–unequal	None 90°

© George V. Kelvin

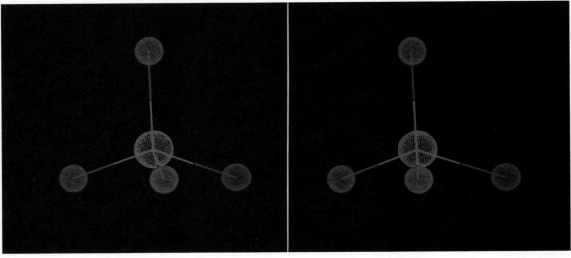

A

Class	Arrangement of SiO_4 tetrahedra	Unit composition–Mineral examples
B Single chains		$(SiO_3)^{2-}$ Pyroxene: The most common pyroxene is augite, $(Ca,Na)(Mg,Fe,Al)(Si,Al)_2O_6$
C Double chains		$(Si_4O_{11})^{6-}$ Amphibole: The most common amphibole is hornblende, $Ca_2(Mg,Fe)_4Al(Si_7Al)O_{22}(OH)_2$
D Sheet silicates		$(Si_2O_5)^{2-}$ Mica, clay minerals E.g., Muscovite: $KAl_2(Si_3Al)O_{10}(OH)_2$
E Framework silicates		SiO_2 Quartz: SiO_2 Potassium feldspar: $KAlSi_3O_8$

© George V. Kelvin

Figure 1–3 (A) Stereofigures of an isolated silica tetrahedron, the fundamental building block of the silicate minerals. (B) Single-chain structure. (C) Double-chain structure. (D) Sheet structure. (E) Framework structure. The structures are distinguished by the spatial arrangements of silica tetrahedra and the number of oxygen atoms which they share.

EXERCISE 1–1 Crystallization of Minerals from a Solution

This exercise will demonstrate the growth of mineral crystals from a solution. It requires a hot plate, a microscope glass slide, and an eyedropper, as well as a solution of saline water (prepared by dissolving 1 to 2 tsp of table salt in 50 to 100 ml of warm water). Turn the hot plate to its lowest setting and put the glass slide on it. Place a few drops of the solution on the slide. Allow some (but not *all*) of it to dry, and then gently add a few more drops. Repeat until there is a buildup of salt on the slide, and then let the solution dry completely. Remove the slide from the hot plate, let it cool, and examine it with your hand lens.

1. Sketch the shapes of the crystals, including their internal structure. Notice that the crystals begin to crowd one another as they grow. What effect does this crowding have on their shapes? Illustrate this.

2. When allowed to grow without interference by other minerals, a mineral such as halite will assume an external shape or **crystal habit** which reflects its internal crystal structure. The number of **crystal faces** in the crystal habit and the angles between them are fixed properties of minerals (Fig. 1–2), and they can be used to distinguish one mineral from another. Examine halite crystals of different sizes on the glass slide. Estimate the angle between the adjacent faces of the crystals (where there is no crowding). _____ Does this angle vary with the size of the crystal? _____

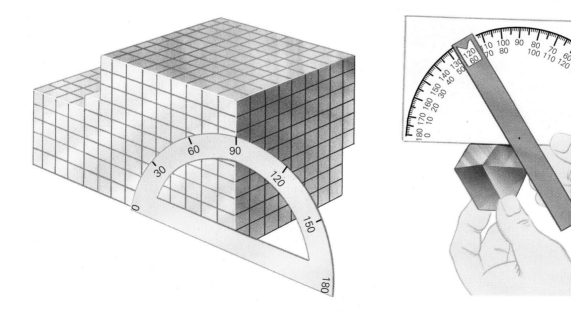

Figure 1–4 Measuring the angles between the crystal faces or cleavage planes of a mineral.

Your instructor will distribute several samples of crystalline fluorite and quartz (or other minerals with good crystal habits) to your class. Fluorite crystals have six crystal faces arrayed in a cubic habit; quartz crystals consist of six-faced prisms capped by multifaceted pyramids.

    1. Measure the angles between the faces of the fluorite and the prismatic faces of the quartz. Illustrate the samples below, and record their interfacial angles.

    2. Compare your data on the minerals to comparable data collected by your classmates. What consistencies (or inconsistencies) do you see in your respective data sets? Your observations provide a demonstration of the law of the constancy of interfacial angles. Explain this law.

Optional Approach to This Exercise: Your instructor will provide you with samples of quartz, calcite, feldspar, and other minerals which display excellent crystal habits, or with crystal models and crystal blocks that represent different structures.

Use the following flow sequence to determine their crystal structures yourself:

A. Using Figure 1–2, identify the principal axes of the crystal. If there is no obvious long axis, the crystal is *probably* **cubic.**

B. If the crystal is not cubic, orient the crystal so that the shorter axes are horizontal and the long axis is vertical or subvertical (see the examples in Figure 1–2). If the angles between the long axis and the two horizontal axes are all equal to 90°, the crystal is either **tetragonal, hexagonal,** or **orthorhombic.** If not, the crystal is either **monoclinic** or **triclinic.**

C. To distinguish between tetragonal, hexagonal, and orthorhombic crystals:
    a. Determine the number of horizontal axes in the crystal; hexagonal crystals have three.
    b. Measure the lengths of the horizontal axes (that is, the widths of the vertical faces); they are equal in tetragonal crystals and unequal in orthorhombic ones.

D. To distinguish between monoclinic and triclinic crystals, examine the angles between the horizontal axes; they are equal (90°) in monoclinic crystals and not equal in triclinic.

EXERCISE 1–3   Silicate Structures

1. Construct a silica tetrahedron with the materials provided by your instructor. With the other members of your laboratory, construct single-chain or double-chain structures with these tetrahedra. You will need to remove some oxygen atoms from the tetrahedra to bond them.

2. Answer the following questions about your silicate chain structure:

a. What is the number of Si atoms in the structure? _____

b. What is the number of O atoms in the structure? _____

c. What is the charge on a Si atom? _____ the cumulative charge of all the Si atoms in the structure?

d. What is the charge on an O atom? _____ the cumulative charge of all the O atoms in the structure?

e. What is the Si-to-O ratio in the structure? _____

f. What is the net charge of the structure? _____

g. Discuss with your classmates the ways in which this charged structure could be electrically neutralized.

EXERCISE 1–4   Structure and Composition of Silicate Minerals

1. The accompanying table lists the chemical formulas of several framework and single and double chain (inosilicate) minerals. The atomic weights, ionic radii, and charges of their constituent elements are also given. Calculate the atomic weight percent of Si in each of the minerals. The calculations for K feldspar ($KAlSi_3O_8$) are shown here as an example:

ELEMENT	ATOMIC WEIGHT	×	NUMBER OF ATOMS		
K	39	×	1	=	39
Al	27	×	1	=	27
Si	28	×	3	=	84 →
O	16	×	8	=	128
					278 ↗

$$\frac{84}{278} \times 100\% = 30\% \text{ Si}$$

2. Examine your data, and answer these questions:

a. Which structures are *generally* characterized by higher Si content: framework or chain? _____

b. What mineral with this same structure is an exception to this generalization? _____

c. This exception is due to the fact that another element (call it X) has substituted for Si in the crystal structure of this mineral. This occurs because X and Si ions have similar ionic radii. Examine the composition of the mineral which you identified in 2b and the ionic radii data in the table. What is Element X? _____ Calculate the atomic weight percent of Si + X, and record it on the table.

d. What is the charge on a Si cation? _____ of an ion of Element X? _____

e. What is the charge on a molecule with the formula Si_4O_8? _____

f. What is its charge after two atoms of Element X are substituted for two Si atoms? _____

g. Consider the charges of other elements in the table: Which elements (or element) could be added to the structure of the altered molecule to neutralize its charge? _____

3. You have just deduced a phenomenon called **coupled substitution.** Describe it.

FRAMEWORK SILICATES			
K feldspar	$KAlSi_3O_8$	___30% Si___	___40% Si + X___
Quartz	SiO_2	_____	_____
Sodic Plagioclase	$NaAlSi_3O_8$	_____	_____
Na-Ca Plagioclase	$Na_{0.5}Ca_{0.5}(Si_{2.5}Al_{0.5})O_8$	_____	_____
Calcic Plagioclase	$CaAl_2Si_2O_8$	_____	_____

CHAIN (INO-) SILICATES			
Hornblende	$NaCa_2Mg_4Al_3Si_6O_{22}(OH)_2$	_____	_____
Tremolite	$Ca_2Mg_5Si_8O_{22}(OH)_2$	_____	_____
Diopside	$CaMgSi_2O_6$	_____	_____
Hedenbergite	$CaFeSi_2O_6$	_____	_____

Element	Atomic wt	Ionic radius (Å)	Ionic charge	Element	Atomic wt	Ionic radius (Å)	Ionic charge
H	1	0.46	+1	Al	27	0.51	+3
O	16	1.40	−2	Si	28	0.39	+4
Na	23	0.97	+1	K	39	1.33	+1
Mg	24	0.66	+2	Ca	40	0.99	+2
				Fe	56	0.64–0.74	+2, +3

ineralogists employ a variety of sophisticated analytical instruments to identify minerals, including the neutron activator (which determines chemical composition) and the X-ray diffractometer (which determines crystal structure). However, it is also possible to identify many minerals on the basis of certain observable or testable **diagnostic properties** which reflect their internal compositions and structures. The diagnostic properties of the most common and economically important minerals are summarized in Table 1–1 and illustrated in Figure 1–5.

Figure 1–5 Some of the diagnostic properties of minerals. (A) Cubic crystal habit and pearly luster of the mineral halite. (B) Cleavage and twinning on the mineral plagioclase feldspar. The twinning striations are seen on the front cleavage plane. The sample rests on its second cleavage plane, which intersects the front cleavage plane at an angle of approximately 90°. (C) Metallic gray luster and cubic cleavage habit of the mineral galena, and the pale golden yellow color of the mineral marcasite (which lies between the galena cubes). (D) Rhombohedral cleavage habit of the mineral calcite, with cleavage in three directions. The sample also exhibits the transparency which is common to calcite. (E) Tabular cleavage habit of the mineral muscovite. Micas tend to cleave into thin transparent sheets, one of which is peeled away by the knife. (F) Conchoidal fracture on the upper right corner of the mineral quartz. The sample also exhibits the translucency and glassy luster which are common in quartz. (G) Purple color and six-sided pyramidal crystal habit of amethyst, a variety of the mineral quartz. The color is due to trace amounts of iron or manganese. (H) Reddish-brown streak and color, dull earthy luster, and uneven fracture habit of the mineral hematite (Jeffrey Scovil).

TABLE 1–1   **Common Rock-Forming Minerals and Their Properties**

MINERAL GROUP OR MINERAL	CHEMICAL COMPOSITION	HABIT, CLEAVAGE, FRACTURE	USUAL COLOR
Amphibole (var. Actinolite)	$Ca_2(MgFe)_5Si_8O_{22}(OH)_2$	Slender crystals, radiating, fibrous	Blackish-green to black, dark green
Amphibole (var. Hornblende)	$Ca_2(Mg,Fe)_4Al(Si_7,Al)$ $O_{22}(OH)_2$	Elongate crystals	Blackish-green to black, dark green
Calcite	$CaCO_3$	Perfect cleavage into rhombs	Usually white, but may be variously tinted
Dolomite	$CaMg(CO_3)_2$	Cleaves into rhombs; granular masses	White, pink, gray, brown
Illite	$K_{0.8}Al_2(Si_{3.2}Al_{0.8})O_{10}(OH)_2$		White
Kaolinite	$Al_2Si_2O_5(OH)_4$		White
Smectite	$(Na,Ca)_{0.3}(Al,Mg)_2Si_4O_{10}$ $(OH)_2 \cdot nH_2O$		White, buff
Albite (sodium feldspar)	$NaAlSi_3O_8$ (sodic plagioclase)	Good cleavage in two directions, nearly 90°	White, gray
Orthoclase (potassium feldspar)	$KAlSi_3O_8$	Good cleavage in two directions at 90°	White, pink, red, yellow-green, gray
Plagioclase (feldspar containing both sodium and calcium)	$(Na,Ca)(Al,Si)_4O_8$	Good cleavage in two directions at 90°	White, gray
Biotite	$K(Mg,Fe)_3AlSi_3O_{10}(OH)_2$	Perfect cleavage into thin sheets	Black, brown, green
Muscovite	$KAl_2(Si_3Al)O_{10}(OH)_2$	Perfect cleavage into thin sheets	Colorless if thin
Olivine	$(Mg,Fe)_2SiO_4$	Uneven fracture, often in granular masses	Various shades of green
Pyroxene (var. Augite)	$(Ca,Na)(Mg,Fe,Al)(Si,Al)_2O_6$	Short, stubby crystals have 4 or 8 sides; prismatic cleavage	Blackish-green to light green
Pyroxene (var. Diopside)	$CaMg(Si_2O_6)$	Usually short, thick prisms; may be granular; prismatic cleavage	White to light green
Pyroxene (var. Orthopyroxene)	$(Mg,Fe)_2(Si_2O_6)$	Cleavage good at 87° and 93°; usually massive	Pale green, brown, gray, or yellowish
Quartz	SiO_2	No cleavage; massive, and as six-sided crystals	Colorless, white, or tinted any color by impurities

HARD-NESS	STREAK	SPECIFIC GRAVITY	OTHER PROPERTIES	TYPE(S) OF ROCK IN WHICH THE MINERAL IS MOST COMMONLY FOUND
5–6	Pale green	3.2–3.6	Vitreous luster.	Low- to medium-grade metamorphic rocks
5–6	Pale green	3.2	Vitreous luster; crystals six-sided with 124° between cleavage faces.	Common in many granitic to basaltic igneous rocks and many metamorphic rocks
3	White	2.7	Transparent to opaque. Rapid effervescence with HCl.	Limestone, marble, cave deposits
3.5–4	White to pale gray	3.9–4.2	Effervesces slightly in cold dilute HCl.	Dolomite
}	The clay minerals are so fine-grained that most physical properties cannot be identified.			Shale
				Shale, weathered bedrock, and soil
				Shale, weathered bedrock, and soil
6–6.5	White	2.6	Vitreous and pearly luster; many show fine striations (twinning lines) on cleavage faces.	Granite, rhyolite, low-grade metamorphic rocks
6	White	2.6	Vitreous to pearly luster.	Granite, rhyolite, metamorphic rocks
6	White	2.6–2.7	Vitreous and pearly luster; may show striations as in albite.	Basalt, andesite, medium- to high-grade metamorphic rocks
2.2–2.5	White, gray	2.7–3.1	Vitreous luster; divides readily into thin flexible sheets.	Granitic to intermediate igneous rocks, many metamorphic rocks
2–2.5	White	2.7–3	Vitreous or pearly; flexible and elastic; splits easily.	Many metamorphic rocks, granite
6.5–7	White	3.2–3.3	Vitreous, glassy luster.	Basalt, peridotite
5.5	Pale green	3.2–3.6	Vitreous, distinguished from hornblende by the 87° angle between cleavage faces.	Basalt, peridotite, andesite, high-grade metamorphic rocks
5.5	White to greenish	3.2–3.6	Vitreous luster.	Medium-grade metamorphic rocks
5.5	White	3.2–3.5	Vitreous luster.	Peridotite, basalt, high-grade metamorphic rocks
7	White	2.6	Includes rock crystal, rose and milky quartz, amethyst, smoky quartz, etc.	Granite, rhyolite, metamorphic rocks of all grades, sandstone, siltstone

Table continued on following page

TABLE 1–1 Common Accessory Minerals and Their Properties

MINERAL GROUP OR MINERAL	CHEMICAL COMPOSITION	HABIT, CLEAVAGE, FRACTURE	USUAL COLOR
Apatite	$Ca_5(OH,F,Cl)(PO_4)_3$	Massive, granular	Green, brown, red
Chlorite	$(Mg,Fe)_6(Si,Al)_4O_{10}(OH)_8$	Perfect cleavage as fine scales	Green
Corundum	Al_2O_3	Short, six-sided barrel-shaped crystals	Gray, light blue, and other colors
Epidote	$Ca_2(Al,Fe)Al_2O(SiO_4)(Si_2O_7)(OH)$	Usually granular masses; also as slender prisms	Yellow-green, olive-green, to nearly black
Fluorite	CaF_2	Octahedral cleavage and cubic crystals	White, yellow, green, purple
Garnet (var. Almadine)	$Fe_3Al_2(SiO_4)_3$	No cleavage; crystals 12- or 24-sided	Deep red
Garnet (var. Grossular)	$Ca_3Al_2(SiO_4)_3$	No cleavage; crystals 12- or 24-sided	White, green, yellow, brown
Graphite	C	Foliated, scaly, or earthy masses	Steel gray to black
Hematite	Fe_2O_3	Granular, massive, or earthy	Brownish-red
Limonite	$FeO(OH)$	Earthy fracture	Brown or yellow
Magnetite	Fe_3O_4	Uneven fracture, granular masses	Iron black
Pyrite	FeS_2	Uneven fracture cubes with striated faces, octahedrons	Pale brass yellow (lighter than chalcopyrite)
Serpentine	$Mg_3Si_2O_5(OH)_4$	Uneven, often splintery fracture	Light and dark green, yellow

HARD-NESS	STREAK	SPECIFIC GRAVITY	OTHER PROPERTIES	TYPE(S) OF ROCK IN WHICH THE MINERAL IS MOST COMMONLY FOUND
4.5–5	Pale red-brown	3.1	Crystals may have a partly melted appearance; glassy.	Common in small amounts in many igneous, metamorphic, and sedimentary rocks
2.0–2.5	Gray, white, pale green	2.8	Pearly to vitreous luster.	Common in low-grade metamorphic rocks
9	None	3.9–4.1	Hardness is distinctive.	Metamorphic rocks, some igneous rocks
6–7	Pale yellow to white	3.3	Vitreous luster.	Low- to medium-grade metamorphic rocks
4	White	3.2	Cleaves easily; vitreous, transparent to translucent.	Hydrothermal veins
6.5–7.5	White	4.2	Vitreous to resinous luster.	The most common garnet in metamorphic rocks
6.5–7.5	White	3.6	Vitreous to resinous luster.	Metamorphosed sandy limestones
1–2	Gray or black	2.2	Feels greasy; marks paper; metallic luster	Metamorphic rocks
5.5–6.5	Reddish-brown	2.5–5	Often earthy; dull appearance.	Common in all types of rocks; can form by weathering of iron minerals and is the source of color in nearly all red rocks
1.5–4	Brownish-yellow	3.6	Earthy masses that resemble clay.	Common in all types of rocks; can form by weathering of iron minerals and is the source of color in most yellow-brown rocks
5.5	Iron black	5.2	Metallic luster; strongly magnetic.	Common in small amounts in most igneous rocks
6–6.5	Greenish-black	5	Metallic luster; brittle.	The most common sulfide mineral: igneous, metamorphic, and sedimentary rocks; hydrothermal veins
2.5	White	2.5	Waxy luster, smooth feel; brittle.	Alteration or metamorphism of basalt, peridotite, and other magnesium-rich rocks

Table continued on following page

TABLE 1–1 Common Economic Minerals and Their Properties

MINERAL GROUP OR MINERAL	CHEMICAL COMPOSITION	HABIT, CLEAVAGE, FRACTURE	USUAL COLOR
Anhydrite	$CaSO_4$	Granular masses, crystals with two good cleavage directions	White, gray, blue-gray
Asbestos	$Mg_3Si_2O_5(OH)_4$	Fibrous	White to pale olive-green
Azurite	$Cu_3(CO_3)_2(OH)_2$	Varied; may have fibrous crystals	Azure blue
Bauxite	$Al(OH)_3$	Earthy masses	Reddish to brown
Chalcopyrite	$CuFeS_2$	Uneven fracture	Brass yellow
Chromite	$FeCr_2O_4$	Massive, granular, compact	Black
Cinnabar	HgS	Compact, granular masses	Scarlet red to red-brown
Galena	PbS	Perfect cubic cleavage	Lead or silver gray
Gypsum	$CaSO_4 \cdot 2H_2O$	Tabular crystals, fibrous, or granular	White, pearly
Halite	$NaCl$	Granular masses, perfect cubic crystals	White; also pale colors and gray
Hematite	Fe_2O_3	Granular, massive, or earthy	Brownish-red, black
Malachite	$CuCO_3 \cdot Cu(OH)_2$	Uneven, splintery fracture	Bright green, dark green
Native copper	Cu	Malleable and ductile	Copper red
Native gold	Au	Malleable and ductile	Yellow
Native silver	Ag	Malleable and ductile	Silver-white
Pyrolusite	MnO_2	Radiating or dendritic coatings on rocks	Black
Sphalerite	ZnS	Perfect cleavage in six directions at 120°	Shades of brown and red
Talc	$Mg_3Si_4O_{10}(OH)_2$	Perfect in one direction	Green, white, gray

HARD-NESS	STREAK	SPECIFIC GRAVITY	OTHER PROPERTIES	TYPE(S) OF ROCK IN WHICH THE MINERAL IS MOST COMMONLY FOUND
3–3.5	White	2.9–3	Brittle; resembles marble but acid has no effect.	Sedimentary evaporite deposits
1–2.5	White	2.6–2.8	Pearly to greasy luster. Flexible, easily separated fibers.	A variety of serpentine, found in the same rock types
4	Pale blue	3.8	Vitreous to earthy; effervesces with HCl.	Weathered copper deposits
1.5–3.5	Pale reddish-brown	2.5	Dull luster; clay-like masses with small round concretions.	Weathering of many rock types
3.5–4.5	Greenish-black	4.2	Metallic luster; softer than pyrite.	The most common copper ore mineral; hydrothermal veins, porphyry copper deposits
5.5	Dark brown	4.4	Metallic to submetallic luster.	Peridotites and other ultramafic igneous rocks
2.5	Scarlet red	8	Color and streak are distinctive.	The most important mercury ore mineral; hydrothermal veins in young volcanic rocks
2.5	Gray	7.6	Metallic luster.	The most important lead ore; commonly also contains silver; hydrothermal veins
1–2.5	White	2.2–2.4	Thin sheets (selenite), fibrous (satinspar), massive (alabaster).	Sedimentary evaporite deposits
2.5–3	White	2.2	Pearly luster, salty taste, soluble in water.	Sedimentary evaporite deposits
2.5	Dark red	2.5–5	Often earthy, dull appearance; sometimes metallic luster.	Huge concentrations occur as sedimentary iron ore; the most important source of iron
3.5–4	Emerald green	4	Effervesces with HCl. Associated with azurite.	Weathered copper deposits
2.5–3	Copper red	8.9	Metallic luster.	Basaltic lavas
2.5–3	Yellow	19.3	Metallic luster.	Hydrothermal quartz-gold veins, sedimentary placer deposits
2.5–3	Silver-white	10.5	Metallic luster.	Hydrothermal veins, weathered silver deposits
1–2	Black	4.7	Sooty appearance.	Black stains on weathered surfaces of many rocks, manganese nodules on the sea floor
3.5	Yellow	4	Resinous luster; may occur with galena, pyrite.	The most important ore mineral of zinc; hydrothermal veins
1–1.5	White	1–2.5	Greasy feel; occurs in foliated masses.	Low-grade metamorphic rocks

DIAGNOSTIC PROPERTIES

There are eight major diagnostic properties of minerals: **crystal habit**, **luster**, **hardness**, **cleavage**, **fracture**, **color**, **streak,** and **specific gravity.** Generally, there is no single diagnostic property which by itself can be used to identify a mineral sample. Rather, we must rely on a number of diagnostic properties to reach this goal.

Crystal Habit

If a mineral crystallizes without any impediments to its growth, it may assume a characteristic shape, or **crystal habit,** which reflects its internal crystal structure. For example, muscovite mica will often display a bookish **tabular** habit, a reflection of its sheet structure, and halite forms nearly perfect cubes with flat square **crystal faces,** a reflection of the **cubic** arrangement of its atoms. Other common crystal habits are illustrated in Figure 1–6. The term **anhedral** is used to describe minerals without well-formed crystal habits (which are vastly more common in rocks).

In addition to a simple crystal habit, some minerals contain intergrowths of two or more crystals or show evidence that their crystals changed the direction and orientation of their growth during crystallization. This appearance is called **twinning**. It is typically manifested as a series of straight parallel lines and grooves called **striations** on the cleavage planes of minerals.

Luster

Luster refers to the way in which light is reflected from the surface of a mineral. The two basic classes are **metallic** luster, which describes the reflection of polished metal surfaces, and **nonmetallic** luster, which includes **pearly, resinous, silky,** vitreous (or **glassy**), and **waxy** lusters. Luster is best described from fresh and unweathered surfaces, and preferably from crystal faces. Weathered minerals are typically described as **earthy** in luster.

Hardness

The **hardness** of a mineral refers to its resistance to scratching. It is a measure of the strength of the bonds between the constituent atoms in a mineral: Minerals with relatively strong chemical bonds have a greater resistance to scratching and are thus harder than minerals with relatively weak chemical bonds.

The relative hardness of a mineral is measured by the Mohs scale of hardness, which was developed by the German mineralogist Friedrich Mohs in the nineteenth century (Table 1–2). This scale contains ten minerals which are arranged in order of their relative hardness, beginning with talc (the softest), with a hardness of 1, and ending with diamond (the hardest) at 10. The scale also includes some common objects, such as glass and fingernails, which could also be used to determine the relative hardness of a mineral.

Cleavage

Cleavage describes the tendency of some minerals to break or split along flat surfaces called **cleavage planes.** Cleavage planes are surfaces of weak chemical bonds in mineral crystals. For example, the chemical bonds in muscovite, which has a sheet structure, are strong within the planes of the silicate sheets but weak between them. This causes muscovite crystals to **cleave** well into thin sheets which can be peeled away like layers of an onion.

Some minerals such as mica, feldspar, and fluorite have good to excellent cleavage in one, two, three, four, or even

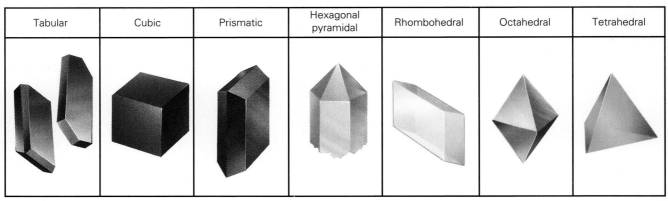

| Tabular | Cubic | Prismatic | Hexagonal pyramidal | Rhombohedral | Octahedral | Tetrahedral |

© George V. Kelvin

Figure 1–6 Common crystal habits of minerals.

TABLE 1–2 Mohs Hardness Scale	
MINERALS OF MOHS SCALE	COMMON OBJECTS
1. ___ Talc	
2. ___ Gypsum	
3. ___ Calcite	Fingernail 2.5 Copper penny 3.0 ___
4. ___ Fluorite	
5. ___ Apatite	
6. ___ Orthoclase	Window glass 5.5
7. ___ Quartz	Steel file 6.5
8. ___ Topaz	
9. ___ Corundum	
10. ___ Diamond	

six different directions, and they cleave cleanly to form regular geometric shapes with smooth cleavage planes (Fig. 1–7). The number of cleavage planes and the angles between them (which you can measure with a protractor) are very useful diagnostic properties for identifying such minerals.

Fracture

Fractures are rough nonplanar breaks which cut randomly through mineral crystals (Fig. 1–7). Minerals fracture rather than cleave whenever the bonds between their constituent atoms are equally strong in all directions, so that there are no preferred planes of weakness in their structures.

Fractures are very common to the framework silicates such as quartz, which exhibits distinct smooth, concentric, dish-shaped **conchoidal** fracture patterns. Other patterns include **fibrous**, which describes a wood-like splintery fracture, and **uneven**, which describes a rough and irregular fracture.

Color and Streak

The **color** of a mineral is one of its most obvious properties, but it is not always diagnostic. The color of a mineral is sometimes controlled by its bulk chemical composition. For example, minerals rich in Si, K, and Al are often light colored, and those rich in Fe and Mg are often dark colored.

However, the color of a mineral can be greatly affected by trace amounts of chemical impurities. For example, **quartz** (SiO_2) is light colored and transparent, but **amethyst** is a variety of quartz which is tinted purple by traces of iron.

Fortunately, there are many minerals (particularly the oxide, sulfide, native metal, and hydroxide minerals) which are not affected in this way and can be recognized on the basis of their colors. The best way to examine the color of such minerals is to grind them against a **streak plate** of unglazed porcelain and examine the color of the powder trace which the mineral leaves behind. (This technique cannot be used for minerals with hardnesses greater than 7, which is the hardness of porcelain.) Minerals with distinctive streaks include **hematite** (which has a reddish-brown streak), **graphite** (grayish-black), and **cinnabar** (scarlet red).

Specific Gravity

Specific gravity is defined as the ratio of the weight of a mineral to the weight of an equal volume of water. It is essentially a measure of the density of a mineral and thus reflects its chemical composition. The specific gravity of a mineral can be measured precisely in a laboratory, and it can be estimated by feeling the heft of the mineral. It is particularly diagnostic of the native metals such as gold (Au), silver (Ag), and copper (Cu), which have specific gravities of 19, 10.5, and 8.9, respectively. It is not very diagnostic of most other minerals, which have specific gravities of about 2.7.

Other Properties

There are several other properties which are diagnostic of specific minerals.

Effervescence describes the reactivity of minerals to dilute hydrochloric acid (HCl). This "acid test" is a diagnostic property of carbonate minerals such as calcite and dolomite. Calcite reacts vigorously to cold HCl, particularly when the acid is dropped on a clean, freshly exposed surface. Dolomite reacts to cold HCl when it is in powdered form, or to warm HCl which is dropped onto a clean surface. (But please do not warm HCl in your laboratory; hot acid is volatile and explosive and requires special equipment for its preparation and handling.)

Fluorescence is the ability of a mineral to absorb ultraviolet light (which is invisible to the human eye) and then emit this absorbed energy in the form of visible light. It is diagnostic of certain varieties of fluorite, calcite, and opal, among other minerals.

Magnetism is attraction to magnets, steel paper clips, and other similar objects. It is diagnostic of the mineral magnetite.

Number of cleavage directions	Angles between cleavage directions	Sketch	Shapes of broken crystal	Sketch
1	180°		Tabular	
2	90°	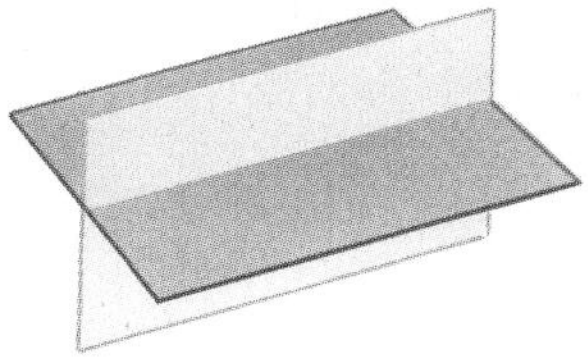	Rectangular prism	
2	Not at 90°	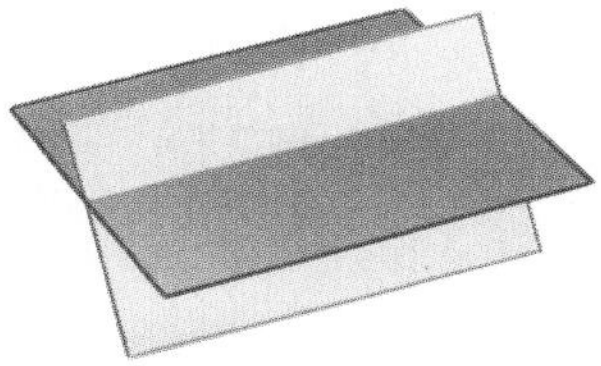	Non rectangular prism	
3	90°	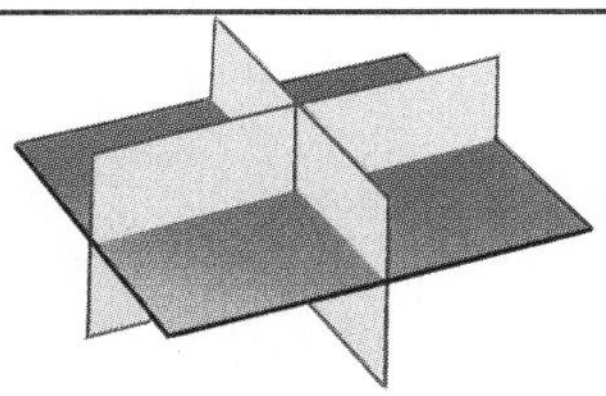	Cubes	
3	Not at 90°	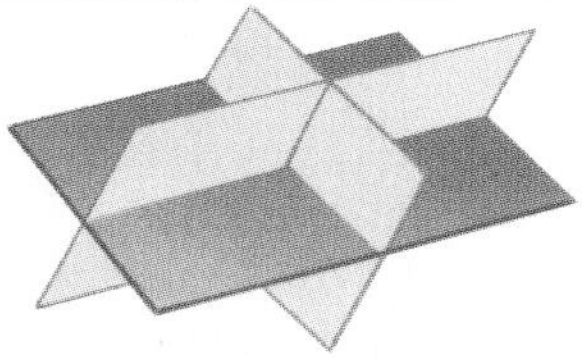	Rhombohedrons	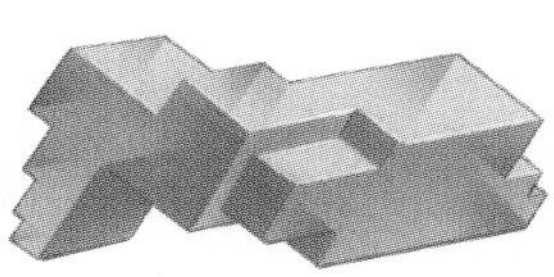
4	Varied	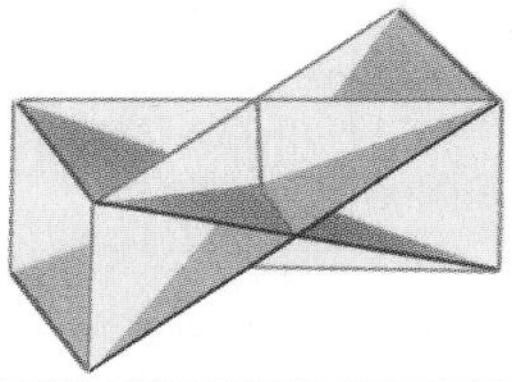	Octahedrons (8 sided)	
6	Varied	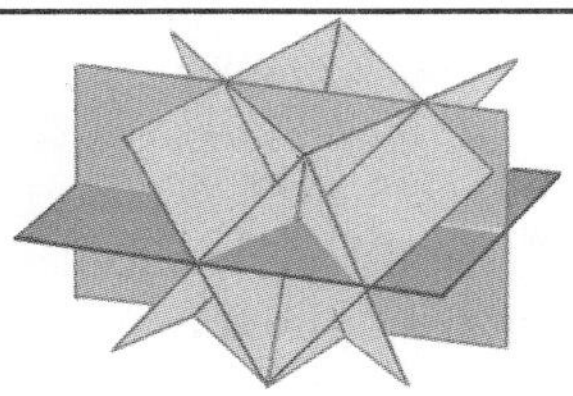	Dodecahedrons (12 sided)	
None			Conchoidal fibrous or uneven	

Figure 1–7 Common cleavage and fracture patterns of minerals.

EXERCISE 1–5   Diagnostic Properties of Known Minerals

This exercise allows you to practice the diagnostic techniques which geologists use to identify minerals. You will be given several known minerals which provide good examples of different diagnostic properties. You should examine and describe each mineral, and record its name and pertinent properties in the space provided. This information will be useful in a later exercise in which you will have to identify unknown minerals by these same diagnostic criteria.

1.   Your instructor will provide (or identify within your sample set) minerals with good crystal habits. Sketch and describe them with the terminology in Figure 1–6.

2.   Your instructor will provide a sample of the mineral plagioclase feldspar, which offers a good example of twinning. Examine the cleavage planes for twinning striations, and sketch them.

3.   Your instructor will provide minerals with different types of lusters. Divide them into separate piles of metallic and nonmetallic minerals. Identify the different types of nonmetallic lusters in the samples.

4.   Your instructor will provide minerals of different degrees of hardness. Measure the hardness of each sample. Begin by scratching each sample with your fingernail (which has a hardness of 2.5), a penny (3.0), glass (5.5), and a steel file (6.5), and then narrow the range of possible hardness by testing one mineral against another.

   When you test for hardness (and later for streak), it is important to be able to distinguish when the mineral has been scratched by the hardness standard (such as a steel file) and when the standard has been scratched by the mineral. You can tell the difference by rubbing the wet tip of your finger across the suspected scratch. When a mineral has been scratched by the standard, the standard will leave a perceptible indented groove on the mineral surface. When the reverse occurs, the standard will leave a powdered streak on the mineral surface which will wipe away.

5. Your instructor will provide samples of the minerals calcite and dolomite, along with a dropper bottle containing dilute hydrochloric acid (HCl), a file, and a watch glass or petri dish. Describe the crystal habit, cleavage, luster and streak of the two minerals. Apply 2 or 3 drops of acid to the minerals and describe their reactions. Use the file to grind some of the dolomite into a powder, letting the powder fall into the watch glass. Apply 1 or 2 drops of acid to it and describe the reaction.

CAUTION *HCl is highly reactive even when it is dilute, and it can cause serious burns if it comes into contact with clothes, skin, eyes or mucous membranes. Handle it with great care, and wash your hands and the minerals after using it.*

6. Your instructor will provide minerals with different breakage patterns. Separate the minerals with cleavage from the minerals without cleavage. Examine the minerals with cleavage. Sketch each sample, identify the number of cleavage directions, measure the angles between them with your protractor or goniometer, and describe their shapes (Fig. 1–7). Examine the minerals without cleavage. Sketch each sample, and describe its fracture pattern as conchoidal, fibrous, or uneven.

7. Your instructor will provide streak plates and minerals with different types of streaks. Describe the streak color of each sample.

8. Your instructor will provide minerals of different specific gravities. Use the heft of the minerals to separate them into two classes: high specific gravity and low specific gravity.

9. Calcite is distinguished by an optical property called **double refraction,** which results from the interplay between light and its crystal structure. Your instructor will provide a sample of optically clear calcite. Place it on this page and read this question through it. What is the effect of double refraction?

Name _____ Section _____ Date _____

Your instructor will now provide you with a set of unknown minerals. Describe the diagnostic properties of each mineral, and record these data in the accompanying table. Use Table 1–1 and the data which you collected in the previous exercise to identify each sample. Remember: Use several diagnostic properties (not just one or two) to identify the minerals.

SAMPLE NO.	CRYSTAL HABIT	LUSTER	HARDNESS	CLEAVAGE AND FRACTURE	COLOR	STREAK	SPECIFIC GRAVITY	MINERAL

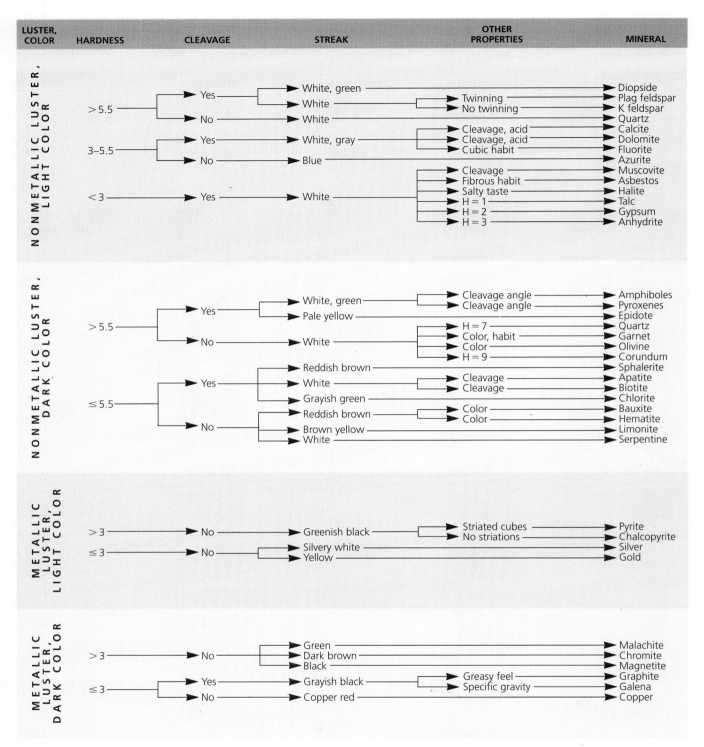

Figure 1–8 A flow chart for the identification of unknown minerals. Given an unknown mineral, you can narrow down its possible identity to a small number of minerals by first describing in this exact sequence its luster and color, hardness, cleavage, and streak. As you describe these properties, follow the arrows from left to right across the flow chart. In some cases, these four properties will be sufficient to identify the mineral. In other cases, you will have to examine or test another diagnostic property in order to identify the mineral. Refer to the mineral tables for more detail about these properties.

2

IGNEOUS ROCKS

Igneous rocks . . . magma, lava . . . plutonic, volcanic rocks . . . phaneritic, aphanitic, glassy, porphyritic, pyroclastic, vesicular textures . . . sialic, intermediate, mafic, ultramafic rocks . . . partial melting, fractional crystallization . . . granite, rhyolite, diorite, andesite, gabbro, basalt, peridotite . . . batholiths, stocks, dikes, sills, laccoliths . . . shield volcanoes, cinder cones, stratovolcanoes, flood basalts . . . viscosity

GRANITE.

Materials Needed *Colored pens or pencils, lead pencils with erasers, hand lens or magnifying glass*

Igneous rocks are formed by the cooling and crystallization of molten rock, or **magma**. Magmas are produced in the lower crust and upper mantle, where high pressures and temperatures result in the melting of large volumes of rock. Molten rock is less dense and more buoyant than solid rock, and so it rises toward the surface of the Earth through fractures and fissures in the crust. Some magmas tend to intrude into the rocks of crust and accumulate in vast subterranean pools, or **magma chambers,** in the upper crust, where they slowly cool and crystallize into **intrusive** or **plutonic rocks**. However, other magmas tend to rise completely through the crust and erupt in **lava** flows on the Earth's surface, where they rapidly cool and crystallize into **extrusive** or **volcanic rocks**.

This chapter examines the formation and nature of igneous rocks and rock bodies. The first part of the chapter discusses the two fundamental properties of igneous rocks, the texture and composition of their mineral grains, and the classification of igneous rocks on the basis of these properties. The second part of the chapter describes plutonic and volcanic rock bodies and then considers the factors which influence their formation and composition.

Igneous rocks are composed of an interlocking network of mineral grains which crystallize during the cooling of magma or lava. The **texture** of the grains—that is, their sizes and shapes—is an important indicator of the rate of cooling of the parent magma (i.e., whether the rocks are intrusive or extrusive in origin). Their **composition**, on the other hand, is an important indicator of the source of their parent magma. For this reason, texture and composition serve as the basis for the description and classification of igneous rocks.

TEXTURE OF IGNEOUS ROCKS

The term **texture** describes the sizes and shapes of the mineral grains in a rock. Six types of textures are commonly seen in igneous rocks: **phaneritic**, **aphanitic**, **glassy**, **porphyritic**, **pyroclastic**, and **vesicular** (Fig. 2–1).

Phaneritic igneous rocks consist of an interlocking mass of coarse, angular mineral grains which are large enough to be seen with the naked eye (that is, they are greater than about 1 millimeter in size). Phaneritic rocks with grains greater than 1 centimeter in size are often called **pegmatites**.

Aphanitic rocks consist of mineral grains which are too small to be seen with the naked eye (that is, less than 1 millimeter). The crystalline nature of aphanitic rocks can be seen with the aid of a microscope, but hand samples are typically massive and featureless in appearance.

Glassy texture describes an igneous rock which is composed of either **amorphous** (noncrystalline) solid matter or mineral grains which are **microcrystalline** and submicroscopic in size—that is, too small to be seen even with a conventional light microscope. Glassy rocks have a shiny, or vitreous, luster to their surfaces and typically fracture in a conchoidal manner. Rocks with glassy textures are often called **obsidian**.

Porphyritic rocks contain a mixture of coarse mineral grains, called **phenocrysts**, embedded in a matrix, or **groundmass,** of finer mineral grains. The rocks may consist of either phaneritic phenocrysts embedded in an aphanitic matrix (**porphyritic-aphanitic texture**) or pegmatitic phenocrysts in a phaneritic matrix (**porphyritic-phaneritic texture**).

Pyroclastic rocks consist of solid fragments (called **pyroclasts**) which are hurled into the atmosphere by volcanic explosions. Pyroclasts are typically composed of varying amounts of mineral grains, igneous rock fragments, and glass, and they range in size from coarse **bombs** to fine **ash** (Table 2–1). Coarse pyroclasts are typically deposited close to volcanic vents, forming steep-sided piles of debris called **cinder cones**. Volcanic ash, which is considerably finer grained, is usually lofted high into the atmosphere and transported many hundreds to thousands of miles downwind from its source. Pyroclastic debris is typically welded together by its own heat to form **pyroclastic rocks** such as **volcanic breccias** and **tuffs**.

A

B

Figure 2–1 (A) Coarse phaneritic texture of a granite. (B) A coarse-grained granite pegmatite with large feldspar crystals. *(William E. Ferguson)* (C) Fine aphanitic texture in a basalt. (D) Glassy texture and typical conchoidal fracture pattern of obsidian. (E) Porphyritic texture in a rhyolite. The rock consists of coarse pink potassium feldspar phenocrysts imbedded in a matrix of very fine-grained minerals. (F) Pyroclastic texture of a volcanic tuff. The rock is composed of pumice fragments and fine volcanic ash. *(Jeffrey Sutton)* (G) Vesicular texture of a basaltic scoria.

C

D

E

F

G

TABLE 2–1	Texture of Pyroclastic Grains and Rocks	
SIZE	**PYROCLAST**	**ROCK NAME**
>64 mm	Blocks and bombs	Volcanic breccia
2–64 mm	Cinders	Volcanic breccia
0.06–2 mm	Coarse ash	Volcanic tuff
<0.06 mm	Fine ash	Volcanic tuff

Vesicles are spherical voids which are formed when magma crystallizes around trapped gas bubbles. Igneous rocks with vesicular textures (**pumice** and **scoria**) can often float on water due to their low density.

Significance of Texture of Igneous Rocks

The texture of an igneous rock is determined by its rate of cooling and crystallization: Slow cooling allows for the crystallization and growth of large mineral crystals, whereas fast cooling results in the rapid crystallization of many small mineral crystals. Consequently, texture is an important indicator of the origins of igneous rocks.

Phaneritic textures are typical of intrusive rocks, which crystallize slowly within the warm crust of the Earth. Aphanitic textures, on the other hand, are typical of extrusive rocks, which crystallize rapidly on the Earth's cold surface. They are also common to the rocks along the margins of igneous intrusives, where magma comes into contact with cooler country rock. Glassy textures are often formed on the surface of an extrusive lava flow, where molten rock comes into contact with the atmosphere and cools so rapidly that crystals have too little time to grow or in some cases even form. They are also common to submarine **pillow lavas**, which crystallize rapidly when they are quenched by contact with the cold waters of the ocean floor.

Porphyritic textures indicate a two-stage cooling history for an igneous rock: a fairly long period of partial cooling in a magma chamber, during which time coarser phenocrysts grow, followed by the sudden extrusion of the remaining magma (and the phenocrysts) and its rapid crystallization into fine groundmass.

Pyroclastic textures are typical of gas-rich and violently explosive volcanic eruptions, such as the eruption of Mount St. Helens in May of 1980. This eruption pulverized the northern face of the mountain and blew off its upper 410 meters, and it sent tons of pyroclastic debris and gas as far as 18 kilometers upward into the atmosphere. There, the westerly winds spread an ash cloud from Washington State to western New York, raining ash across the upper Mississippi and Missouri valleys and darkening the skies of Washington, Idaho, and Montana for several days.

Vesicular textures are common to gas-rich extrusive flows. Generally, gas tends to bubble upward through lava flows and escape into the atmosphere (a process called **degassing**). Vesicular textures form when gas bubbles become trapped by the very rapid cooling of lava.

COMPOSITION OF IGNEOUS ROCKS

The crust of the Earth is almost entirely (i.e., 98 percent) composed of eight elements: silicon and oxygen (which together make up about 74 percent of the crust by weight), aluminum, magnesium, calcium, sodium, potassium, and iron. Therefore, it should be no surprise that igneous rocks are composed primarily of silicate minerals, and particularly **olivine** (Fe-Mg silicate), **pyroxene** (Ca-Mg-Fe-Al silicate), **amphibole** (Na-Ca-Mg-Fe-Al silicate), **biotite** (K-Mg-Fe-Al silicate), **muscovite** and **potassium** (or **K-**) **feldspar** (both K-Al silicates), **plagioclase feldspar** (Na-Ca-Al silicate), and **quartz** (SiO_2).

Four major classes of igneous rocks can be defined on the basis of composition (Fig. 2–2): **sialic**, **mafic**, **ultramafic**, and **intermediate rocks**.

Sialic rocks (also called **felsic rocks**) contain more silica (72 percent on average) than all other classes of igneous rocks, and they are consequently rich in the mineral quartz. They also contain abundant K, Na, and Al, which are present in the forms of K-feldspar and Na-rich plagioclase. Sialic rocks are typically white, gray, red, and pink due to the abundance of these light-colored minerals.

Mafic rocks are silica-poor (50 percent) but rich in Ca, Fe, and Mg, which are present in the forms of Ca-rich plagioclase, olivine, and some pyroxene. Ultramafic rocks are especially poor (45 percent) in silica but rich in Fe and Mg, which are present in the forms of olivine and pyroxene. Mafic and ultramafic rocks are typically colored dark gray, brown, and black by olivine and pyroxene grains. They are also denser than sialic rocks due to their higher content of Fe and Mg.

Intermediate rocks are, as their name indicates, intermediate in composition between sialic and mafic rocks. They contain high amounts of silica (59 percent on average) and fairly high percentages of Fe and Mg, all of which are present in the forms of plagioclase feldspar, pyroxene, and amphibole. Intermediate rocks are generally gray and green.

Factors Affecting Composition of Igneous Rocks

The composition of an igneous rock can reflect several factors, including the source of its parent magma, the interaction between its parent magma and the rocks of the crust, and the timing of its crystallization.

Magmas can be generated by the melting of the rocks of the upper asthenosphere at divergent margins (Fig. 2–3). The upper asthenosphere is largely composed of **peridotite**, a silica-poor igneous rock which is composed of olivine, pyroxene, and some plagioclase feldspars. When a crustal rift

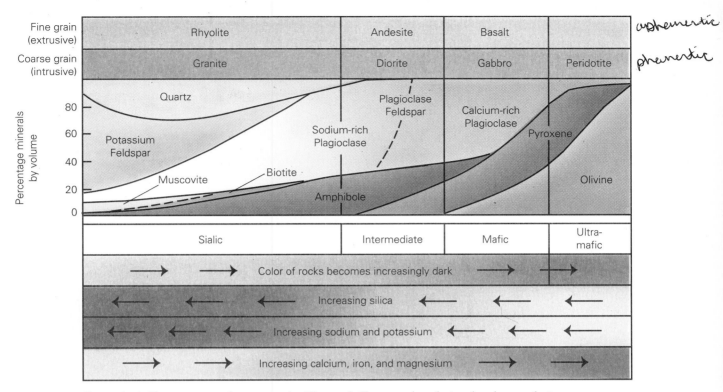

Figure 2–2 The properties of igneous rocks. The graph illustrates the relative abundances of minerals in sialic, intermediate, mafic, and ultramafic rocks; the two bars at the top of the figure show the textures of the common igneous rocks, and the four bars at the bottom show relative trends in their colors and chemical compositions.

opens between two diverging plate margins, the decreased pressure on the asthenosphere causes the melting of the minerals within the peridotite. This melting occurs in a sequence which can be predicted from Bowen's reaction series (Fig. 2–4): The plagioclase feldspars and pyroxene melt first because they have lower melting temperatures, and olivine melts last because it has a higher melting temperature. This sequential process is called **partial melting**, and it results in a gradual evolution in the compositions of the resultant magmas: The earliest ones are Ca-, Fe-, and Mg-rich mafic magmas, whereas the later ones are Fe- and Mg-rich ultramafic magmas.

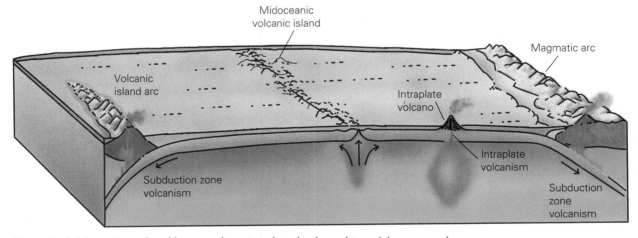

Figure 2–3 Magma is produced between diverging plates by the melting of the upper asthenosphere, and it erupts on the surface to form the submerged volcanoes of the mid-ocean ridges. Magma is also produced between converging plates by the melting of subducted crust. This magma rises into the crust to form plutons and erupts on the surface to form volcanic mountain ranges, or *magmatic arcs*, and volcanic island chains, or *island arcs*.

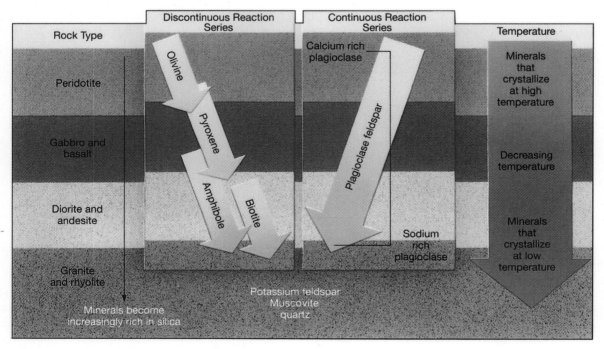

Figure 2–4 The crystallization of minerals from a magma occurs in an orderly sequence called Bowen's reaction series.

Magmas can also be generated by the melting of the rocks of the crust at subduction margins (Fig. 2–3). The crust is often composed of both silica-rich and silica-poor rocks. When crustal rocks are subducted, the downward increase in temperature results in a similar sequence of partial melting and the generation of magmas of different compositions. The first minerals to melt are silica-rich ones such as quartz and potassium feldspar, which generate sialic magmas. This is followed by the gradual melting of rocks with less and less silica and the production of intermediate and eventually mafic magmas.

Magmas may also become "contaminated" with minerals from other rocks as they rise through the crust of the Earth. One of the most common examples of this contamination occurs when basaltic magma, which is generated at temperatures of 1100° to 1400°C, rises upward and comes into contact with the granitic rocks of the lower continental crust, which begin to melt at temperatures of 700° to 900°C. The melting granite enriches the basaltic magma with silica-rich minerals, and the magma becomes more intermediate or even sialic in its composition.

Finally, igneous rocks of different compositions can crystallize from a single magma by the process of **fractional crystallization**, which also can be predicted from Bowen's reaction series. When a magma cools, the first minerals to crystallize are olivine and pyroxene. Once they have crystallized, these minerals often settle to the bottom of the magma chamber, where they accumulate as ultramafic rocks. As cooling continues, the next minerals to crystallize are the Ca-rich plagioclase feldspars and then amphibole, which ac-

cumulate as mafic and then intermediate rocks. The last minerals to crystallize are quartz, K-feldspar, and Na-rich plagioclase feldspar, which accumulate as sialic rocks (Fig. 2–5). By this process, four different types of rocks are crystallized at different times during the cooling history of a single magma.

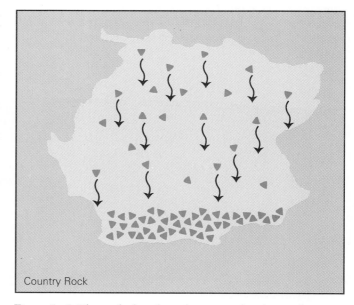

Figure 2–5 The gradual cooling of a magma chamber produces a succession of igneous rock types over time by the process of fractional crystallization.

CLASSIFICATION OF IGNEOUS ROCKS

Igneous rocks are classified on the basis of their texture and mineral content (Fig. 2–2). The common igneous rock types are **granite**, **rhyolite**, **diorite**, **andesite**, **gabbro**, **basalt**, and **peridotite** (Fig. 2–6 and Table 2–2).

Granite and rhyolite are both sialic rocks. They are both composed largely of quartz, potassium feldspar, and Na-rich plagioclase feldspar, and they are typically white, gray, red, and pink. They are differentiated on the basis of their texture and genesis: Granite is a medium- to coarse-grained plutonic rock, whereas rhyolite is an aphanitic or porphyritic-aphanitic volcanic rock. Granite is the most abundant rock type in the continental crust. It is often called the **basement rock** because it often underlies the sedimentary rocks that blanket the interiors of the continents.

Gabbro and basalt are both dark-colored mafic rocks composed largely of Ca-rich plagioclase feldspar, pyroxene, and olivine. However, gabbro is a phaneritic plutonic rock, whereas basalt is an aphanitic, porphyritic-aphanitic, or glassy volcanic rock. Basalt is also the most abundant rock type in the oceanic crust as well as in continental flood basalts, such as those of the Columbia Plateau.

Diorite and andesite are both intermediate in composition between sialic and mafic rocks: They are both medium- to dark-colored rocks that are composed largely of amphibole, Ca- and Na-rich plagioclase feldspars, and pyroxene. However, diorite is a phaneritic plutonic rock, and andesite is its aphanitic or porphyritic-aphanitic volcanic equivalent. Andesite is abundant in the volcanic rocks of the Andes Mountains, from which it acquired its name.

Peridotite is a green phaneritic plutonic rock that is composed largely of olivine and pyroxene, and it is believed to be the most common rock type in the upper mantle. It has no volcanic equivalent.

Other terms which are commonly used to describe igneous rocks are **obsidian**, **pumice** and **scoria**, and **volcanic breccias** and **tuffs**.

Obsidian is a general term which describes a massive volcanic rock with a glassy texture. Pumice and scoria are rocks which are rich in vesicles; **pumice** describes a light-colored sialic or intermediate rock, whereas **scoria** describes a dark-colored mafic or ultramafic rock. Volcanic breccias and tuffs are both pyroclastic rocks (Table 2–1). Volcanic breccias are composed primarily of blocks, bombs, and cinders, whereas tuffs are composed largely of fine volcanic ash.

TABLE 2–2 Common Igneous Rocks and Their Properties

ROCK PROPERTIES	ROCK NAME
Composed largely of quartz, Na plagioclase, K-feldspar, and some mica and amphibole; phaneritic intrusive rock; light color: red to pink when rich in K-feldspar, white to gray when rich in plagioclase	Granite
Composed largely of quartz, Na plagioclase, K-feldspar, and some mica and amphibole; extrusive rock, either aphanitic or porphyritic-aphanitic with light quartz and feldspar phenocrysts; light color, typically pink, white, and gray	Rhyolite
Composed largely of Na-Ca plagioclase, amphibole, and pyroxene; phaneritic intrusive rock; typically speckled with light plagioclase and dark ferromagnesian minerals	Diorite
Composed largely of amphibole, Na-Ca plagioclase, and pyroxene; extrusive rock, either aphanitic and glassy or porphyritic-aphanitic with phenocrysts of amphibole and plagioclase; medium-dark color, typically gray, green, and brown	Andesite
Composed largely of Ca plagioclase, pyroxene, and some olivine; phaneritic intrusive rock; dark color, typically gray, green, and black, and fairly dense; plagioclase can form long dark crystals	Gabbro
Composed largely of Ca plagioclase, pyroxene, and some olivine and amphibole; aphanitic, porphyritic-aphanitic, and glassy extrusive rock; dark color, typically gray, brown, and black; dense and massive; glass and vesicles common	Basalt
Composed largely of olivine and pyroxene; phaneritic intrusive rock; green color from olivine; rarely exposed at the surface, with no extrusive equivalent	Peridotite
Composed of massive volcanic glass; black with red, yellow, and brown streaks of oxidized minerals; glassy luster; breaks with conchoidal fracture pattern	Obsidian
Composed of porous extrusive rock with many vesicles; pumice is rhyolitic and andesitic in composition, light in color; scoria is mafic and dark in color	Pumice and Scoria
Composed of volcanic pyroclasts, largely coarse blocks, bombs, cinders, and some coarse ash; medium to dark, typically brown and grayish-brown	Volcanic Breccia
Composed of volcanic pyroclasts, largely ash and some coarse glass and mineral fragments; lightweight, and soft and friable; dull, light earthy color, typically white, gray, and yellow	Volcanic Tuff

A

B

C

D

E

F

Figure 2–6 Common igneous rocks. (A) Granite. (B) Rhyolite. (C) Diorite. (D) Andesite. (E) Gabbro. (F) Basalt. (G) Peridotite. (H) Obsidian. (I) Volcanic breccia. (J) Tuff. *(Jeffrey Scovil)*

EXERCISE 2–1   Common Igneous Rock-Forming Minerals

Refer to Table 1–1 in the previous chapter. Summarize the color, crystal habit, cleavage, and fracture of the common igneous rock-forming minerals on the following table.

MINERAL	COLOR	CRYSTAL HABIT	CLEAVAGE AND FRACTURE	HARDNESS
Olivine				
Pyroxene				
Amphibole				
Biotite				
Muscovite				
K-feldspar				
Plagioclase				
Quartz				

EXERCISE 2–2   Texture and Composition of Igneous Rocks

Rhyolite is composed of quartz, potassium feldspar, and minor amounts of sodium-rich plagioclase, biotite, and amphibole.

1. Which of these minerals would most likely constitute the phenocrysts in a porphyritic rhyolite? Which would likely constitute the groundmass? Explain your answers.

2. Speculate on the crystallization history of this igneous rock type and the possible origins of its parent magma.

1.   Examine the samples of igneous rocks which are provided by your instructor. Describe their physical properties, and summarize this information on the following table.

   a. If a rock sample is phaneritic or porphyritic-phaneritic in texture, you will be able to see individual mineral grains with your naked eye, a hand lens, or a binocular microscope. Refer to the table in the previous exercise, and identify the minerals in such samples. In addition, indicate whether each mineral is abundant (A), common (C), or rare (R).

   - *Abundant* indicates that a mineral makes up at least 50 percent of the rock sample (at least one out of every two grains).

   - *Common* indicates that it makes up between 10 percent (one out of ten grains) and 50 percent of the sample.

   - *Rare* indicates that it is present in very small amounts, less than 10 percent of the sample.

   b. If a rock sample is porphyritic-aphanitic in texture, determine the composition of the phenocrysts and estimate the abundance of each mineral in the sample.

   c. If a rock sample is aphanitic in texture, examine its color carefully. Use the color to indicate the likely mineral composition of such rocks. Do the same for the groundmass in porphyritic-aphanitic rocks.

2.   Classify each sample as either sialic (S), intermediate (I), mafic (M), or ultramafic (U), and as either intrusive (Int) or extrusive (Ext).

3.   Name each sample on the basis of the information in Table 2 – 2.

SAMPLE	TEXTURE	COMPOSITION	COLOR	OTHER PROPERTIES	CLASSIFICATION	NAME

The principal types of igneous rock bodies are crustal **plutons** and surficial **volcanoes** and **lava plateaus**.

PLUTONS

Plutons are intrusive rock bodies which have crystallized within the crust of the Earth. They can be further classified as **batholiths, stocks, dikes, sills,** and **laccoliths** on the basis of their size, shape, and orientation relative to the **country rocks** into which they intrude. Two types of orientations can exist between a pluton and its surrounding country rocks: **concordance**, in which a pluton has intruded parallel to the bedding or structure of the country rock, and **discordance**, in which a pluton cuts across the bedding or structure of the country rock (Fig. 2–7).

Batholiths and stocks are the largest types of plutons. Batholiths are greater than 100 square kilometers in surface area and average about 10 km in thickness, and they have bulbous shapes with steeply dipping walls and deep floors. Batholiths are formed deep within the crust of the Earth, but they are often uplifted to the surface by tectonic processes. When they are exposed, they form massive mountains with sheer, vertical cliffs which are popular among rock climbers. Stocks are smaller intrusive rock bodies with surface areas of less than 100 square kilometers. They may be separate and distinct rock bodies, although in many cases what appears to be a stock is actually a small segment of a larger batholith which has been only partially exposed by erosion. Batholiths and stocks are both discordant rock bodies, and they are usually composed of coarse-grained sialic rock (granite).

Dikes and sills are tabular, sheetlike plutons which range in thickness from a few centimeters to a few kilometers. They differ in orientation relative to their country rocks. Dikes are discordant plutons which crystallize from magmas that intrude into fractures, faults, and joints in the country rock. They are often found clustered in parallel or radiating sets called **dike swarms**. Sills are concordant plutons which crystallize from magmas that intrude parallel to the bedding or structure of the country rock. Laccoliths are mushroom-shaped sills with flat bases and dome-shaped tops which push against and fold the overlying country rock.

VOLCANOES AND FLOOD BASALTS

Volcanoes are composed of extrusive igneous rocks which have crystallized on the surface of the Earth. There are three principal types: **shield volcanoes, cinder cones,** and **stratovolcanoes**.

Shield volcanoes are broad, dome-shaped volcanic mountains that are largely composed of layer upon layer of basaltic lava flows. They have gentle slopes of less than 12° and can reach a height of 9000 meters. The most prominent set of shield volcanoes in the world is the Hawaiian Island chain of the central Pacific Ocean.

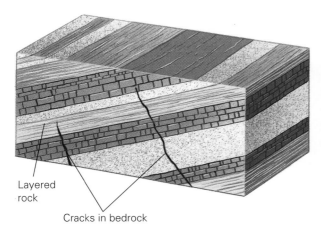
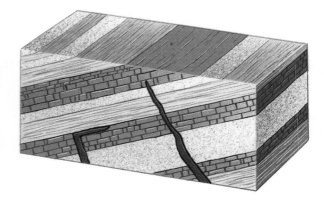

Layered rock

Cracks in bedrock

Figure 2–7 Sills are concordant plutons which form parallel to the bedding, or layering, of the country rock. Dikes are discordant plutons which cut across the structure of the country bedrock.

Cinder cones are piles of pyroclasts which accumulate around volcanic vents. This debris is ejected into the air during violent volcanic eruptions and then accumulates around the vent in small (100- to 400-meter) piles with steep (up to 30°) slopes. The most famous cinder cone is Parícutin, which began to form in a Mexican farm field in 1943 and grew to a height of 400 meters within 2 years.

Stratovolcanoes (also known as **composite cones**) are typically composed of interlayered andesitic lava flows and pyroclastic debris. They have moderate slopes but steep, often snowcapped summits, and they range in height from 100 to 3500 meters. Stratovolcanoes are the most beautiful of all volcanoes, but also the most dangerous because of their erratic and unpredictable eruptive style. The Cascade Range of western North America is a string of stratovolcanoes which includes Mount St. Helens and Mount Rainier.

Lava plateaus are broad, flat-lying plains which are formed of stacked horizontal layers of extrusive rocks called **flood basalts**. Flood basalts crystallize from highly fluid lavas which flow gently from linear **fissures** in the Earth's crust and spread over large areas. One of the largest lava plateaus in the world is the Columbia River Plateau, which extends over 200,000 square kilometers of the northwestern United States and reaches a maximum thickness of 3000 meters.

COMPOSITION OF PLUTONIC AND VOLCANIC ROCKS

Magma is generated by the melting of the rocks of the upper asthenosphere and lower crust, and once formed, it rises upward through the crust to the surface of the Earth. However, geologists have noted that the behavior of magma varies with its composition: Silica-poor magmas appear to be able to rise completely through the crust to the surface, where they erupt and crystallize into basalt and andesite, whereas silica-rich magmas tend to "stall" and crystallize into granite within the crust itself.

The primary cause of this difference in the behavior of magmas of different compositions is their **viscosity**. Viscosity is defined as the resistance of a fluid to flow, and it is inversely proportional to the flow rate of that fluid. For example, a high-viscosity fluid (such as molasses) flows very slowly from a jar or bottle, whereas a low-viscosity fluid (such as water) flows very rapidly.

The viscosity of a fluid is largely controlled by two factors: its molecular composition and structure, and its temperature. For example, molasses is composed of complex organic molecules which are linked in long chains called **polymers**. When molasses flows, these polymers become tangled and knotted, making the molasses stiff and viscous. Water, on the other hand, is composed of simple H_2O molecules, which do not form such long polymers and thus do not impede the flow. However, molasses can be made to flow more readily by heating it: The heat breaks the long chains of polymers into shorter chains, which are less likely to tangle and knot during flow.

The same laws of fluid flow apply to magmas of different compositions. Silica-rich magmas are relatively viscous because they contain many long, complex chains of silicate polymers. Consequently, they tend to rise through the crust slowly, which gives them sufficient time to cool and solidify into plutonic rocks. Silica-poor magmas, on the other hand, are less viscous because they contain shorter silicate polymers. Consequently, they tend to rise more rapidly, and there is little time for them to cool and solidify before they reach the surface.

The viscosity of lavas also affects their eruptive styles. Violent eruptions and the production of large volumes of pyroclastic debris are caused by the build-up of pressure from gases which are dissolved within lavas. Such pressure build-ups and explosions are common in relatively viscous lavas, which do not degas rapidly. On the other hand, gentle lava flows, which spread like sheets over large areas of land, are more typical of less viscous lavas.

Mount St. Helens prior to the 1980 eruption.
(Courtesy of Larry Davis)

Name ______________________ Section ____________ Date ____________

EXERCISE 2-4 Structure, Viscosity, and Igneous Rock Bodies

The following table lists the most abundant minerals in sialic, intermediate, and mafic igneous rocks and magmas, their compositions, and their crystal structures.

ROCK/ MAGMA	COMMON MINERALS	CRYSTAL STRUCTURE	SILICON CONTENT Mineral	SILICON CONTENT Average
SIALIC	Quartz SiO_2	Framework		
	Potassium feldspar $KAlSi_3O_8$	Framework		
	Na-Plagioclase feldspar $NaAlSi_3O_8$	Framework		
INTERMEDIATE	Plagioclase feldspars $Na_{0.5}Ca_{0.5}(Si_{2.5}Al_{0.5})O_8$	Framework		
	Amphibole (var. Hornblende) $NaCa_2Mg_4Al_3Si_6O_{22}(OH)_2$	Double chain		
	Pyroxene (var. Diopside) $CaMgSi_2O_6$	Single chain		
MAFIC	Ca-Plagioclase feldspar $CaAl_2Si_2O_8$	Framework		
	Pyroxene (var. Diopside) $CaMgSi_2O_6$	Single chain		
	Olivine (var. Forsterite) Mg_2SiO_4	Isolated tetrahedra		

1. Calculate the atomic-weight percent of Si in each mineral on the table. The atomic weights of all the elements in the minerals are:

H: 1 O: 16 Na: 23 Mg: 24 Al: 27 Si: 28 K: 39 Ca: 40

Record your data on the table under the "Mineral" side of the "Silicon Content" column on the table. The calculations for quartz (SiO_2) are shown here as an example:

$$\text{Atomic weight of } SiO_2 = \text{Atomic weights of 1 Si atom} + 2 \text{ O atoms} = 28 + (16 \times 2) = 60$$

$$\text{Weight percent of Si in } SiO_2 = (28 \div 60) \times 100\% = 46.7\%$$

2. What mineral has the highest Si content? _____ the lowest? _____

3. What kinds of crystal structures are generally found in the minerals with the highest Si contents? _____
_____

What mineral is the exception to this generalization? _____
_____

What kinds of crystal structures are generally associated with the lowest Si contents? _____
_____

4. Calculate the average Si content of the three magma/rock types. Assume that each rock contains equal proportions of its three constituent minerals. Record your results on the table.

5. What is the general relationship between the Si content of a magma and its viscosity? Why does it occur?

6. Examine the average silicon content values of the three rock/magma types. Explain why plutons are generally composed of granite and why volcanoes are often composed of basalt.

7. Which magma type would *most likely* erupt violently and construct cinder cones? Explain your answer.

8. Why are lava plateaus typically composed of basalt rather than andesite?

9. The slopes of stratovolcanoes are typically composed of andesitic lava flows separated by layers of pyroclastic debris. What does this indicate about the composition of the source of their magmas?

| EXERCISE 2–5 | Igneous Rock Bodies |

1. Figure 2–8 illustrates a hypothetical cross section through the upper crust of the Earth, showing a variety of igneous rock bodies. Locate and label examples of batholiths, stocks, dikes, sills, laccoliths, flood plateaus, and volcanoes.

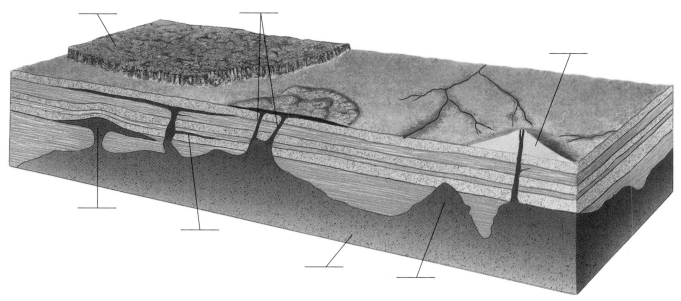

Figure 2–8 Intrusive and extrusive igneous rock bodies.

2. Figure 2–9 illustrates two possible ways in which a tabular igneous rock body can come to be sandwiched between two layers of sedimentary rocks:

- By the concordant intrusion of a magma between two older layers of sedimentary rock; or
- By an extrusive lava flow over the older sedimentary rock layer, followed by the deposition of the younger rock layer over the lava flow.

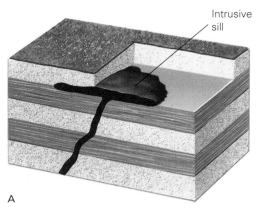

Intrusive
sill

A

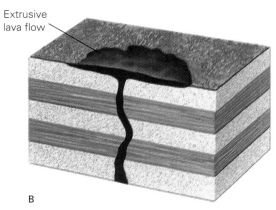

Extrusive
lava flow

B

Figure 2–9 Two mechanisms for forming a tabular igneous rock body between two sedimentary rock layers. (A) Concordant intrusion of a sill. (B) Extrusion of lava over sedimentary rock, followed by its burial by younger sediments.

a. What textural features might indicate whether the tabular rock body in Figure 2–9 is intrusive or extrusive in origin? Explain your answer.

b. Where would you look for such diagnostic features?

c. What compositional features might indicate whether the tabular rock body is intrusive or extrusive in origin? Explain your answer.

d. Where would you look for such diagnostic features?

e. What kind of igneous rocks would be highly indicative of an *intrusive* origin for the tabular rock body?

f. What kind of igneous rocks would be highly indicative of an *extrusive* origin for the tabular rock body?

3

SEDIMENTARY ROCKS

KEY WORDS

Weathering . . . detrital sediments, rock fragments, and clay minerals . . . erosion and transport . . . size, sorting, and composition . . . clastic, biogenic, and chemical sediments . . . lithification

SANDSTONE ROCK, BRYCE CANYON NATIONAL PARK

(© Tony Stone Worldwide/David Schultz)

Materials Needed Colored pens or pencils, lead pencils with erasers, hand lens or magnifying glass

Sedimentary rocks are composed of mineral fragments, rock fragments, and organic debris that are bound together by natural mineral cements. They are not a particularly abundant rock type compared to igneous and metamorphic rocks: They constitute a mere 5 percent of the total volume of the crust, forming only a thin veneer (thickened slightly at the edges of continents and absent in other places) over the surfaces of the continents and the floors of the oceans. However, they are an important rock type for two reasons.

First, sedimentary rocks are formed on the surface of the Earth by many different types of physical, chemical, and biogenic processes, and their texture and composition reflect these processes and the surficial conditions of the time and place of their formation. In addition, they may contain **fossils**, the physical remains of ancient animals and plants. Fossils are a very important part of the geologic record: They document the origin and evolution of life on this planet, and they reveal information about the climate, topography, and vegetation of the continents and the temperature, salinity, and currents of the oceans in the geologic past. In short, sedimentary rocks contain a record of the **paleotectonic** history of the crust, the **paleoclimatic** history of the atmosphere, and the **paleogeography** and **paleooceanography** of the continents and oceans, as well as the most complete record of the evolution of life on Earth.

Second, sedimentary rocks are economically valuable. Petroleum, natural gas, and water are typically found in subsurface **reservoirs** composed of sedimentary rocks, and bituminous coal is both a sedimentary rock and a major fossil fuel. Some sedimentary rocks are sources of raw material, such as building stone and cement (limestones and gypsums), fertilizer (evaporites), spices (salt), ceramic (shales), and glass (sandstones), whereas others contain sizable economic deposits of metallic ores, including aluminum, manganese, copper, lead, zinc, iron, uranium, gold, and silver.

43

WEATHERING

The first step in the formation of a sedimentary rock is the breakdown of exposed sediment sources or **parent rocks** by chemical and mechanical weathering processes.

Chemical weathering is the result of the interaction between the atmosphere and the minerals in a parent rock and in soil water. Some minerals (particularly those rich in Si and Al) are resistant to chemical weathering reactions to some degree, but most minerals (particularly those rich in Mg, Ca, Fe, and Na) are either altered into new **sedimentary minerals** (the most important of which are the **clay minerals**) or are completely dissolved.

Mechanical weathering is the process by which a parent rock is broken into coarse **rock fragments** by purely physical means. One of the most important mechanical weathering processes is **frost wedging**, which occurs when water percolates into cracks in a rock, freezes, and expands in volume. This expansion generates sufficient force to fracture the rock further and liberate large rock fragments from it.

The products of weathering are **detrital sediment** and **soluble matter**. The sediment consists of **detrital grains** of various sizes and compositions, including mineral and rock fragments which survived weathering with no significant alteration in their compositions, and clay mineral grains, which were produced during weathering. The soluble matter consists largely of dissolved Ca, Na, Mg, Cl, K, Si, HCO_3 (bicarbonate), and SO_4 (sulfate). It is released into streams and the ground-water system, and as we shall see later, it constitutes the basic building blocks for carbonate and chemical sedimentary rocks.

Composition of Detrital Grains

There are hundreds of different minerals in igneous and metamorphic rocks, but the vast majority of them are highly susceptible to either chemical weathering or the rigor and wear of transport. Consequently, geologists have found that the detrital grains in sedimentary rocks are largely composed of four **detrital species**: quartz, feldspar, rock fragments, and clay minerals.

Quartz is the most abundant detrital species in sedimentary rocks for several reasons: It is common in many different types of igneous and metamorphic rocks; its chemical composition makes it highly resistant to chemical weathering; and its physical hardness (7) and lack of cleavage enable it to endure the rigors of transport. For this reason, quartz is considered an index of the **maturity** of detrital grains—that is, it is a measure of the intensity of weathering and mechanical wear to which the grains were subjected at some time in their past.

Feldspar is the most abundant mineral in igneous and metamorphic parent rocks. However, it is less common than quartz in sedimentary rocks because it is very susceptible to chemical weathering, particularly in warm, humid climates where the rate and efficacy of chemical weathering are high. Consequently, feldspar is considered an index of the **climate** of the source region of the detrital grains: High amounts of feldspar indicate an arid climate where chemical weathering is at a minimum due to the lack of moisture, whereas low amounts indicate a warm, humid climate.

Rock fragments are largely the product of mechanical weathering, which produces abundant coarse polymineralic grains. The most common varieties are metamorphic rock fragments, particularly schist, phyllite, and metaquartzite fragments, and sedimentary rock fragments, particularly shale fragments. Rock fragments are produced in greatest abundance in mountainous source regions, where the steep slopes and high relief promote rapid mechanical weathering (by gravity) and the production of coarse polymineralic grains. For this reason, rock fragments are considered an index of the **topography** of the source region.

Clay minerals are silicate minerals that are produced by the chemical weathering and alteration of minerals. For example, the exposure of orthoclase feldspar to slightly acidic soil water will alter this mineral to the clay mineral **kaolinite** and release dissolved Si (silicon) and K (potassium) into the water, as described by the following equation:

$$2KAlSi_3O_8 + 2H^+ + 9H_2O \longrightarrow$$
$$\text{(orthoclase)} \qquad \text{(acid)} \qquad \text{(water)}$$

$$Al_2Si_2O_5(OH)_4 + 4H_4SiO_4 + 2K^+$$
$$\text{(kaolinite)} \qquad \text{(silicic acid)} \qquad \text{(dissolved potassium)}$$

There are literally hundreds of different types of clay minerals in modern soils. However, most clay minerals are stable only at the surface of the Earth, and they cannot survive the temperatures and pressures of burial in sedimentary basins. Consequently, most clay minerals in sedimentary rocks are composed of either **kaolinite**, **chlorite**, **smectite**, or **illite**.

The composition of detrital grains is graphically represented on a **ternary diagram** (Fig. 3–1). The three apices or **end members** in this diagram are quartz, feldspar, and rock fragments. Each sample is represented by a single **sample point** that defines its relative proportions (totaling 100 percent) of these three end members.

The position of a sample point within a ternary diagram reveals important information about the genesis of detrital grains, particularly the composition of their parent rock and their weathering and erosion history. A sample point near the quartz apex, for example, indicates that the grains were

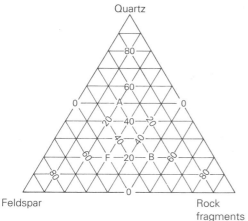

Examples

Sample A contains 50% quartz, 30% feldspar and 20% rock fragments
Sample B contains 20% quartz, 30% feldspar and 50% rock fragments

Figure 3–1 Ternary diagram for the graphic representation of the composition of detrital grains. Every point within the triangle can be defined by its position relative to the three apices of the triangle, which represent 100 percent of quartz, feldspar, and rock fragments. The compositions of samples A and B are given as examples. Note that the percentages of quartz, feldspar, and rock fragments total 100 percent in both samples.

either (1) derived from a quartz-rich parent rock such as a quartzose sandstone or (2) intensely weathered in a warm, humid climate. A sample point near the feldspar apex indicates that the detrital grains were either rapidly weathered or weathered in an arid climate, so that the feldspars survived the weathering process. A sample near the rock-fragment apex indicates high relief and high rates of erosion (and little time for chemical weathering) in the source area.

Texture of Detrital Grains

Geologists use the **Wentworth–Udden scale** (Table 3–1) to describe the sizes of detrital grains. This scale defines four major size classes (and several subclasses): **gravel**, **sand**, **silt**, and **clay**.

Gravel grains are larger than 2 millimeters in diameter, whereas sand grains are smaller but still visible to the naked eye. Silt and clay grains are too fine to be seen with the naked eye. However, silt grains, which are the larger of the two, feel gritty when rubbed between your fingertips or against your teeth, whereas clay grains feel smooth and slick.

The variability in the sizes of grains is described with terms such as **well sorted** and **poorly sorted**. For example, a sediment composed largely (>90 percent) of sand might be described as well sorted, whereas one composed of 60 percent gravel and 40 percent sand might be described as moderately or poorly sorted.

EROSION AND TRANSPORT

The next step in the formation of sedimentary rocks is the erosion of detrital grains from their sources and their transport into basins. This can be accomplished by both gravity-driven processes, such as rock falls and debris flows, and fluid-flow processes, such as streams, glaciers, and wind.

When detrital grains are transported by streams and wind, the coarser gravel and sand grains are usually slowly dragged, rolled, and bounced over the stream bed or desert floor, but the finer silt and clay grains are transported much more rapidly downcurrent because they are suspended in (and move as fast as) the water column or the atmosphere.

SIZE CLASS	SUBCLASSES	DIAMETER (mm)	ROCK NAME	
Gravel	Boulder	> 256	Conglomerate (rounded) or breccia (angular)	
	Cobble	64–256		
	Pebble	4–64		
	Granule	2–4		
Sand	Very coarse	1–2	Sandstone	
	Coarse	0.500–1		
	Medium	0.250–0.500		
	Fine	0.125–0.250		
	Very fine	0.063–0.125		
Silt		0.004–0.063	Siltstone	Shale (fissile)
Clay		< 0.004	Claystone	

TABLE 3–1 Wentworth–Udden Grain-Size Scale

(Silt and Clay grouped as MUD; Siltstone, Claystone, and Shale grouped as MUDSTONE)

This difference in **transport mode** segregates, or **sorts**, detrital grains on the basis of their sizes as the silt and clay grains outrace the gravel and sand grains, and it produces a net decrease in grain size in the downstream or downwind direction.

Sorting is also a consequence of the different flow velocities that are necessary to erode and transport grains of different sizes (Fig. 3–2). For example, gravel grains require high flow velocities for transport, which confines them to stream channels, tidal inlets, and other high-energy environments, but clay can be transported by low-velocity flows into flood plains, lakes, swamps, and other low-energy environments. In addition, gravel grains are transported only during the peaks of floods, when flow velocities are highest, whereas clay can be transported for longer periods of time by the everyday movement of sluggish streams and moderate breezes.

DEPOSITION

The next step in the formation of sedimentary rocks is the deposition of detrital grains and dissolved matter. This deposition is accomplished by physical, organic, and chemical means, each of which produces a different class of sediment.

Physical Deposition

Detrital grains are deposited whenever their transport agent slows in velocity and loses its capacity to transport them. When a stream empties into an ocean, for example, its flow velocity decreases rapidly due to the friction between the stream current and the ocean, and the stream gradually deposits its sediment load at its mouth to form a **delta deposit**. A current will first deposit its coarsest sediment grains and then progressively finer grains as it slows in velocity. This physical deposition of the detrital grains produces the first of the three major classes of sediments, the **clastic sediments**.

Organic Deposition

There are many species of marine and lake organisms that can absorb dissolved matter from water, precipitate it into sedimentary minerals such as calcite, opal, and apatite, and then incorporate these new minerals into their bodies. When these organisms die, their hard body parts, called **skeletal grains**, accumulate on the sea floor or lake bed to form the second major class of sediments, the **biogenic sediments**.

The most common group of biogenic sediments is the **carbonates**, which consist of the **calcareous** skeletal grains of many different marine plants and animals. These grains range from coarse gravel-sized coral and shell fragments to fine, clay-sized **tests** of algae, and they are composed of the minerals **calcite**, **aragonite** (two forms of $CaCO_3$), and **dolomite**. The grains are often found mixed with calcareous **nonskeletal grains** such as **ooids** (spherical sand-sized grains), **intraclasts** (carbonate rock fragments), and **micrite** (clay-sized calcareous grains).

A second important type of biogenic sediments is the **siliceous sediments** such as diatomaceous ooze. Such sediments consist of the remains of **radiolarians**, **diatoms**, and

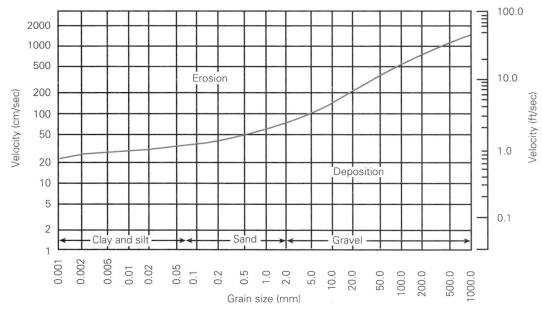

Figure 3–2 Hjulstrom diagram. Flow velocities needed to erode and deposit detrital grains of different sizes. For any given size class, grains at rest will be eroded when the velocity rises above the line, and grains in transport will be deposited when it falls below this line.

other microscopic organisms which construct skeletal parts with the mineral **opal** and other varieties of SiO_2.

A third important type of biogenic sediments is the **carbonaceous sediments**, the **coal group**. Coals are composed of terrestrial woody plant debris (**peat**) that accumulates in the backwaters of swamps and river flood plains, where the **anoxic** (oxygen-deficient) water conditions allow for the preservation of organic matter.

Chemical Deposition

Soluble matter can also be precipitated from surface and ground waters by nonorganic chemical reactions to form chemical sediments. The most common reaction of this type occurs during the evaporation of a saline lake or isolated sea, when minerals such as **halite** (NaCl), **gypsum** ($CaSO_4 \cdot 2H_2O$), and **anhydrite** ($CaSO_4$) are precipitated on the lake bed or sea floor. This process forms a type of chemical sediment known as the **evaporites**.

Evaporites are often deposited in flat layers, or **salt beds**, that extend over large areas to form natural pavements. Examples of such salt beds can be found at Edwards Air Force Base in California, where they form the landing strip for the U.S. space shuttles. Evaporites can also be deposited within clastic and carbonate sediments in the form of discrete **nodules**.

Carbonate minerals can also precipitate by nonorganic processes to form nonorganic carbonate rocks such as **travertine** and **dolomite**. Several quartz-related minerals, including opal, chalcedony, jasper, and flint, can also precipitate from silica-rich waters to produce an inorganic variety of **chert**.

Chemical sediments differ from clastic and biogenic sediments in that they are not generally composed of discrete mineral grains. Rather, they are more similar to igneous and metamorphic rocks in that they have **crystalline textures** composed of interlocking mineral crystals.

Sediments and Depositional Environments

Sedimentary rocks are deposited in various continental, coastal, and marine environments. Every depositional environment can be characterized by a specific set of physical, organic, and chemical depositional processes. The nature of these depositional processes controls the types of sediments which accumulate there (Table 3–2).

Generally, clastic sediments accumulate in many different depositional environments ranging from alluvial fans at the bases of mountains to the bottom of the sea floor. Sand and gravel accumulate wherever currents are strongest: in the channels of alluvial fans, streams, and deltas; in windy deserts; on beaches and barrier islands; in tidal inlets; and in submarine canyons. On the other hand, silt and clay accumulate under quieter conditions: in flood plains, lakes, and swamps; in coastal lagoons, tidal flats, and bays; across conti-

nental shelves and slopes; and on the deep sea floor. Carbonaceous sediments also accumulate in the fertile backwaters of flood plains and swamps, where vegetation is dense and oxygen is scarce.

Carbonate and siliceous sediments are most often deposited in marine environments, but they are found in lakes as well. Coarse-grained carbonates such as fossiliferous and oolitic limestones accumulate around wave-battered reefs (which are often called **carbonate factories**), along beaches and barriers, in tidal inlets, and in submarine canyons. Fine carbonates such as micrite and chalk, on the other hand, accumulate in quiet waters: in coastal lagoons, tidal flats, and bays; across continental shelves and slopes; and on the deep ocean floor. Fine-grained siliceous detritus is also an important sediment component on the ocean floor.

Chemical sediments are common to aqueous environments that are prone to evaporation and hypersalinity, particularly desert playa lakes (such as the Great Salt Lake) and large bodies of marine water with poor connections to the open sea (such as the Persian Gulf).

LITHIFICATION

The last step in the formation of sedimentary rocks is the burial of unconsolidated sediments and their **lithification** into sedimentary rock. Sediments can be lithified by several different processes, including **compaction**, **cementation**, and **recrystallization**.

Compaction is the result of the downward pressure which is exerted on buried sediment by overlying sediments. When a sediment is deposited, its grains are usually separated by significant amounts of empty space called **pores**. When it is buried, compaction rotates, bends, and breaks the grains to force them into a denser interlocking configuration, and it reduces the volume of pore space by 20 to 80 percent or even more. Clay deposits are particularly susceptible to compaction: The clay particles are rotated and realigned by compaction so that they are parallel, which produces the thinly layered **fissile** structure which characterizes shales.

Cementation is the result of the crystallization of mineral cements in the pores of sediment. The cements are precipitated from surface and ground waters which are rich in dissolved ions and which circulate through the pores of sediment. There are many minerals which can bind sediments together in this fashion, but the most common are quartz, calcite, dolomite, and the clay minerals.

Recrystallization involves the rearrangement of the crystal structures of sedimentary minerals and/or the incorporation of new elements into the crystal structures. The most common reactions of this type are the recrystallization of calcareous grains to form an interlocking network of coarse calcite crystals, and the introduction of magnesium (Mg) into the structure of calcite or aragonite grains to form a new calcareous mineral, dolomite.

TABLE 3–2 **Sedimentary Processes and Products**

		SEDIMENTARY PROCESSES	MAJOR SEDIMENT/ROCK TYPE
CONTINENTAL	Alluvial Fans	High-velocity flow of water through distributary channels, deposition of bedload on channel bottom and channel bars during waning flood	Gc and SSc: moderately sorted; rounded gravel grains; may be graded
		Debris flows and rock slides from adjacent mountain slopes	mGb and mSS: very poorly sorted; angular gravel grains
	Streams	High-velocity flow of water through channels, deposition of bedload on channel bottom and channel bars during waning flood	SS: moderately to well sorted; may be graded
		Low-velocity sheetflow across vegetated floodplains, deposition of suspended load during waning flood	Sh: typically red to green in color; may be graded; may contain carbonaceous debris
	Deserts	Migration of windblown dunes across desert floor	SS: very well sorted and rounded grains; commonly quartz-rich
	Glaciers	Melting of glacial ice, deposition of till and moraines	mGb, mSS: very poorly sorted; angular gravel grains
	Lakes	Deposition of suspended load by waning fluvial currents	Slt, Cly, Sh: moderately to well sorted; may be graded; black when organic-rich
		Evaporation of lake water	Ev
COASTAL	Deltas	High-velocity flow of water through distributary channels, deposition of bedload at distributary-mouth bars, reworking of bars by waves	SS: moderately to well sorted; may be graded and gravelly
		Low-velocity sheetflow across marshes and bays, deposition of suspended load during waning flood	Sh: thin graded deposits; typically dark gray to black; may contain nodules of pyrite and other sulfide minerals
		Accumulation of organic (plant) debris in swamps	Co
		Deposition of suspended load on continental shelf	Sh: typically dark gray to black; may be graded; may contain fossils
	Beaches and Barriers	Reworking and sorting of sediment by waves	SS: well sorted; may contain fossil fragments
		Wind reworking of beach sand and accumulation in coastal dunes	SSf: very well sorted and rounded grains; often quartz-rich
	Tidal Flats	High-velocity flow of water through tidal channels	SS: moderately to well sorted; may contain fossil fragments
		Low-velocity sheetflow of water across tidal flats and coastal marshes	Sh: typically dark gray to black; may contain fossils or carbonaceous debris
MARINE	Reefs and Lagoons	Formation of calcareous skeletal and nonskeletal grains in reefs, reworking and sorting by waves and tides	Ls: composed of coarse skeletal, ooid, intraclast, and pellet grains in younger rocks, but typically crystalline in older rocks; moderately to well sorted
		Deposition of fine suspended grains in quiet waters of back-reef lagoon	Ls: micritic, but may contain coarser skeletal and nonskeletal grains
	Submarine Canyons and Fans	High-velocity flow of turbidity currents through submarine canyons and fan distributary channels, deposition of bedload during waning flow	Gc and SSc: moderately to well sorted; may be graded (turbidites); may be composed of clastic or carbonate grains, depending on source
		Low-velocity sheetflow across interdistributary areas of submarine fans, deposition of suspended load during waning flow	Sh: typically dark gray to black; may be graded
		Debris flows and rock slides	Gb and SSc: very poorly sorted; angular gravel grains; may be composed of clastic or carbonate grains, depending on source
	Abyssal Plain	Deposition of clay and fine biogenic grains suspended in seawater and air	Cly, Ck, and Di: may contain recognizable microfossils and chemical nodules
		Chemical deposition	Ch: chert and manganese nodules

KEY:
Gc	Gravel and Conglomerate	mSS	Muddy Sandstone	Cly	Clay and Claystone	Di	Diatomite
Gb	Gravel and Breccia	SSc	Coarse Sand and Sandstone	Sh	Shale and Mudstone	Co	Peat, Lignite, and Coal
mG	Muddy Gravel	SSf	Fine Sand and Sandstone	Ls	Limestone and Dolomite	Ev	Evaporites
SS	Sand and Sandstone	Slt	Silt and Siltstone	Ck	Ooze and Chalk	Ch	Other Chemical Rocks

EXERCISE 3-1   Composition of Detrital Sediments

The following table lists the compositions of the sand fractions in two samples of recently weathered granite. The two granite parent rocks are similar in mineral composition, but they were weathered in different climates: One was weathered in the hot and humid Appalachian Mountains of Georgia, whereas the other was weathered in the cool and dry Rocky Mountains of Colorado.

1. Plot the compositional data on the ternary diagram in Figure 3-1. Which sample was probably taken from the humid Appalachians? the arid Rockies? Explain your answer, and record it on the table.

2. The ternary diagram also contains a third point F, which represents a sample from the Fountain Formation. The Fountain is a sandstone that was eroded from granitic parent rocks in the central Rocky Mountains and deposited by streams across the western-central United States during the Pennsylvanian period (approximately 300 million years ago). This part of the United States presently has an arid climate.
    a. Determine the quartz, feldspar, and rock fragment contents of the Fountain sample from the ternary diagram. Complete the table with this information.
    b. Compare the composition of the Fountain to those of the modern weathered granites. Which weathered granite is more similar in composition to the Fountain?
    c. What does this indicate about the climate of the central Rocky Mountains during the Pennsylvanian period? Explain your answer, and record it on the table.

SAMPLE	QUARTZ	FELDSPAR	ROCK FRAGMENTS	CLIMATE
Weathered granite #1	59%	29%	12%	
Weathered granite #2	33%	38%	29%	
Fountain sandstone				

1.   Refer back to Figure 2–4, the illustration of Bowen's reaction series in the Igneous Rocks chapter. Which three minerals in Bowen's reaction series should be most resistant to chemical weathering? least resistant? Explain your answer.

2.   What is the relationship between temperature of mineral crystallization and resistance to chemical weathering?

3.   Which igneous rocks would be most resistant to chemical weathering? least resistant? Explain your answer.

EXERCISE 3-3   Constructing a Grain-Size Comparator

The sizes and sorting of detrital grains can be visually estimated with the use of comparators, which contain detrital grains of a standard size or sorting. Construct your own grain-size comparator by following these steps:

1.   Take two pieces of white cardboard or posterboard, approximately 5 × 7 inches or smaller. On one piece, mark off six equal-sized squares with a ruler and pencil, and cut a circular window from the center of each square.

2.   Place strips of double-sided tape across the uncut piece of cardboard and mount the cut piece over it. Cut any extra tape from the edges with a scissor.

3.   Your instructor will provide samples of detrital grains of different sizes. Label a window with the name of each grain size provided, and then sprinkle a small amount of each size fraction into the appropriate hole.

4.   Your instructor will provide a set of detrital sediment. Use your grain-size comparator to determine the grain size of each sample. In addition, use the sorting comparator (Fig. 3–3) to describe their sorting. Record these data in the space below.

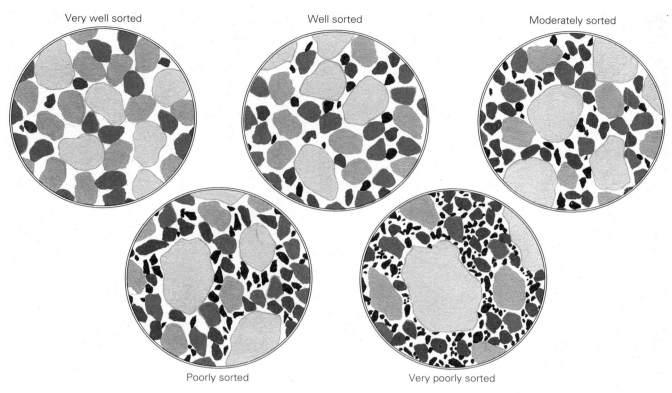

Figure 3–3 A comparator for estimating the sorting of sediments and sedimentary rocks. Notice that the grain size becomes more variable with decreased sorting.

The Tuscarora Sandstone was deposited by streams that flowed across the mideastern part of the United States during the Silurian period, approximately 400 million years ago. Figure 3–4 shows the locations of exposures (**outcrops**) of this sandstone and the maximum size of its detrital grains at each of these outcrops. What was the general direction of stream flow in this area during the Silurian period? Explain your answer.

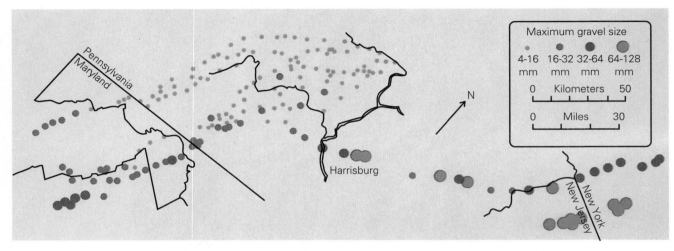

Figure 3–4 Maximum size of quartz gravel clasts at outcrops of the Tuscarora Sandstone in the central Appalachian Mountains.

This simple laboratory experiment will allow you to observe the formation of evaporites. It requires a hot plate, a 50 or 100 ml beaker of water, table salt, a hand lens, eyedropper, and a microscope glass slide.

1.   Warm the beaker of water on the hot plate. Place 1 to 2 tsp of table salt in the water, and stir until it is dissolved.

2.   Turn the hot plate to its lowest setting, and put the glass slide on it. Place a few drops of the solution on the slide. Allow some (but *not all*) of it to dry, and then gently add a few more drops. Repeat until there is a buildup of salt on the slide, and then let the solution dry completely. Remove the slide from the hot plate, and let it cool.

3.   Examine the glass slide with your hand lens, and illustrate what you see.

Name _____ Section _____ Date _____

1. Your instructor will provide you with poorly sorted sediment. Use your comparator to describe the ranges of grain sizes in this sample. Place the sample in a glass jar, fill it with water, and seal its top. Shake the jar vigorously until all the grains are suspended in the water column, and then place the jar on the table and allow the grains to settle. After a few minutes, examine the sediment column at the bottom of the jar with your grain-size comparator. The sediment will be graded. Describe and illustrate this feature.

2. When you shook the jar, you created a turbulent current. What happened to the velocity of this current when you set the jar at rest? Why is the sediment graded?

3. Figure 3–5 shows a map of a hypothetical river mouth on a seacoast. The flow velocity of the river current is indicated and contoured in 10-cm/sec intervals.
  a. What happens to the flow velocity of current as you move seaward from the river mouth?

  b. Use the Hjulstrom diagram (Fig. 3–2) to determine the grain sizes that would be deposited in the areas between the contour lines. Illustrate the areal pattern of size variation that would result from the river discharge on the figure.

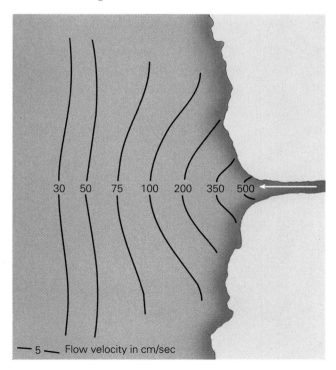

30   50   75   100   200   350   500

— 5 — Flow velocity in cm/sec

Figure 3–5 Hypothetical model for the flow of a river current into an ocean. The velocity of the river current is contoured.

The following minerals are some of the most common grains and cements in sedimentary rocks. Refer back to Table 1–1 and complete the following table by summarizing their diagnostic properties.

MINERAL	DIAGNOSTIC PROPERTIES	MINERAL	DIAGNOSTIC PROPERTIES
Quartz		Calcite	
Potassium feldspar		Gypsum	
Plagioclase feldspar		Anhydrite	
Biotite mica		Halite	
Muscovite mica		Clay minerals	
Dolomite		Hematite	

Figure 3–6 on pages 56 and 57 and Table 3–3 summarize the classification of some of the most common sedimentary rocks. Generally, there are separate classification schemes for clastic, biogenic, and evaporitic sedimentary rocks, because they differ significantly in their origins and lithology.

CLASTIC ROCKS

Clastic sedimentary rocks are generally classified on the basis of their **texture** (particularly the sizes of their detrital grains) because this is often a good indicator of their transport history and their depositional environment. For example, stream channels, which are characterized by strong currents and high flow velocities, are typically distinguished by the coarse, gravelly texture of their deposits, whereas flood plains are distinguished by the silt- and clay-rich deposits of sluggish overbank flows.

Sandstones are often further classified on the basis of their detrital grain composition, because this property reveals much information about the tectonic and climatic setting of the parent rock and its composition. Three major classes of sandstones can be defined on the basis of their relative abundance of quartz, feldspar, clay minerals, and rock fragments (Fig. 3–7): the **quartz arenites, arkoses,** and **graywackes**.

Quartz arenites are composed largely of well-sorted, well-rounded quartz sand grains that are cemented by quartz or calcite. They contain very little or no mud, and they are usually white to buff-colored. Quartz arenites can be derived from igneous and metamorphic parent rocks that are subjected to intense chemical weathering in a warm, humid climate. However, they are more often produced by the erosion, or **recycling**, of older quartz arenites.

Arkoses (or *arkosic arenites*) are composed largely of moderately sorted and highly angular quartz and feldspar grains. The grains are typically cemented by calcite and small amounts of hematite (iron oxide), which gives arkoses their distinctive red color. Arkoses are produced by the weathering and erosion of granites and gneisses in arid climates, where the rates of chemical weathering are fairly low.

Figure 3–7 Sandstones can be classified as quartz arenites, arkoses, and graywackes on the basis of their quartz, feldspar, and rock fragment contents.

Graywackes (or *lithic arenites*) are composed largely of rock fragments and clay minerals. They are usually poorly sorted, and they have a distinctive black-and-white or salt-and-pepper appearance. Graywackes are most often derived from shales, slate, phyllites, and schists that are exposed in mountain areas where the climate, relief, and rate of erosion promote mechanical weathering and the production of rock fragments.

BIOGENIC SEDIMENTARY ROCKS

Carbonate biogenic rocks are classified primarily on the basis of their mineralogy; they are called **limestones** if they are composed of the mineral calcite, and **dolomites** if they are composed of the mineral dolomite.

It is also common to modify the terms limestone and dolomite with descriptions of the types of skeletal and nonskeletal grains which are present in the rocks. Some common terms for this are **fossiliferous**, which describes rocks rich in the skeletal remains of calcareous organisms; **micritic**, which describes rocks rich in fine carbonate mud; and **oolitic**, which describes rocks rich in ooids. In addition, the terms **crystalline limestone** and **crystalline dolomite** describe rocks which have been recrystallized to the point that they no longer contain any recognizable skeletal or nonskeletal grains.

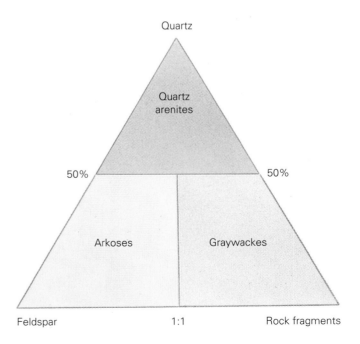

Figure 3–6 Common sedimentary rocks. *(Jeffrey Scovil)*

(B) Conglomerate

(E) Graywacke

(H) Shale

(A) Breccia

(D) Arkose

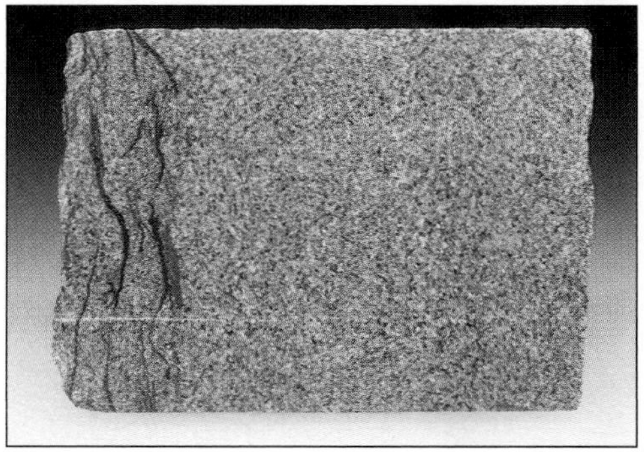

(G) Siltstone

(C) Quartz arenite

(F) Claystone

(K) Coquina

(N) Bituminous coal

(Q) Rock gypsum (light) and calcite (dark)

(J) Dolomite

(M) Lignite

(P) Rock salt (halite)

(I) Fossiliferous limestone

(L) Chalk

(O) Chert

Other terms which are used to describe carbonate rocks include **coquina** and **chalk**. Coquina is a colloquial term for a poorly lithified and highly porous shelly limestone. Chalk is a rock which is composed of the silt- and clay-sized calcareous skeletal parts of algae, foraminifera, and other microscopic marine organisms.

Siliceous rocks are classified on the basis of composition and grain type as well as degree of crystallization, which is reflected by their hardness and density. For example, **diatomite** is used to describe a soft, poorly lithified rock composed of diatom fragments, whereas **diatom chert** describes a diatom-rich rock which has been partly recrystallized and welded together into a harder, denser mass.

Carbonaceous rocks are classified on the basis of **rank**, which describes their contents of carbon, water, and gas. The rank is usually apparent from the rock's degree of compaction and crystallinity—that is, whether it is composed of loosely bound debris with clearly recognizable plant fragments (**lignite**) or highly compacted crystalline carbon (**coal**). **Bituminous coal** is the most abundant rank of coal. It has a very high (80 to 85 percent) carbon content and yields many BTUs of energy when it is burned, which makes it an important energy resource around the world.

CHEMICAL ROCKS

Chemical rocks are generally classified on the basis of mineral composition. The most common varieties are the **evaporites** such as **rock salt** and **rock gypsum**, the carbonate rocks **travertine** and **dolomite**, and the siliceous **cherts**. The textures of chemical rocks may also be described as **coarsely crystalline** (containing large visible mineral grains) and **finely crystalline** (grains are too small to be seen with the naked eye).

TABLE 3–3 Common Sedimentary Rocks and Their Properties

CLASTIC SEDIMENTARY ROCKS	
Rock Properties	**Rock Name**
Composed largely of angular, poorly sorted gravel grains; grains are typically supported in a matrix of sand, mud, and cement; rich in rock fragments, quartz, and chert grains	Breccia
Composed largely of rounded, moderately sorted gravel grains; typically grain supported; pores filled with sand, mud, and cement; rich in rock fragments, quartz, and chert grains	Conglomerate
Composed largely of well-rounded and well-sorted quartz sand grains; white, buff, tan, pink, or brown	Quartz Arenite
Composed largely of angular, moderately sorted feldspar and quartz sand grains; often reddish-brown (redbed) due to the presence of hematite cement	Arkose
Composed largely of poorly sorted sand-sized rock fragments and quartz grains, as well as abundant mud; typical salt-and-pepper appearance	Graywacke
Composed largely of clay minerals and clay-sized quartz and mica; massive and nonfissile; soft and scratchable with fingernails, but tends to fracture conchoidally; color varies from tan and light greenish-gray to reddish-brown and black, depending on composition; rock feels smooth when rubbed across fingers or teeth	Claystone
Composed largely of silt-sized detrital grains, such as quartz and mica; nonfissile but often contains thin layers (laminae) visible to the naked eye; typically hard due to quartz content, and tends to break along laminae; white, tan, and buff in color; rocks feel gritty when rubbed across fingers or teeth	Siltstone
Composed of silt and clay-sized grains such as clay minerals, quartz, and mica; fissile and very finely laminated; soft and scratchable with fingernails, and tends to break cleanly between laminae; variable color, but often gray to black due to presence of organic matter	Shale

BIOGENIC SEDIMENTARY ROCKS	
Rock Properties	**Rock Name**
Composed largely of skeletal and nonskeletal grains, cements, and intergrown crystals of calcite (limestone) and dolomite; generally white, tan, and buff, but can be tinted dark brown, gray, or black by organic matter; limestone effervesces strongly in cold HCl; dolomite effervesces modestly when acid is applied to fresh or scratched surface; varieties include: • *fossiliferous*: composed largely of skeletal remains of calcareous invertebrates such as forams, corals, oysters, and molluscs; • *oolitic*: composed largely of well-sorted spherical sand-sized ooids; • *micritic*: composed largely of fine carbonate mud; hard, dense, and tends to fracture conchoidally; • *pelletal*: composed largely of sand-sized fecal pellets and other unrecognizable skeletal and nonskeletal grains; • *intraclastic*: composed largely of carbonate rock fragments; • *crystalline*: composed largely of interlocking coarse or fine crystals of calcite or dolomite; recognizable skeletal and nonskeletal grains are not common	Limestone and Dolomite
Colloquial term for a poorly lithified and highly porous limestone which is composed of moderately to well-sorted shells and shell fragments; common to beaches in Florida and the Bahamas	Coquina
Composed largely of the silt- and clay-sized remains of calcareous algae and forams; soft and scratchable with fingernails, and very porous; effervesces strongly in cold HCl; white, light gray, or buff	Chalk
Composed largely of the silt- and clay-sized remains of diatoms and other siliceous organisms; well lithified with silica cement but fairly friable and breakable; does not react to HCl; white to gray	Diatomite
Composed largely of the silt- and clay-sized remains of siliceous organisms; grains have partly recrystallized and intergrown into a denser, harder form, but skeletal types are still recognizable; white, gray, salt-and-pepper to black; sometimes thinly banded; varieties include *radiolarite* and *diatom cherts*	Chert
Composed largely of partly decomposed and moderately compacted (but still recognizable) plant debris; low density, poorly lithified, and very friable; generally reddish-brown or yellowish-brown; modest BTU levels when burned due to significant gas and water content	Lignite
Composed largely of highly compacted recrystallized plant debris; few recognizable particles; dark gray to black with streaks of mud, pyrite, and other matter; burns with smoky flame, yields high BTU levels due to high carbon content; also called *soft coal* because it can be broken by hand	Bituminous Coal

CHEMICAL SEDIMENTARY ROCKS	
Rock Properties	**Rock Name**
Composed of interlocking coarse to fine cubic crystals of halite (NaCl); salty taste; soft and scratchable with fingernail; clear, white to gray, but sometimes tinted by impurities such as iron oxide or clay	Rock Salt
Composed of interlocking coarse to fine elongate crystals of gypsum ($CaSO_4-2H_2O$); soft and scratchable with fingernail; typically white, buff, or pink ~~clear also~~	Rock Gypsum
Composed of interlocking coarse to fine crystals of calcite ($CaCO_3$); generally appears as alternating bands of light yellowish-brown pure calcite and dark reddish-brown to gray calcite with traces of iron oxide	Travertine
Composed of interlocking coarse to fine crystals of dolomite with no recognizable skeletal or nonskeletal grains; effervesces weakly in cold HCl, but vigorously in warm HCl; white to gray; not distinguishable from biogenic dolomite	Dolomite
Composed of interlocking very fine crystals of quartz (SiO_2); hardness of 7 and dense; conchoidal breakage pattern; white, gray, blue, yellow, or red; commonly forms nodules in limestones	Chert

---

EXERCISE 3 – 8   Description and Classification of Sedimentary Rocks

---

Examine the tray of sedimentary rocks provided by your instructor. Describe and classify each sample, and summarize your data on the accompanying table. Follow these steps in your analysis:

1. Separate rocks which are composed of discrete grains from rocks which are crystalline.

2. Examine the granular rocks carefully.
   a. Identify granular limestones and dolomites with the dilute hydrochloric acid (HCl) provided by your instructor. Limestones fizz vigorously with cold HCl; dolomites fizz modestly if the acid is applied to a freshly exposed or scratched surface. Record your answers in the column labeled "Grain Composition."

      CAUTION   *HCl is highly reactive even when it is dilute, and it can cause serious burns if it comes into contact with clothes, skin, eyes, or mucous membranes. Handle it with great care, and wash your hands after its use.*

   b. Identify any skeletal or nonskeletal grain types which are apparent in any limestones and dolomites. Record your observations in the column labeled "Grain Type." If the rocks are recrystallized, indicate this condition in the same column.
   c. Describe the texture of any clastic rocks. Record your observations in the columns marked "Grain Size" and "Grain Sorting."
   d. Refer to the Mineral Identification Tables in the first chapter. Identify the detrital grain types in the clastic rocks and estimate their abundances. Record your observations in the column marked "Grain Composition."
   e. Identify any carbonaceous rocks by examining the samples for recognizable plant debris.

3. Examine the crystalline rocks carefully.
   a. Identify any travertine, dolomite, and crystalline limestone with the dilute HCl. Record your findings in the column labeled "Mineralogy." You may also be able to distinguish travertine from crystalline limestone by looking for color banding, which is common in travertine.
   b. Identify any bituminous coal on the basis of its color and hardness. Record your findings in the column labeled "Mineralogy."
   c. Refer to the Mineral Identification Tables. Identify the compositions of the remaining samples, and record your findings in the column labeled "Mineralogy."
   d. Describe the textures of the crystalline rocks as either coarse or fine, and record this information in the column labeled "Texture."

4. Name each sample, and record your answer on the table.

5. For each sample, list all of the depositional environments in which it possibly could have accumulated.

GRANULAR ROCKS				CRYSTALLINE ROCKS			
Grain Size	Grain Sorting	Grain Composition	Grain Type	Mineralogy	Texture	Rock Name	Depositional Environments

---

EXERCISE 3-9 Composition of Sandstones

Plot the following compositional data on the ternary diagram in Figure 3–7 and classify each sandstone.

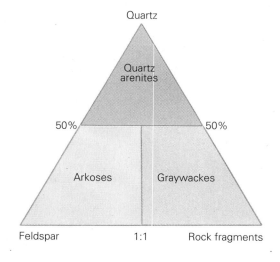

SANDSTONE NO.	QUARTZ	FELDSPAR	ROCK FRAGMENTS	COMPOSITIONAL CLASS
1	65%	15%	20%	
2	10%	75%	15%	
3	10%	10%	80%	
4	30%	30%	40%	

4

METAMORPHIC ROCKS

KEY WORDS

Metamorphism . . . recrystallization, neomorphism, and metasomatism . . . foliation and metamorphic minerals . . . metamorphic grade . . . slate, phyllite, schist, and gneiss . . . contact, regional, burial, and hydrothermal metamorphism . . . juvenile and metamorphic waters

FOLIATED MICA SCHIST

Materials Needed *Colored pens or pencils, lead pencils with erasers, hand lens or magnifying glass*

Metamorphic rocks are the products of the alteration (**metamorphism**) of rocks by elevated temperatures, pressures, and directed stresses. Metamorphism can occur throughout the Earth's crust, mantle, and core, and it can alter igneous, sedimentary, and even older metamorphic rocks. It is an important process along convergent plate margins (both collision and subduction types), where vast volumes of rock are metamorphosed by the high temperatures and stresses of tectonic deformation. It is also common to deeply buried rocks in sedimentary basins, to country rocks around hot magma chambers, and to rocks beneath surficial lava flows.

Metamorphism is a **solid-state reaction** which usually occurs at temperatures of 200° to 800°C and pressures of 2 to 15 kilobars. Minerals generally do not melt under such conditions, but their crystal structures can be rearranged and their compositions changed by chemical reactions with

other minerals and fluids in their parent rocks. Such metamorphic reactions alter the textures of rocks and produce a distinctive set of **metamorphic minerals** that are more stable under higher temperature and pressure conditions. Metamorphic reactions are often assisted by warm and chemically rich ground waters and other types of fluids which migrate through the many fractures, faults, and pores in crustal rocks and mobilize and exchange elements with them.

Geologists study the textures and compositions of metamorphic rocks because they reflect the original lithology of their parent rocks and the physical and chemical conditions of their metamorphic environment. Thus, they provide important evidence of the origin and tectonic evolution of the Earth's crust, and they serve as "paleo-thermometers" and "paleo-barometers" which record the temperatures and pressures of the crust in the geologic past. In addition, metamorphic rocks are important economic and energy resources. Metamorphism of bituminous coal, for instance, yields **anthracite coal**, a fossil fuel with high BTUs, and the mineral **graphite** (C), a natural lubricant and the "lead" in pencils. The metamorphic mineral **garnet** is a common abrasive in sandpaper and is a semi-precious stone. Gold, silver, and other metals form rich ore deposits in some metamorphic rocks.

This chapter begins with a review of metamorphic reactions and their products, and then continues with the description and classification of metamorphic rocks. It concludes with a consideration of the origins of metamorphic rocks, particularly the tectonic and other geologic processes which lead to metamorphism.

Metamorphism has profound effects on the mineral grains in a parent rock. It can recrystallize them, and change their sizes and shapes; it can strain and deform them, and reorient them in a new direction; and it can change their chemical composition and mineralogy. Consequently, texture and composition are the most important descriptive properties of metamorphic rocks.

METAMORPHISM AND ROCK TEXTURE

When a rock is exposed to the temperature and pressure conditions of metamorphism, the chemical bonds in its constituent minerals can be broken easily and rearranged into configurations which are more stable under those conditions. This rearrangement is generally accompanied by the growth of the mineral grains, as smaller grains are incorporated into the crystal structures of larger ones. The effect of this process of **recrystallization** is to coarsen the texture of the parent rock and create a dense network of interlocking grains within it.

Recrystallization is a common reaction during the metamorphosis of quartz arenite. Quartz arenites are typically composed of rounded quartz sand grains which come into contact with one another at **point contacts** along their edges (Fig. 4–1A). When quartz arenites are metamorphosed into **quartzites**, their grains recrystallize and become coarser and more angular, and they develop welded or **sutured** contacts with one another (Fig. 4–1B). These same textural changes occur in limestones when they metamorphose into **marble**.

Whenever mineral grains are recrystallized under conditions of high temperature but low pressure, they grow in random directions. When they are simultaneously heated and stressed by tectonic deformation, on the other hand, their growth is oriented perpendicular to the direction of the principal stress (Fig. 4–2). This parallel orientation of mineral grains produces a layered or banded appearance called **foliation**, which is a distinctive textural feature of many metamorphic rocks. Foliation is most obvious in rocks which contain platy and elongate grains such as biotite, chlorite, and hornblende. It is particularly common in metamorphosed shales and mudstones, which are rich in platy clay minerals.

Texture of Metamorphic Rocks

Two major classes of metamorphic rocks can be distinguished on the basis of texture: **nonfoliated** and **foliated** (Fig. 4–3).

Nonfoliated rocks are characterized by lack of foliation and by generally massive, structureless appearances. They are composed largely of interlocking grains of minerals such as quartz and calcite which are **equigranular** (that is, about the same size) and blocky and angular in shape. However, they may also contain stretched and deformed rock fragments and fossil clasts.

Foliated rocks are characterized by their layered and banded appearances. Four types of foliation textures can be distinguished on the basis of grain size and thickness of foliation: **slaty**, **phyllitic**, **schistose**, and **gneissic** (Fig. 4–4).

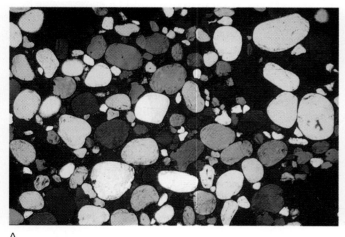

A

B

Figure 4–1 (A) Quartz grains in a quartzose sandstone. Note the roundness of the grains and the point contacts between them. (B) Quartz crystals in a quartzite. Note the blocky, angular shapes of the crystals and the sutured contacts between them.

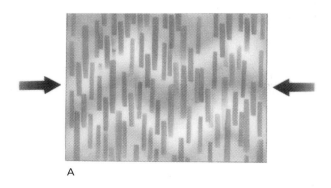

A

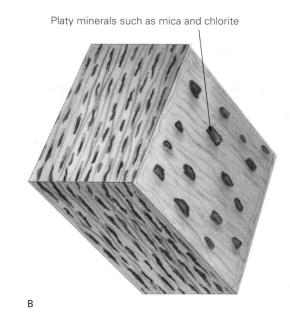

Platy minerals such as mica and chlorite

B

Figure 4–2 When platy and elongate minerals recrystallize under conditions of high stress, they grow perpendicular to the direction of the principal stress and parallel to one another to produce the layered texture called foliation.

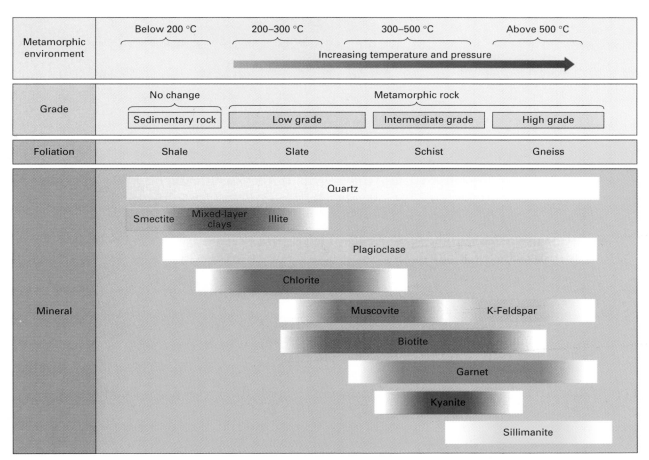

Figure 4–3 The foliation textures and compositions of metamorphic rocks of different grades.

A

B

C

D

E

F

Figure 4–4 (A) Slate. (B) Phyllite. (C) Schist. (D) Gneiss. (E) Quartzite. (F) Marble. (G) Hornfels with porphyroblasts. (H) Metaconglomerate. (I) Anthracite coal. *(Jeffrey Scovil)*

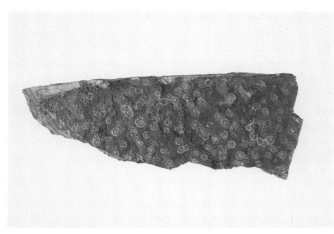

G

H

I

- Slaty textures are characterized by very fine-grained minerals which generally are not visible to the naked eye, and extremely thin, planar, and parallel foliation bands. Rocks with slaty textures (called **slates**) make excellent

blackboards and tabletops because they tend to split into very thin and perfectly flat slabs. This tendency is called **slaty cleavage**.

- Phyllitic textures are characterized by fine- to medium-grained minerals which are visible (but often barely) to the naked eye and by thin, wavy, and parallel foliation bands. Rocks with phyllitic textures (called **phyllites**) often have a distinctly silky and shiny luster due to the reflection of light from mica grains.

- Schistose textures are characterized by medium- to coarse-grained minerals which are clearly visible to the naked eye and by moderately thick and planar foliation bands. Rocks with schistose textures (**schists**) also have a shiny and silky luster due to their mica content.

- Gneissic textures are characterized by coarse-grained minerals and thick foliation bands. The foliation bands typically alternate in composition and color, with minerals such as quartz and feldspar forming the lighter-colored bands and minerals such as biotite and hornblende forming the darker-colored bands. The foliation bands of rocks with gneissic textures (**gneisses**) are sometimes deformed into broad, large **folds** and small (< 1 cm) **crenulations** by the intense pressures of metamorphism.

Foliated and nonfoliated rocks may also contain unusually large crystals called **porphyroblasts**. Porphyroblasts can be composed of garnet, staurolite, and other minerals, and they often have euhedral crystal shapes.

METAMORPHISM AND ROCK COMPOSITION

The compositions of metamorphic rocks are largely determined by the compositions of their parent rocks. Consequently, the most common minerals in metamorphic rocks are also the most common minerals in igneous and sedimentary rocks: quartz, feldspar, amphibole, pyroxene, muscovite, biotite, calcite, and so on.

However, metamorphism can also cause profound changes in the mineralogy of a parent rock. This alteration can be the result of two processes: **neomorphism** and **metasomatism**.

Neomorphism is the rearrangement of the existing elements in a mineral and the formation of a new mineral with the same chemical composition but a different crystal structure. Neomorphism is the typical metamorphic reaction of clay minerals, which are abundant in sedimentary parent rocks. Clay minerals tend to recrystallize into the mica minerals **chlorite**, **muscovite**, and **biotite micas**, which are among the most common minerals in metamorphic rocks.

Metasomatism is a chemical interaction between minerals and crustal fluids in which the compositions of the minerals are altered by the exchange or substitution of ele-

67

ments. It results in the formation of a fairly distinctive suite of **metamorphic minerals** including **staurolite**, **graphite**, **garnet**, and **wollastonite**.

CLASSIFICATION OF METAMORPHIC ROCKS

Foliated metamorphic rocks are named primarily on the basis of texture, using the terms **slate**, **phyllite**, **schist**, and **gneiss**. The names of coarser-grained foliated rocks (schists and gneisses) are often qualified by descriptions of the most abundant minerals in the rocks (e.g., **muscovite schist**) or the composition of less common but distinctive metamorphic or porphyroblastic minerals (e.g., **garnet schist**) (Table 4–1).

Nonfoliated metamorphic rocks are named primarily on the basis of composition, using terms such as **marble**, **amphibolite**, and **quartzite**. They may also be described further on the basis of texture, using terms such as **coarse grained** or **fine grained** (Table 4–1).

TABLE 4–1 Common Metamorphic Rocks and Their Properties

	TEXTURE	ROCK PROPERTIES	ROCK NAME
FOLIATED	Slaty	Very fine grained; minerals too fine to be seen with the unaided eye; very thinly foliated, and tends to split into thin planar slabs (*slaty cleavage*); dull, earthy luster; typically gray and black, sometimes red and green; product of low-grade metamorphism of shale and mudstone	Slate
FOLIATED	Phyllitic	Fine to medium grained; minerals barely visible to the eye; thin, wavy, parallel foliation; silky to shiny luster due to reflection from mica flakes; product of low- to intermediate-grade metamorphism of shale, mudstone, and slate	Phyllite
FOLIATED	Schistose	Medium to coarse grained with plainly visible grains of mica, garnet, staurolite, and other metamorphic minerals; moderately thick planar foliation; typically has silky to shiny luster due to reflection from mica flakes; product of intermediate- to high-grade metamorphism of siltstone and silty sandstone, shale, slate, phyllite, basalt, and granite; varieties include: • *serpentinite*: composed largely of the mineral serpentine and some talc, magnetite, and chlorite; serpentine may be in the fibrous form of asbestos; dark green and black snake-skin color and slippery feel; product of metamorphism of peridotite • *soapstone*: composed largely of the mineral talc and possibly chlorite and mica; may also have massive nonfoliated texture; very soft rock with soapy feel; typically light colored with a dull, earthy luster; also a product of metamorphism of peridotite	Schist
FOLIATED	Gneissic	Coarse grained with thick foliation bands which may be folded; alternation of light bands of quartz and feldspar with dark bands of biotite and ferromagnesian minerals is common; product of high-grade metamorphism of shale, schist, granite, and many other rocks	Gneiss
NONFOLIATED		Composed of elongate, angular quartz grains which have intergrown and welded together into a hard, dense, massive crystalline rock; recrystallized mica flakes are also common; product of intermediate- to high-grade metamorphism of quartz arenite	Quartzite
NONFOLIATED		Contains gravel clasts which have been stretched, deformed, and welded together into a hard, dense rock; product of low- to intermediate-grade metamorphism of conglomerates, particularly pebbly rocks rich in soft, pliable rock fragments	Metaconglomerate
NONFOLIATED		Composed of coarse, angular crystals of calcite or dolomite which have intergrown and welded together into a dense crystalline rock; texture often described as *sugary*; soft rock, and reactive to HCl; white, gray, pink, and brown and sometimes contains dark organic-rich bands and streaks; product of metamorphism of limestone and dolomite	Marble
NONFOLIATED		Composed of crystalline carbon and traces of clay and pyrite; blue-black with shiny and glassy luster; hard, dense rock which fractures conchoidally; product of low-grade metamorphism of bituminous coal and lignite	Anthracite Coal
NONFOLIATED		Composed of crystalline carbon; steel gray to iron black with dull to shiny luster; very soft with greasy feel; leaves streak on paper; product of intermediate- to high-grade metamorphism of anthracite coal	Graphite
NONFOLIATED		Composed of fine-grained quartz, plagioclase, pyroxene, and various metamorphic minerals; hard and dense; dark gray, green, and black; product of low- to intermediate-grade *contact* metamorphism of shale, basalt, and gabbro	Hornfels
NONFOLIATED		Composed of coarse-grained amphibole, plagioclase, and various metamorphic minerals; hard and dense; dark gray to black; product of intermediate- to high-grade metamorphism of basalt and gabbro	Amphibolite

EXERCISE 4–1   Metamorphism of Sedimentary Rocks

Your instructor will provide you with samples of either (a) marble and its parent rock, limestone, or (b) quartzite and its parent rock, quartz arenite.

1. Examine the samples with a hand lens or binocular microscope. Describe and illustrate their textures, including the sizes and shapes of their grains and the contacts between them.

2. What changes occur in the texture of a limestone or quartz arenite when it is metamorphosed?

EXERCISE 4–2   Metamorphic Minerals

The following table lists the names of some of the most common or most important minerals in metamorphic rocks. Describe the hardness, crystal habit, cleavage or breakage pattern, color, and any other important diagnostic properties of each mineral. Refer to Table 1–1 for this information.

MINERAL	HARDNESS	CRYSTAL HABIT	CLEAVAGE OR FRACTURE	COLOR AND LUSTER	OTHER DIAGNOSTIC PROPERTIES
Chlorite					
Muscovite					
Biotite					
Garnet					
Staurolite					
Epidote					
Andalusite					
Talc					
Graphite					
Wollastonite					
Actinolite					
Sillimanite					
Tremolite					

Examine the metamorphic rock samples provided by your instructor. Describe and classify each sample using the information in Table 4 – 1. Follow these steps in your analysis:

1. Separate foliated from nonfoliated rocks.

2. Determine the composition of the nonfoliated rocks, and classify them:
   a. Separate the dark- and light-colored nonfoliated rocks.
   b. The dark-colored rocks are either anthracite coal, amphibolite, or hornfels. Coal is distinguished by its color, luster, fracture, and crystallinity; amphibolite and hornfels are distinguished by their crystal size and composition.
   c. The light-colored rocks are either quartzite, metaconglomerate, or marble. Marble can be identified by its color and reactivity to HCl; quartzite and metaconglomerate can be identified by their hardness and grain composition.
   d. Describe the mineralogy of any porphyroblasts in any of the samples.

3. Examine the foliated rocks.
   a. Describe their texture and luster. Classify them as either slaty, phyllitic, schistose, or gneissic.
   b. Describe the color and composition of the major minerals in any schist and gneiss samples. Add this information to their classification.
   c. Describe the mineralogy of any porphyroblasts in any of the samples.

4. List all the possible parent rocks of each sample.

SAMPLE	TEXTURE	COMPOSITION	COLOR AND LUSTER	OTHER PROPERTIES	ROCK NAME	METAMORPHIC GRADE	PARENT ROCK

METAMORPHIC GRADE

Metamorphic reactions can occur in many different parts of the crust and upper mantle, and under many different temperature and pressure (T-P) conditions. This variability in the T-P conditions of different metamorphic environments leads to different intensities of metamorphism, ranging from **low grade** to **high grade** (Fig. 4–5).

Low-grade metamorphism occurs in the upper 10 km of the Earth's crust, where temperatures are generally less than 300°C and pressures are less than 3 kb. High-grade metamorphism occurs in the lower crust and upper mantle, approximately 20 to 55 km below the surface, where temperatures range from 500° to 800°C and pressures reach 12 kb or more. It also occurs in shallower parts of the crust around the margins of igneous intrusions, where temperatures are high but pressures are fairly low.

The textures of metamorphic rocks are good indicators of the T-P conditions in their metamorphic environment and thus function as "paleo-thermometers" and "paleo-barometers" for the Earth's crust. For example, slate is usually indicative of low-grade metamorphism; phyllite and schist, intermediate-grade metamorphism; and gneiss, high-grade metamorphism. Metamorphic rocks also contain many minerals, including **chlorite**, **muscovite**, **biotite**, **garnet**,

staurolite, kyanite, and sillimanite, which recrystallize under very limited T-P conditions. Such **index minerals** are also reliable indicators of metamorphic grade (Fig. 4–3). For example, chlorite is formed and stable only under low-grade T-P conditions, and it decomposes and recrystallizes into muscovite or biotite when the metamorphic grade rises to higher levels. Consequently, the presence of chlorite in a metamorphic rock is indicative of metamorphism in a fairly shallow crustal setting, where temperatures are less than 300°C and pressures are usually less than 3 kb.

METAMORPHIC PROCESSES

There are three types of geologic processes which produce metamorphic rocks: **contact**, **regional**, and **hydrothermal metamorphism**. These three processes are distinguished on the basis of their origins, the types of metamorphic reactions which occur during them, and the volumes of rock which are metamorphosed by them.

Contact Metamorphism

Contact metamorphism describes the localized alteration of rock by hot magma and lava. It is a common phenomenon beneath the volcanic arcs of subduction zones, where large volumes of molten rock are constantly being generated (Fig. 4–6). It also affects the surficial rocks which lie directly below extrusive lava flows.

Contact metamorphism is accomplished by two agents: the heat of magmas and lavas, and the crustal fluids (**juvenile waters**) which they release as they cool. High stress is not an especially important agent of contact metamorphism, for it usually occurs in the upper crust or on the surface of the Earth. Consequently, contact-metamorphosed rocks are typically nonfoliated. They are also quite varied in composition, because they can be formed from any type of rock—igneous, sedimentary, or metamorphic—that happens to be in contact with magma or lava.

Contact metamorphism affects a relatively small volume of rock, for it occurs only in the country rocks around igneous intrusions and the rocks directly beneath lava flows. The intensity of contact metamorphism is always highest at a rock–magma or rock–lava contact and diminishes rapidly from that contact, because rocks are generally poor conductors of heat. As a result, a **metamorphic aureole** or halo of metamorphic rocks of different grades forms around igneous rock bodies (Fig. 4–7). The thickness of this halo depends on several factors, including the volume and temperature of the magma or lava and the volume and composition of their juvenile waters.

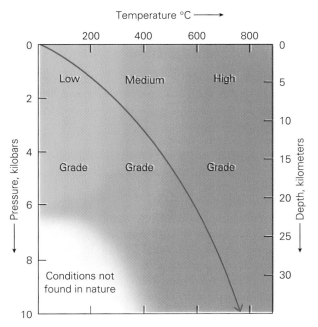

Figure 4–5 Temperatures, pressures, and depth ranges of the three metamorphic grades.

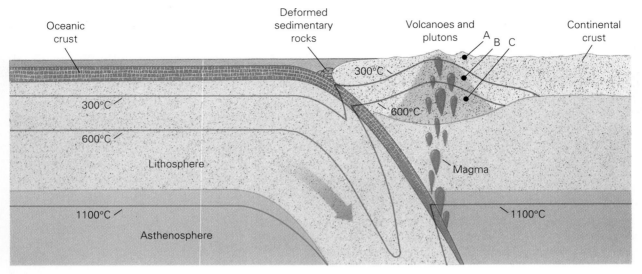

Figure 4–6 Metamorphic environments at a subduction zone. The heat of magmatism and the compressive stress of convergence result in regional dynamothermal and contact metamorphism in the shaded part of the overriding plate.

Regional Metamorphism

Regional metamorphism describes the alteration of broad regions of the Earth's crust and large volumes of rocks during tectonic deformation and rock burial. It can be the result of one of two processes: **regional dynamothermal metamorphism** or **burial metamorphism**.

Regional dynamothermal metamorphism occurs in and along the lengths of subduction zones and fold–thrust belts in convergent plate boundary settings (Fig. 4–6). It results from the high temperatures and pressures which develop when batholiths and other igneous intrusions are emplaced into the crustal rocks around subduction zones, and when

crustal rocks are deformed into mountain belts between two converging plates. Given that it results from the high temperatures *and* directed stresses of tectonic deformation, regional dynamothermal metamorphism produces high- to low-grade metamorphic rocks which are usually foliated.

Burial metamorphism occurs in the deepest parts of basins, where sedimentary rocks are metamorphosed by the heat of the Earth's interior. It usually occurs at depths greater than 2 km, where temperatures range from 50° to 300°C, and it is usually not accompanied by any tectonic deformation. Consequently, burial metamorphism typically produces low-grade nonfoliated metamorphic rocks.

Hydrothermal Metamorphism

Hydrothermal metamorphism is primarily the product of the chemical reactions between crustal rocks and fluids. The fractures, faults, joints, and pores of crustal rocks are often permeated by warm, chemically rich fluids which are derived from three sources: the downward percolation of ground water from the surface and upper crust; the release of **juvenile waters** from magmas as they cool; and the release of **metamorphic waters** from crustal rocks as they are heated and pressured during metamorphism.

Hydrothermal metamorphism often results in the dissolution of feldspar grains and other minerals which are rich in soluble elements such as K, Na, Ca, and Mg. The exchange of these dissolved elements with more stable minerals can also result in the neomorphic formation of clay minerals, biotite and muscovite micas, feldspars, and other types of metamorphic minerals. The dissolution of silicate minerals also releases large volumes of silica, which is often precipitated in fractures and joints to form **quartz veins**. These quartz veins are often important ore deposits, because they can also contain significant concentrations of precious metals such as gold, silver, copper, lead, and zinc.

Figure 4–7 A contact metamorphic aureole around a pluton.

EXERCISE 4-4   Metamorphic Grade

1. Return to the table in Exercise 4–3. Characterize the metamorphic grade of each of the rock samples as low, intermediate, or high.

2. Figure 4–8 is a simplified geologic map of an ancient orogenic belt in the Scottish Highlands. The bedrock between the two faults consists of late Precambrian and Cambrian mudstones which were folded and regionally metamorphosed during the late Cambrian and Ordovician periods. The metamorphic minerals in these formations are shown on the map.

   a. Locate the highest-grade metamorphic zone.
      1. Locate the approximate center of this zone, and write the letter H there.
      2. Estimate the maximum temperature of metamorphism in this zone.
         Record your answer here _____.
   b. Locate the lowest-grade metamorphic zone.
      1. Locate the approximate center of this zone, and write the letter L there.
      2. Estimate the maximum temperature of metamorphism in this zone.
         Record your answer here _____.
   c. What was the direction of the tectonic stresses which deformed and metamorphosed these rocks? Explain your answer.

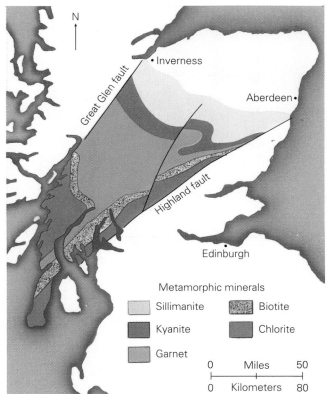

Figure 4–8 Zonation of metamorphic minerals in the late Precambrian and Cambrian metamorphic rocks of the Scottish Highlands.

   d. The thermal gradient is a measure of the change in the temperature of metamorphism across a metamorphic belt. Calculate the thermal gradient for the metamorphic rocks of Scotland by following these steps:
      1. Measure the distance in kilometers between points H and L.
      2. Measure the difference in estimated maximum temperature between the high-grade and low-grade metamorphic zones.
      3. Divide the temperature difference by the distance between the two points.
         Express your answer in degrees Celsius per kilometer here: _____

1. Examine the illustration of a subduction zone in Figure 4–6.
    a. Use different colored pencils to shade the zones of low-grade, intermediate-grade, and high-grade regional metamorphism in the continental plate.
    b. Locate point A. Complete the following table by describing (1) the types of metamorphic rocks which could form at this location and (2) the metamorphic minerals which they could contain, given the following parent rocks:

PARENT ROCK	METAMORPHIC ROCKS	METAMORPHIC MINERALS
Shale		
Limestone		
Bituminous coal		

    c. Locate point B, and complete the following table in a similar fashion:

PARENT ROCK	METAMORPHIC ROCKS	METAMORPHIC MINERALS
Shale		
Peridotite		
Conglomerate		
Basalt		

    d. Locate point C, and complete the following table:

PARENT ROCK	METAMORPHIC ROCKS	METAMORPHIC MINERALS
Shale		
Quartz arenite		
Basalt		
Granite		

2. Examine the illustration of a contact metamorphism halo in Figure 4–7.
    a. Locate point D. Complete the following table by describing (1) the types of metamorphic rocks which could form at this location and (2) the metamorphic minerals which they could contain, given the following parent rocks:

PARENT ROCK	METAMORPHIC ROCKS	METAMORPHIC MINERALS
Shale		
Quartz arenite		
Basalt		

    b. Locate point E. Complete the following table in a similar fashion:

PARENT ROCK	METAMORPHIC ROCKS	METAMORPHIC MINERALS
Shale		
Quartz arenite		
Basalt		

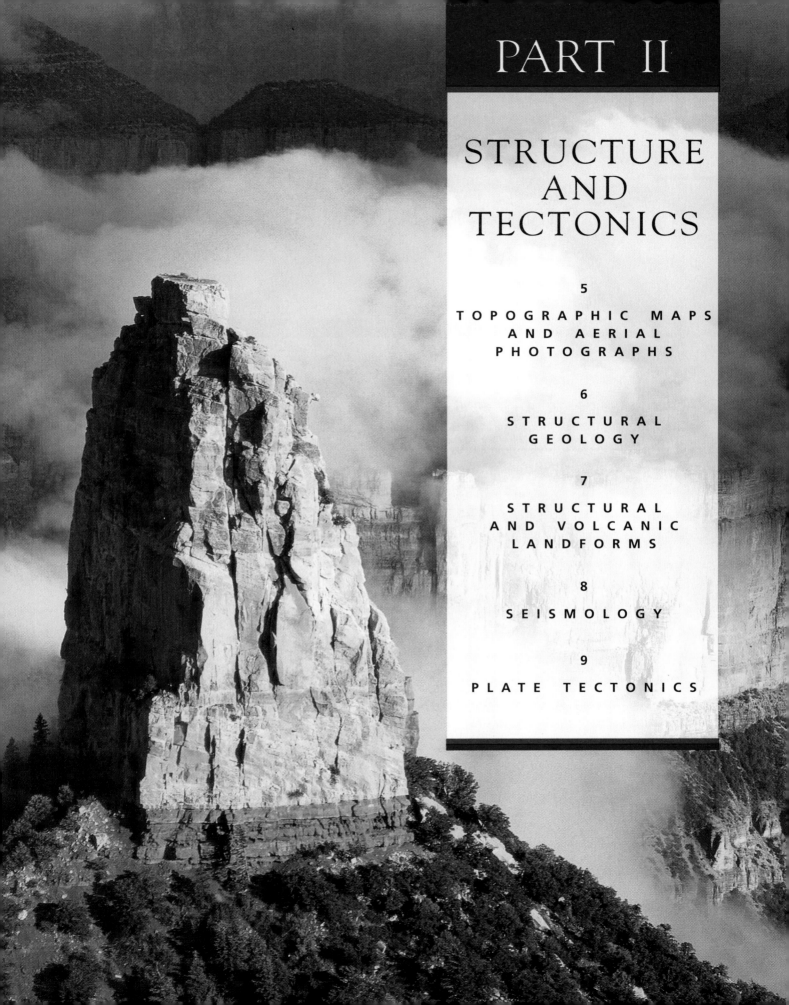

PART II

STRUCTURE AND TECTONICS

5

TOPOGRAPHIC MAPS AND AERIAL PHOTOGRAPHS

KEY WORDS

Topography . . . plan view and quadrangles . . . fractional, verbal, and graphic scales . . . bearings, true north, and magnetic north . . . latitude, longitude, equator, and prime meridian . . . townships, ranges, sections, quarters, and tracts . . . elevation, relief, and slope . . . contour lines, contour intervals, and index contours . . . topographic profiles and vertical exaggeration . . . stereophotographs and stereoscopes

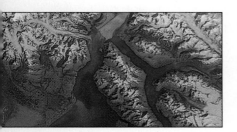

HUBBARD GLACIER

Materials Needed *Colored pens or pencils, lead pencils with erasers, ruler, protractor, calculator, stereoscope*

Topographic maps and aerial photographs are reduced images of the surface of the Earth which show the shape and configuration (**topography**) of the land, the density of its vegetation cover, and the locations of beaches, streams, lakes, buildings, roads, bridges, and other natural and cultural features. Geologists use these images extensively to study landforms and to construct geologic maps of the bedrock lithology and structure. However, they also serve many other purposes. They are used, for example, by engi-neers to plan construction; by agricultural scientists to survey soils and by farmers to estimate crop yields; by military forces to plan troop movements on the ground and along the coasts; and by hikers, bikers, hunters, and camp-ers to locate themselves in the field, among other things.

This laboratory manual utilizes many topographic maps and aerial photographs of different parts of the United States to illustrate landforms and their origins. The maps were pro-duced by the United States Geologic Survey (USGS), which has been constructing them according to uniform standards since 1882. The aerial photographs were provided by the Earth Resources Observation System (EROS) Data Center, which is the national repository for aerial pho-tographs of the United States.

This chapter instructs students on the basic skills which are necessary to read topographic maps and aerial pho-tographs. It begins with an examination of the components of topographic maps and the construction of topographic profiles, and concludes with an introduction to aerial pho-tographs and the study of stereophotographs. Students will gain the experience which is necessary to interpret topo-graphic maps and aerial photographs in subsequent chapters in this laboratory manual.

Topographic maps are graphic representations of the three-dimensional shape and configuration of the rough surface of the Earth. Like many other types of maps, they provide a **plan view** of the land (i.e., the perspective of someone who is looking straight down at the ground from an airplane or satellite). In addition, they employ a standard system of **scales** to measure ground distances, **bearings** to indicate directions, **coordinate systems** to locate points and features, and symbols to represent natural and cultural features (Table 5–1). Topographic maps differ from other maps in their use of **contour lines** to show the **elevation, slope,** and **relief** of the landscape.

SCALE

A topographic map shows a reduced image of a square or rectangular parcel of land called a **quadrangle.** The degree of reduction is indicated by the **scale,** which is the ratio between a distance on a map and the corresponding distance on the ground. Three types of scales are commonly used on topographic maps: <u>**fractional, verbal,** and **graphic.**</u>

A <u>fractional scale</u> is a numerical expression of the ratio between distances on the map and distances on the ground. A fractional scale of 1:1000, for example, indicates that 1 unit of distance (English or metric) on the map represents 1000 equal units of distance on the ground. For instance, a map distance of 1 in. would represent 1000 in. (approximately one fifth of a mile) on the ground, whereas a map distance of 5 cm would represent 5000 cm (or 50 m) on the ground.

A <u>verbal scale</u> is a written description of the relationship between map distances and ground distances (in that order). For example, it may read "1 cm to 1 km"—that is, 1 centimeter on the map equals 1 kilometer on the ground.

A graphic scale is a calibrated bar or line which is divided into units representing ground distances. This is perhaps the most convenient form of a scale, and it is used on most of the maps (topographic and others) in this laboratory manual.

For practical purposes, the scales of topographic maps vary with the actual size of the map area (Fig. 5–1). Topographic maps of the small island of Puerto Rico, for example, are produced at **1:20,000** scale (i.e., 1 cm = 200 m), whereas maps of the 48 conterminous states and Hawaii are produced with scales of **1:24,000** (1 in. = 2000 ft) and **1:62,500** (1 in. = 1 mile approximately). Topographic maps of Alaska, Antarctica, and large sections of the United States, on the other hand, are produced at scales which range from 1:62,500 to **1:1,000,000** (1 in. = 16 miles).

There are advantages and disadvantages to these different scales of presentation. A **large-scale map**—for example, a map with a scale of 1:24,000—illustrates a fairly small area of land, but it provides a great amount of detailed information about its topography. On the other hand, a **small-scale map**—for example, a map with a scale of 1:62,500 or higher—shows a considerably greater area of land but offers less detailed information about its topography.

BEARINGS

A **bearing** is the direction between two points. Two important bearings are usually indicated on a topographic map: **true north** and **magnetic north**.

True north is the bearing within the map area toward the geographic **North Pole**, where the axis of the Earth intersects its surface (Fig. 5–2). Topographic maps are always oriented with true north toward their tops. True north can also be indicated with an arrow marked True North (TN) or more simply North (N).

Magnetic north is the bearing within the map area toward the northern pole of the Earth's magnetic field, which is presently located about 700 kilometers from the geographic North Pole in northern Canada. It is an important bearing because the magnetic north pole attracts the needles of compasses. Field geologists use this phenomenon to determine directions and measure the orientations of rock bodies and geologic structures.

There is an angular difference, or **magnetic declination**, between true and magnetic north because they are not located in the same place on the globe. This angle varies in direction and magnitude depending on the relative position of a given map area with respect to the magnetic north pole. It also varies temporally because the magnetic north field is not permanently fixed in position, but rather migrates slowly throughout geologic time. The directions of true and magnetic north for a given area, the angle between them, and the date on which these directions were measured are often indicated on topographic maps so that "true" or geographic directions can be calculated from magnetic-compass bearings (Fig. 5–2).

COORDINATE SYSTEMS

There are two coordinate systems which can be used to indicate the locations of specific points and features on the surface of the Earth: the **latitude–longitude system** and the **township-range system**.

TABLE 5-1 Topographic Map Symbols

Primary highway, hard surface . . . .

Secondary highway, hard surface . . . .

Light-duty road, hard or improved surface . . . .

Unimproved road . . . .

Road under construction, alinement known . . . .

Proposed road . . . .

Dual highway, dividing strip 25 feet or less . . . .

Dual highway, dividing strip exceeding 25 feet . . . .

Trail . . . .

Railroad: single track and multiple track . . . .

Railroads in juxtaposition . . . .

Narrow gage: single track and multiple track . . . .

Railroad in street and carline . . . .

Bridge: road and railroad . . . .

Drawbridge: road and railroad . . . .

Footbridge . . . .

Tunnel: road and railroad . . . .

Overpass and underpass . . . .

Small masonry or concrete dam . . . .

Dam with lock . . . .

Dam with road . . . .

Canal with lock . . . .

Buildings (dwelling, place of employment, etc.) . . . .

School, church, and cemetery . . . .

Buildings (barn, warehouse, etc.) . . . .

Power transmission line with located metal tower . . . .

Telephone line, pipeline, etc. (labeled as to type) . . . .

Wells other than water (labeled as to type) . . . . o Oil . . . . . o Gas

Tanks: oil, water, etc. (labeled only if water) . . . . ● ● ● Ⓦ Water

Located or landmark object; windmill . . . . o . . . . ¥

Open pit, mine, or quarry; prospect . . . . ✕ x

Shaft and tunnel entrance . . . . ▪ . . . . Y

Horizontal and vertical control station:

   Tablet, spirit level elevation . . . . BM △ 5653

   Other recoverable mark, spirit level elevation . . . . △ 5455

Horizontal control station: tablet, vertical angle elevation VABM △ 95l9

   Any recoverable mark, vertical angle or checked elevation △3775

Vertical control station: tablet, spirit level elevation . . . . . BM ✕ 957

   Other recoverable mark, spirit level elevation . . . . ✕ 954

Spot elevation . . . . ✕ 7369   ✕ 7369

Water elevation 670   670

Boundaries: National . . . .

   State . . . .

   County, parish, municipio . . . .

   Civil township, precinct, town, barrio . . . .

   Incorporated city, village, town, hamlet . . . .

   Reservation, National or State . . . .

   Small park, cemetery, airport, etc. . . . .

   Land grant . . . .

Township or range line, United States land survey . . . . . .

Township or range line, approximate location . . . .

Section line, United States land survey . . . .

Section line, approximate location . . . .

Township line, not United States land survey . . . .

Section line, not United States land survey . . . .

Found corner: section and closing . . . .

Boundary monument: land grant and other . . . . □ . . . . □

Fence or field line . . . .

Index contour	⌇	Intermediate contour	
Supplementary contour		Depression contours	
Fill		Cut	
Levee		Levee with road	
Mine dump		Wash	
Tailings		Tailings pond	
Shifting sand or dunes		Intricate surface	
Sand area		Gravel beach	

Perennial streams		Intermittent streams	
Elevated aqueduct		Aqueduct tunnel	
Water well and spring	o ⌒	Glacier	
Small rapids		Small falls	
Large rapids		Large falls	
Intermittent lake		Dry lake bed	
Foreshore flat		Rock or coral reef	
Sounding, depth curve	10	Piling or dolphin	o
Exposed wreck		Sunken wreck	
Rock, bare or awash; dangerous to navigation			

Marsh (swamp)		Submerged marsh	
Wooded marsh		Mangrove	
Woods or brushwood		Orchard	
Vineyard		Scrub	
Land subject to controlled inundation		Urban area	

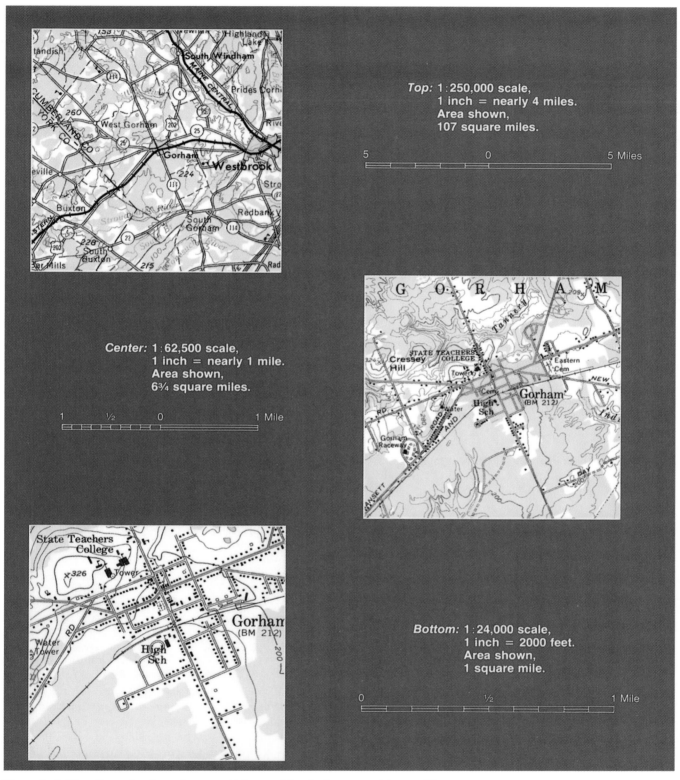

Top: 1:250,000 scale,
1 inch = nearly 4 miles.
Area shown,
107 square miles.

5   0   5 Miles

Center: 1:62,500 scale,
1 inch = nearly 1 mile.
Area shown,
6¾ square miles.

1   ½   0   1 Mile

Bottom: 1:24,000 scale,
1 inch = 2000 feet.
Area shown,
1 square mile.

0   ½   1 Mile

Figure 5–1 The effects of scale on the area and degree of detail of a topographic map. The 1:250,000 scale map covers the largest geographic area but provides relatively little detail about its topography. The 1:62,500 map shows the topography of a smaller area in more detail, and the 24,000 scale map shows the smallest geographic area but provides great detail about its topography.

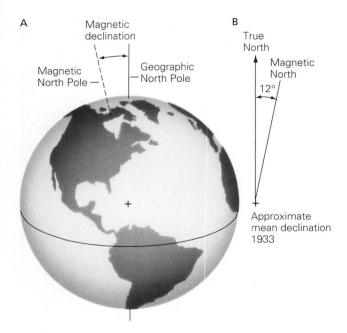

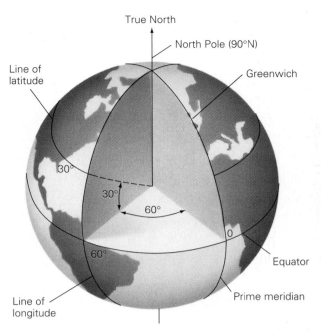

Figure 5–2 True and magnetic north poles and their declination. (A) True geographic north and magnetic north do not presently coincide. The angle between the Earth's true and magnetic axes is the magnetic declination. (B) The symbol used to show the directions of true and magnetic north and the magnetic declination on a topographic map.

Figure 5–3 The latitude–longitude coordinate system. Latitude indicates the vertical angle (0° to 90°) and relative location (north or south) between any point on the surface and the equator. Longitude indicates the horizontal angle (0° to 180°) and relative location (east or west) between any point and the prime meridian.

Latitude and Longitude

The surface of the Earth is divided by an imaginary coordinate grid consisting of lines of **latitude** and **longitude**. The lines of latitude (or **parallels**) run in the east–west direction, parallel to the Earth's equator. The lines of longitude (or **meridians**) run in the north–south direction from the North Pole to the South Pole (Fig. 5–3). The locations of specific points and features can be indicated accurately by their **coordinates** within this grid of imaginary lines.

Latitude and longitude are both measured in units of **degrees** (°), because they are angular measurements of the distances between a given point and two lines of reference: the **equator** and the **prime meridian**, which runs through the city of Greenwich, England. Latitude measures the angular distance to the north or south between any point and the equator, whereas longitude measures its angular distance to the east or west from the prime meridian. For greater accuracy of location, a single degree can be subdivided into 60 equal units called **minutes** ('), whereas a single minute can be subdivided into 60 equal units called **seconds** (").

There are 180° of latitude: 90° between the equator (0°) and the North Pole (90°N), and 90° between the equator and the South Pole (90°S). There are 360° of longitude: 180° to the east of the prime meridian (0°), and 180° to its west. The 180°W and 180°E meridians are one and the same line and mark the International Date Line.

Topographic maps are always bordered on the top (north) and bottom (south) by lines of latitude and on the left (west) and right (east) by lines of longitude. The latitude and longitude coordinates are usually inscribed at the four corners and along the edges of the maps. They are also constructed at various sizes, but the most common are **7.5' quadrangles** (which extend 7.5' of latitude from north to south and 7.5' of longitude from east to west) and **15' quadrangles** (which represent an area 15' of latitude by 15' of longitude).

Township-Range System

The **township-range system** (or Public Land Survey System) was established by the Land Ordinance Act of Congress in 1785 to subdivide the land west of the Ohio and Mississippi rivers. This system consists of a coordinate system of **township lines** and **range lines**, which partition the western states into 6-mile-square parcels of land called **townships** (Fig. 5–4). The township and range lines are numbered according to their distances from specified lines of latitude, called **base lines**, and lines of longitude, called **principal meridians**, which have been established in every western state. The location of a township can thus be indicated by a **township number** (such as T2S), which defines its relative position north or south of the nearest base line,

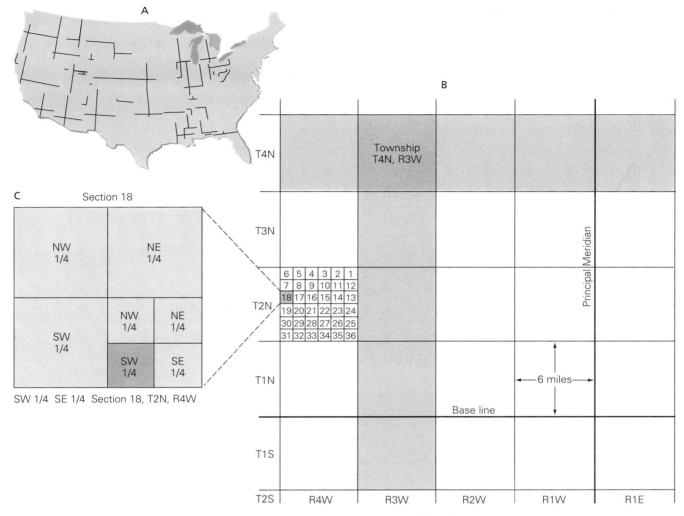

Figure 5–4 The township-range coordinate system. (A) The principal meridians and base lines in the United States. (B) The basic structure of the township-range system consists of 6-mile-square townships each of which is divided into 36 1-mile-square sections. The location of a township is indicated by its position north or south of a base line (T4N in the example) and east or west of a meridian (R3W in the example). (C) Division of a section into quarters and tracts.

and a **range number** (such as R3W), which defines its relative position east or west of the nearest principal meridian.

Each township is further divided into 36 **sections**, each of which is 1 mile square and contains 640 acres of land. Sections are numbered from 1 to 36 in a zig-zag pattern which begins in the northeast corner of the township and ends in its southeast corner.

Each section can be further divided into four **quarters** (the NW¼, NE¼, SW¼, and SE¼ quarters) of land, each of which consists of 160 acres. Similarly, each quarter can be subdivided into four **tracts** of land, each of which contains 40 acres.

The location of any feature with a township-range coordinate is typically given as a long string of terms such as SW¼, SE¼, Sect. 18, T2N, R4W. This particular string locates the shaded tract in Figure 5–4, in the southwest

quarter of the southeast quarter of Section 18, in the second row of townships north of the local base line and the fourth column west of the local principal meridian.

MAP SYMBOLS

Topographic maps use standard symbols to show the locations of physiographic features such as rivers, beaches, and levees; cultural features such as roads and buildings; and the type and density of vegetation. Generally, different colors are used to indicate different classes of features. For example, contour lines are usually printed in brown, water bodies in blue, vegetation in green, roads in red, and cultural features in black (Table 5–1).

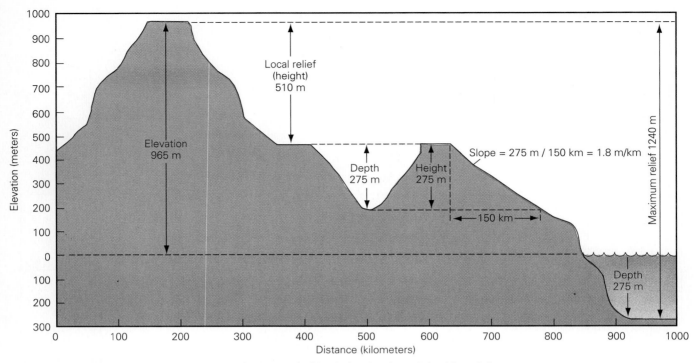

Figure 5–5 Definition and measurement of elevation, relief (including height and depth), and slope.

CONTOUR LINES

Topographic maps are distinguished from other types of maps in that they also show the elevation, relief, and slope of the land surface (Fig. 5–5) with **contour lines**.

Elevation (or **altitude**) is defined as the vertical distance between a given point and mean sea level, which is the standard reference level for all elevation measurements. It is expressed in English units (feet) on older (pre-1980s) topographic maps, and English and metric units on recent maps.

Relief is the difference in elevation between any two points. The term **local relief** describes the height or depth of a topographic feature (such as a peak or valley) relative to its immediate surroundings. *Maximum relief* refers to the difference in elevation between the highest and lowest points in a map area.

Slope is defined as the change in elevation over a given horizontal distance. It is calculated by dividing the relief between two points by the horizontal distance between them, and it is typically expressed in units of either feet per mile (ft/mi) or meters per kilometer (m/km).

Contour lines are imaginary lines which connect points of equal elevation (Fig. 5–6). A contour line marks the intersection between an imaginary horizontal plane with a specific elevation and the land surface. Topographic maps use a series of contour lines of different elevations to illus-

trate the elevation and shape of the land. The difference in elevation between successive lines is called the **contour interval** (CI), and it is generally a constant value throughout a given topographic map. Every contour line is therefore a multiple of the contour interval. For instance, if the CI is 25 ft, the contour lines mark elevations of 0′, 25′, 50′, 75′, and so on.

The contour interval which is used on a particular topographic map depends on the maximum relief within the map area. Where the maximum relief is high (for example, in mountain ranges), the contour interval may be 50 to 100 feet or more; where the relief is low (for example, in coastal plains), it may be 5, 10, or 20 feet.

Every fifth line in a series of contour lines is an **index contour** printed darker and thicker and marked along its length with its elevation (Fig. 5–7A). The elevation of any contour line can be determined by counting up or down from the closest index contour. The contour interval is usually recorded below the scale of a topographic map, but it can also be determined by calculating the difference in elevation between two successive index contours and dividing this value by 5.

Topographic maps also indicate the elevations of **benchmarks** (Fig. 5–7B) and prominent natural and cultural features such as mountain peaks, lake surfaces, and bridges (Fig. 5–7C). Benchmarks, represented by the letters BM,

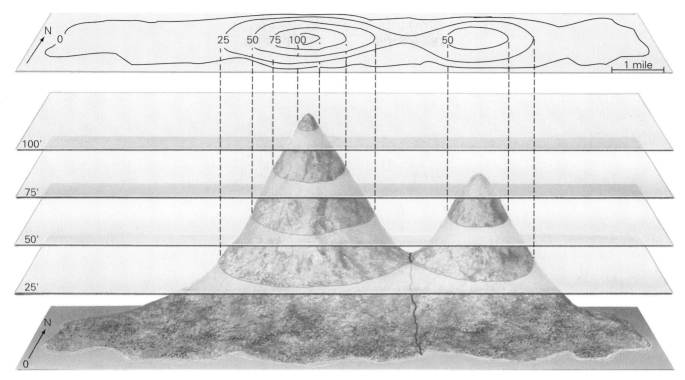

Figure 5–6 Construction of a topographic map. A contour line defines the intersection between a landform and a horizontal plane which lies at a specific elevation. Four such horizontal planes, spaced at a contour interval of 25 feet, are used to contour the island in this figure. The line of intersection on each plane is then projected upward from the land surface to the horizontal map surface.

are permanent metal markers which have been placed throughout the country by the U.S. Geological Survey and the Bureau of Land Management. They are periodically surveyed by ground crews and remote sensing techniques, so they can be used as reference points for elevation measurements in their immediate areas.

Some of the rules governing the reading and interpretation of contour lines are summarized in Table 5–2 and illustrated in Figure 5–7.

Note the extreme detail of this false-color imagery of Glacier Bay National Park, Alaska. The Brady Icefield covers much of the Fairweather Range and is the source of many glaciers. Large amounts of sediment are deposited in the sea by glacial meltwaters. *(NASA, LANDSAT 1057-19542)*

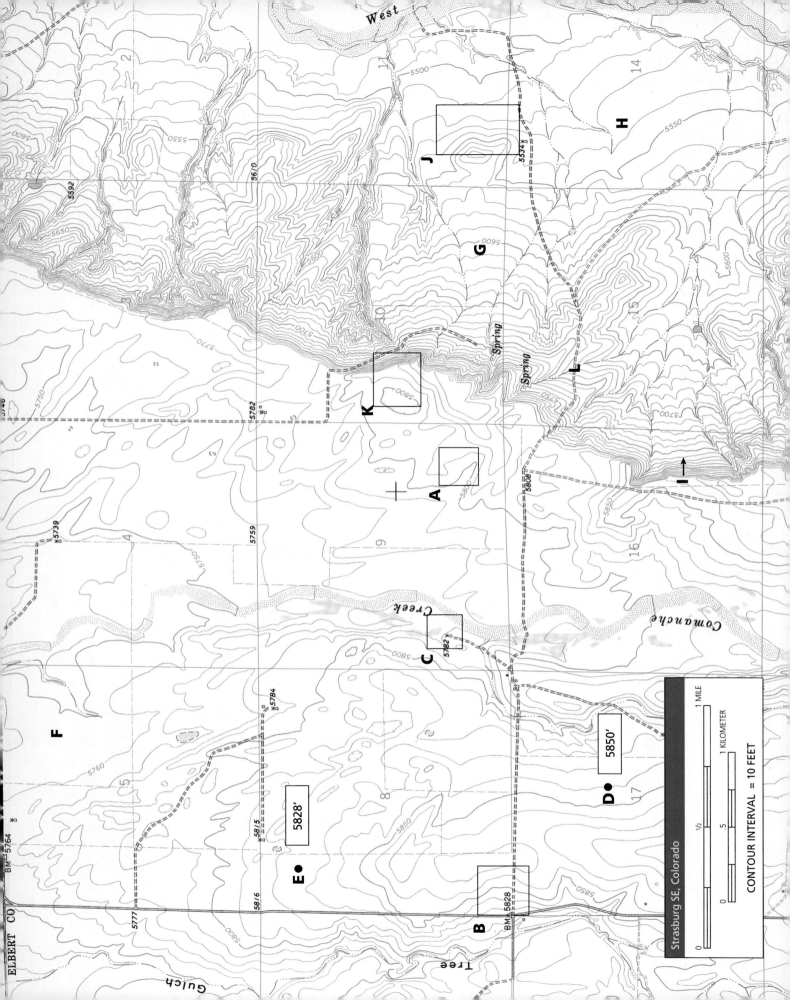

Strasburg SE, Colorado

CONTOUR INTERVAL = 10 FEET

TABLE 5-2 Reading Contour Lines

1. The elevation of any point which lies on a contour line is equal to the elevation of that contour line (Fig. 5-7D). The elevation of any point which lies between two contour lines must be estimated on the basis of its relative distances from the two lines (Fig. 7E).

2. The spacing between contour lines on a topographic map reflects the **slope** or **gradient** of the land—that is, the change in elevation over a given distance.
   - Level ground is not crossed by any contour lines (5-7F).
   - A uniform slope is represented by equally spaced contour lines (5-7G).
   - A gentle slope is represented by widely spaced contour lines (5-7H).
   - A steep slope is represented by closely spaced contour lines (5-7I).
   - A vertical cliff is represented by merged contour lines.

3. Contour lines do not usually cross, except at overhanging cliffs. There, the topography of the cleft beneath the cliff is indicated by **dashed** contour lines.

4. Contour lines always come to closure with themselves, forming rough circles and ellipses. However, this closure may not be apparent within the map area of a single topographic quadrangle, and many contour lines end abruptly at a map's edge.

5. Every point enclosed by a **solid** contour line is topographically higher than the line itself. Thus, solid contour lines enclose **topographic highs** such as mountain ridges and hills (5-7J).

6. Hachured lines are used to mark the contours of closed depressions with no drainage outlets, such as sinkholes and ponds (5-7K).
   - Every point enclosed by a **hachured** contour line is topographically lower than the line itself.
   - The outer hachured contour line around a topographic low has the same elevation as the closest solid contour line.

7. Contour lines which cross rivers and valleys form V's which point in the upstream direction (5-7L).

TABLE 5-3 Conversion Tables

LENGTHS

1 centimeter (cm)	=	0.01 m	=	0.394 in.
1 meter (m)	=	100 cm	=	3.28 ft
1 kilometer (km)	=	1000 m	=	0.621 mi
1 inch (in.)	=	2.54 cm	=	25.4 mm
1 foot (ft)	=	30.48 cm	=	0.305 m
1 mile (mi)	=	5280 ft	=	1.6 km
1 fathom	=	6 ft	=	1.8 m

AREAS

1 sq meter	=	10,000 sq cm	=	10.76 sq ft	=	1550 sq in.
1 sq cm	=	0.0001 sq m	=	0.001076 sq ft	=	0.1550 sq in.
1 sq ft	=	0.09290 sq m	=	929 sq cm	=	144 sq in.
1 sq in.	=	0.0006452 sq m	=	6.452 sq cm	=	0.00694 sq ft

ANGLES

1 degree (°)	=	60 minutes (')
1 minute (')	=	60 seconds (")

◀ **Figure 5-7** Topographic map of Strasburg SE, Colorado, with examples of contour lines and other map symbols.

Topographic Profiles

A topographic map is an overhead, or plan-view, perspective of the elevation, relief, and slope of the surface of the Earth. However, a different perspective on the topography of the surface can be gained by constructing a cross-sectional view, or **topographic profile**, through the surface itself.

A topographic profile is easily constructed by following these steps (Fig. 5–8):

1. Draw a straight line between the two points between which the profile is to be constructed (Fig. 5–8A).

2. Place the edge of a piece of paper along this line, and label its endpoints. Place tick marks on the paper to mark the intersections of contour lines and the profile line. Record the elevations of the contour lines at every tick mark (Fig. 5–8B).

3. Record the horizontal scale of the map in Figure 5–8A at the base of a piece of graph paper (Fig. 5–8C). In addition, construct an elevation scale along the vertical axis of the graph paper. The lowest elevation on this scale should be slightly less than the elevation of the lowest point along the profile line; the highest elevation should be slightly greater than the elevation of the highest point.

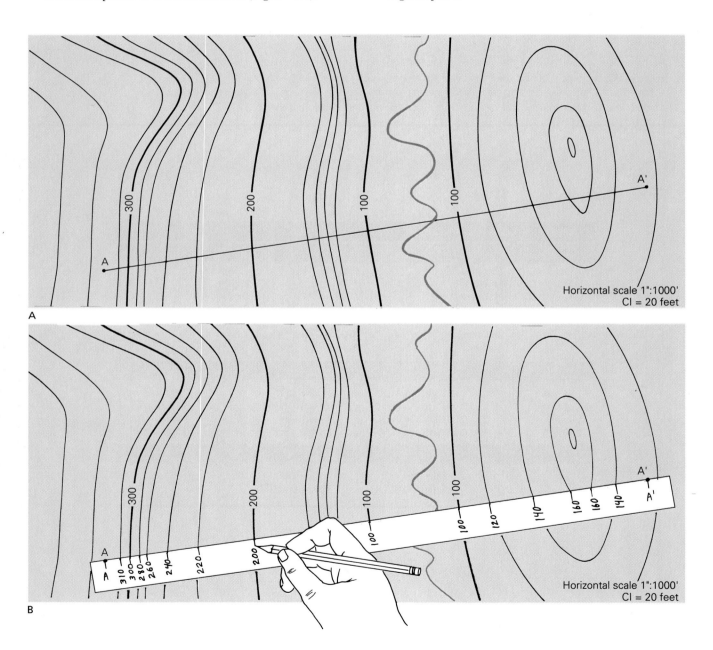

4. Lay the edge of the paper along the horizontal axis of the graph paper. Mark off the endpoints of the profile line and the tick marks, and record their elevations below them. Then project each tick mark straight up to its elevation level, and place a dot there (Fig. 5–8C).

5. Connect the dots with a smooth, curving line to form the topographic profile (Fig. 5–8C). You will have to estimate the slopes of the line segments between each pair of points on the topographic profile on the basis of the general shape of the land along the profile line.

The horizontal scale of a topographic profile is always equal to the horizontal scale on the topographic map. However, the difficult part of constructing a topographic profile is the choice of the elevation scale for the vertical axis: If the vertical scale is too small (see Fig. 5–8D), it will subdue the relief and slope of the land; if it is too large, it will exaggerate them.

Generally, it is necessary to have some degree of **vertical exaggeration** along a topographic profile to accentuate the relief and slope of the land surface, but it is also necessary to calculate and record the vertical exaggeration so that it is understood by a viewer that the topographic profile is indeed distorted. The vertical exaggeration (abbreviated **VE**) can be calculated by dividing the vertical scale by the horizontal scale. For instance, the vertical exaggeration of Figure 5–8C is calculated in the following manner:

$$VE = \frac{1''/100'}{1''/1000'} = \frac{1''}{100'} \times \frac{1000'}{1''} = \frac{1000}{100} = 10\times$$

Figure 5–8 Construction of a topographic profile from a topographic map. See text for instructions.

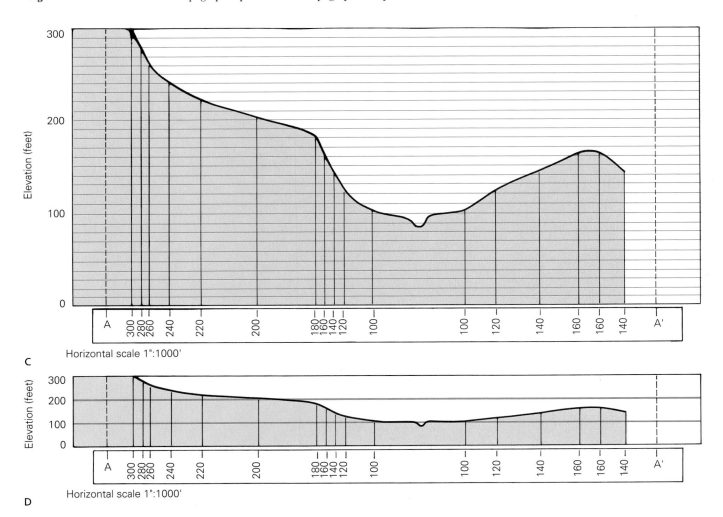

E X E R C I S E   5 – 1   Components of a Topographic Map

Topographic map sheets are large (several square feet), and they cannot be reproduced in their entireties on the pages of this laboratory manual. Rather, this book uses page-size or smaller segments of topographic sheets, most of which have (for the sake of space and clarity) been stripped bare of their latitude and longitude coordinates, true and magnetic north arrows, quadrangle titles, and other information except for their scales and contour intervals.

However, your instructor will provide you with an example of a full-size topographic map so that you can examine all its components. Answer the following questions about this map:

1. What is its scale (in verbal and fractional forms)? _____

2. How many feet are represented by 1 in.? _____ by 3 in.? _____ by 6 in.? _____

3. What lines of latitude form its northern and southern borders? _____

4. What lines of longitude form its eastern and western borders? _____

5. What is its size (in degrees, minutes, and seconds)? _____

6. What are its dimensions in miles in the north–south direction? _____ in the east–west direction? _____

7. Refer to the conversion charts in Table 5–3. What are its dimensions in kilometers in the north–south direction? _____ in the east–west direction? _____

8. What is the total area (in square miles) which is represented by this map? _____

9. Locate the title. What geographic area (city and state) does the map represent? _____

10. What is the magnetic declination? _____ What year was it measured? _____

11. What is the contour interval? _____

12. What is the highest point (in feet) in the map area? _____ the lowest point? _____

13. What is the highest point (in meters) in the map area? _____ the lowest? _____

14. What is the maximum relief in the study area, in feet? _____ in meters? _____

The map in Figure 5–9 contains several spot elevations (measured in feet) and the path of a stream. Contour these data at a contour interval of 10 ft.

To begin, study the map carefully. Locate the highest and lowest points in the map area and circle them. In addition, determine the general direction of slope of the land relative to these points.

The lowest contour line (110 ft) has already been traced on the map. There are no points which lie exactly at this elevation, but there are several points which lie slightly above and below it. Note that the 110-ft contour line has been placed closer to the 111-ft elevation mark than to the 106-ft mark, and that it forms a V when it crosses the stream.

Continue by adding the 120-ft contour and progressively higher contour lines in a similar fashion. Trace each line lightly; contouring is a trial-and-error process, and you will probably have to erase and change them as you go along.

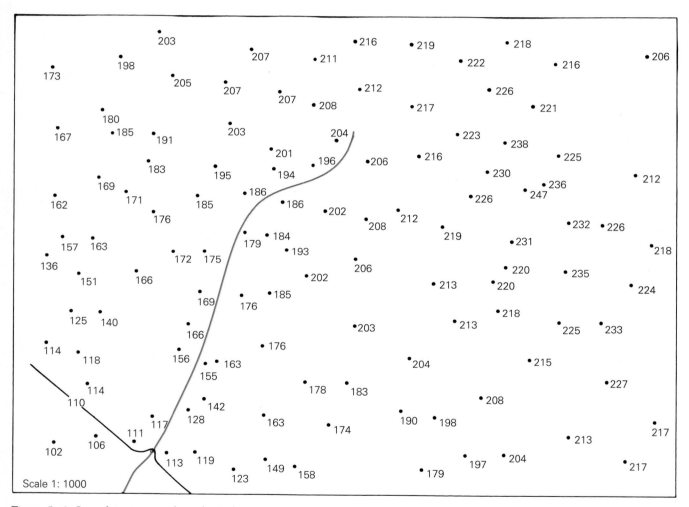

Figure 5–9 Spot elevations in a hypothetical map area.

EXERCISE 5–3   Topography and Topographic Profiles

The map in Figure 5–10 contains several contour lines and one benchmark elevation.

1. What is the contour interval of the lines? Label the elevations of the index contours in the spaces provided.

2. What is the lowest elevation in the map area? the highest? the maximum relief?

3. Construct a topographic profile along line ABCD. Use a vertical elevation scale of 1 in. = 100 ft.

4. Calculate the following:
   a. The vertical exaggeration of the profile
   b. The slope of the land between points B and C
   c. The slope of the land between points C and D

5. Describe the relationship between the slope and the spacing of contour lines.

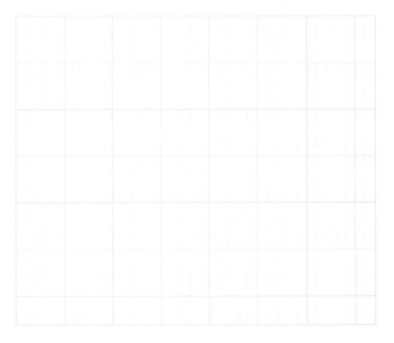

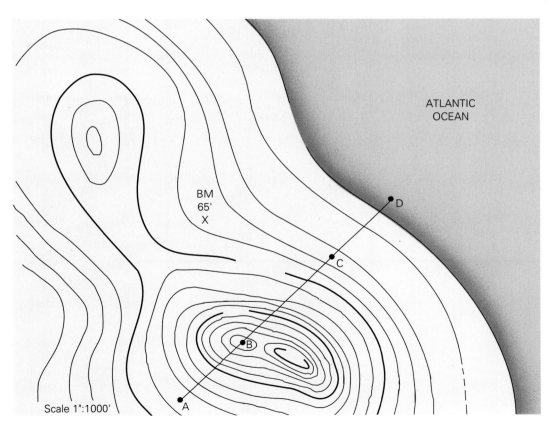

Scale 1":1000'

Figure 5–10 Contour lines and benchmark elevations in a hypothetical map area.

Examine the borders around topographic Map 5-1. Note that the latitude and longitude coordinates and township and range lines are inscribed along its margins.

1. What lines of latitude form the map's northern and southern borders? _____

2. What lines of longitude form its eastern and western borders? _____

3. What is the size (in degrees, minutes, and seconds) of this quadrangle? _____

4. What are the northernmost and southernmost township lines in the map area? _____

5. What are the easternmost and westernmost range lines in the map area? _____

6. Locate and shade with a colored pencil an example of a section. What are its township and range numbers? _____

7. Draw a scale in the lower-center part of the map.

8. What is the distance in feet and miles between the Evergreen Cemetery and the Loup City radio tower? _____

   What is this same distance in meters and kilometers? _____
   What is the compass direction between them? _____

9. What is the contour interval for this map? _____

10. Locate points A, B, C, and D on the map. Record the following information about each point.

	Elevation	Latitude–Longitude	Township–Range—Section—Quarter—Tract
A			
B			
C			
D			

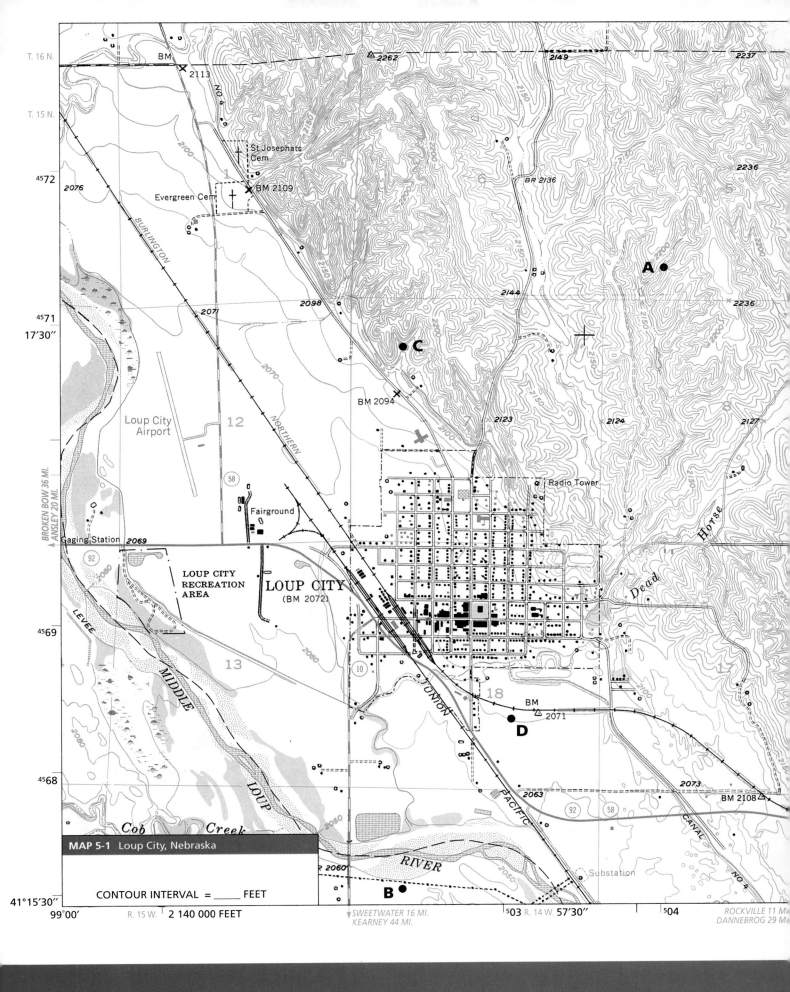

MAP 5-1 Loup City, Nebraska

CONTOUR INTERVAL = _____ FEET

11. Locate (by latitude–longitude and township-range) examples of the following:

	Latitude–Longitude	Township–Range—Section—Quarter—Tract
Railroad station		
Cemetery		
Lake		
Town		
Flood plain marsh		
School		
Highway intersection		
Benchmark		
Steep cliff		
Church		

12. What is the highest elevation in the map area? _____ the lowest? _____ the maximum relief? _____

13. Construct a topographic profile along a line which runs northeast from Loup City Airport to the center of section 6. Use a vertical scale of 1 in. = 100 ft. Calculate its vertical exaggeration, and indicate it on the profile.

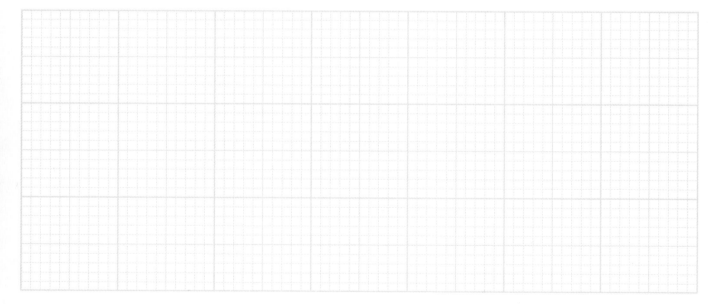

14. Locate and indicate the steepest-sloping segment along this transect. Calculate its slope.

The majority of the aerial photographs in this lab manual come from the National High Altitude Photography (NHAP) and National Aerial Photography Program (NAPP) collections at the EROS Data Center. These two series of images were taken from airplanes flying at altitudes of 20,000 to 40,000 ft along transects which systematically criss-crossed the entire United States. The spacing between exposures along these transects was sufficiently close so that there is approximately 60 percent overlap between adjacent images. The images are all **vertical** images, in which the center of the photograph is directly beneath the plane rather than at an **oblique** angle to it (Fig. 5–11). The remaining photographs were taken from the space shuttle and the Landsat satellites during Earth orbit.

Some of the color aerial photographs in this laboratory manual were taken with special lens filters and films which are sensitive to ultraviolet and infrared wavelengths of light (which are not visible to the human eye). This nonvisible radiation is recorded in **false-color images**, in which the true colors of surficial features are replaced with other colors. For example, the type and density of vegetation are usually indicated by various shades of red; water bodies are colored various shades of blue (light blue when they are muddy, dark blue when they are clear); and deserts and cities appear in dark gray and black shades. False-color images are also useful for identifying and mapping soil and rock types, prospecting for water and mineral resources, measuring areas of urbanization and deforestation, and other similar purposes.

STEREOPHOTOGRAPHS

The aerial photographs in this laboratory manual are usually presented in the form of **stereophotographs**, an example of which is shown in Figure 5–12. A stereophotograph consists of a pair or triplet of overlapping aerial photographs taken at the same altitude. When such overlapping images are viewed through a specially designed pair of glasses called a **stereoscope**, the slight difference in their angles of perspective (Fig. 5–12) allows the eyes to sense their depth and perceive the relief of the surface. The brain processes this information by merging the two separate flat images into a single three-dimensional image, giving the viewer the amazing perspective of looking straight down at the ground from an airplane.

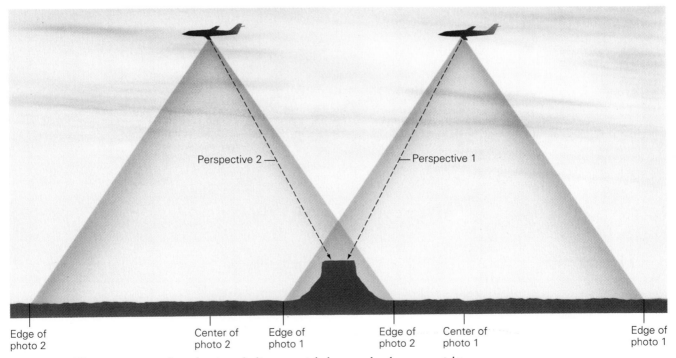

Figure 5–11 The perspective and overlapping of adjacent aerial photographs along an aerial transect. The two different perspectives on the feature of interest will produce a three-dimensional image when the photographs are viewed with a stereoscope.

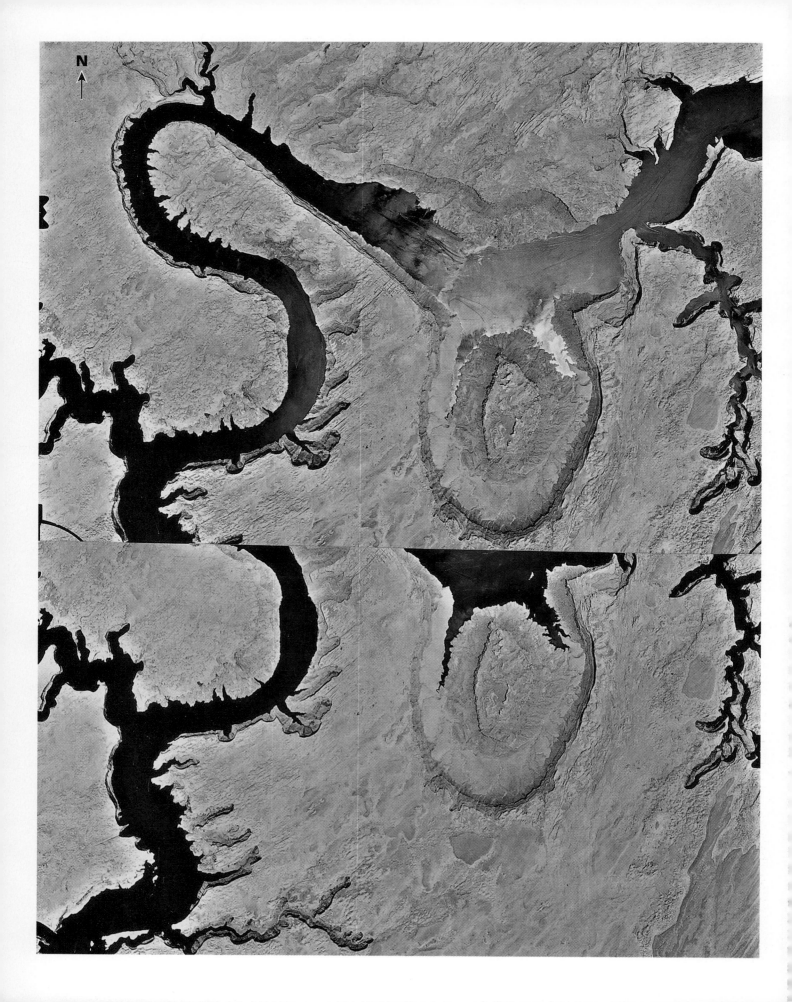

Your instructor will provide you with a stereoscope so that you can examine the stereophotographs in Figure 5–12. Follow these procedures:

1. Examine the photographs, and locate a prominent point or feature which is present in both images.
2. Place the center of the stereoscope directly over the line separating the two photographs, and adjust the two lenses so that they are both directly over the same point in each photograph.
3. Look through the stereoscope, relax your eyes, and allow the two images to merge. If the images do not merge immediately, twist the stereoscope slightly until a three-dimensional image appears. If the images still do not merge, check to make sure that the lenses are properly placed over the same point.

Once you have become proficient at the use of stereophotographs, you may want to try to achieve the same effect without the stereoscope. Simply hold the images at arm's length, fix your eyes on the same point in the two photographs, and move the images in and out slowly until they merge.

Scales

The NAPP and NHAP aerial photographs are printed and enlarged so that they have an approximate scale of 1:40,000. However, the exact scale varies with the focal properties of the camera lens (particularly its focal length) and the precise altitude at which the photographs were taken. Wherever the land is divided by the township-range system, the scale can usually be determined more precisely from the sizes of sections, which are 1 mile square. It can also be determined by comparing the distance between two topographic or cultural features on an aerial photograph to the distance between them on a topographic map of the same area.

The scale of a stereophotograph is slightly distorted in both the horizontal and vertical directions. The horizontal scale is slightly larger at the center of an image than at its edges, which are more distant from the camera and viewed from an oblique angle. The vertical scale is also exaggerated, so mountains, hills, trees, and buildings appear higher and slopes appear steeper than they actually are.

Examine the stereophotographs in Figure 5–12 again, and note the horizontal and vertical distortions.

◀ **Figure 5–12** The Rincon of the Colorado River.

6

STRUCTURAL GEOLOGY

KEY WORDS

Geologic structures . . . tension, shearing, and compression . . . geologic maps and cross sections . . . outcrop patterns, stratigraphic sections, and attitudes . . . Brunton compass, strike, and dip . . . folds, limbs, axes, and axial planes . . . anticlines, synclines, domes, and basins . . . normal, reverse, thrust, strike-slip, and oblique faults . . . joints . . . angular unconformities, disconformities, and nonconformities . . . Nicholas Steno and the principles of horizontality, superposition, and cross-cutting relationship

THE ANDES HAVE RISEN ALONG THE ACTIVE CONTINENTAL MARGIN OF WESTERN SOUTH AMERICA.

(Gray Thompson)

Materials Needed *Colored pens or pencils, lead pencils with erasers, ruler, protractor*

The crust of the Earth is a complex assemblage of many different bedrock bodies or **formations** that can be distinguished from one another on the basis of their lithology, color, and fossil content. The bedrock also contains **folds, faults,** and other types of **geologic structures** which are the products of **crustal deformation** during tectonic and other geologic processes. **Structural geology** is the study of the distribution, geometry, and orientation of the bedrock bodies in the Earth's crust, the types of structures which they contain, and the origins of these structures.

Rock deformation is a common process along the margins of the tectonic plates. For example, the crust is extensively fractured and faulted in rift zones, where it is stretched by **tension,** and along transform margins, where it is **sheared** between two plates. In addition, the crust is greatly faulted *and* folded in orogenic belts and subduction zones, where it is squeezed and **compressed** between two converging plates. Geologic structures are therefore important evidence of the tectonic history of the Earth's crust, and particularly the movement and interaction of its tectonic plates. It is the study of geologic structures which enables geologists to reconstruct the tectonic origin and evolution of the crust, including the history of the formation and growth of continents and mountain belts, the openings and closings of oceans, and the tectonic configuration of the lithosphere in the geologic past.

This chapter begins with an introduction to geologic maps and cross sections, which are the principal media for illustrating the arrangement and structures of rock bodies in the Earth's crust. This is followed by a description of the geometry and outcrop patterns of three major classes of geologic structures: **folds, faults** and **joints,** and **unconformities.** This chapter concludes with a discussion of the principles of relative dating of rocks and geologic events, and some exercises which demonstrate the interpretation of geologic maps and cross sections and the reconstruction of the geologic history of the crust from them.

G eologic maps illustrate the **outcrop patterns**, spatial orientation or **attitudes**, and structures of rock formations in the upper crust. They are the products of many long hours of field work, describing bedrock outcrops, measuring the attitudes of rock layers and geologic structures, and surveying the outcrops of formations and mapping the **contacts** between them. This field work is often accompanied by the study and interpretation of topographic maps and aerial photographs, because the composition and structure of the surficial bedrock are often reflected in its landforms and vegetation patterns (see Chapter 7).

A typical geologic map illustrates the outcrop patterns of rock formations with different colors and alphanumeric characters representing their relative ages and names (Table 6–1). The lithologies, thicknesses, and relative ages of the rock formations and their representative colors and symbols are summarized in the map key or legend. The key may also contain a typical **stratigraphic section** for the map area which shows the formations in proper stratigraphic order—that is, with the oldest at the base and the youngest at the top—and their relative resistance to erosion (i.e., whether they are cliff formers or slope formers).

TABLE 6–1 Geologic Map Symbols

Attitudes of Strata

	Strike and dip of inclined strata with dip angle
	Strike and dip of overturned strata with dip angle
	Horizontal strata
	Vertical strata

Attitudes of Folds

	Anticline
	Syncline
	Overturned anticline
	Overturned syncline

Ages of Rocks

Q	Quaternary	**P**	Pennsylvanian
T	Tertiary	**M**	Mississippian
K	Cretaceous	D	Devonian
J	Jurassic	S	Silurian
TR	Triassic	O	Ordovician
P	Permian	$\in$	Cambrian
		P$\in$	Precambrian

Attitudes of Faults

	Uplifted and downthrown blocks of dip-slip fault
	Inclined fault plane with dip direction and angle
	Thrust fault with teeth pointing toward the thrust sheet
	Strike-slip fault

Plunging anticline with trend direction and angle

Plunging syncline with trend direction and angle

Plunging overturned anticline with trend direction and angle

Plunging overturned syncline with trend direction and angle

STRIKE AND DIP

A typical geologic map also records field measurements of the attitudes of stratified rock formations and other geologic features with various symbols (Table 6–1). The attitudes of rock strata are defined by their **strikes** and **dips**, which are measured at field outcrops with a **Brunton magnetic compass** (Fig. 6–1).

Strike is the compass bearing of a line formed by the intersection of an inclined rock stratum with an imaginary horizontal plane. It measures the direction or trend of the surface outcrop of the stratum relative to the magnetic north pole. For example, it may be given as *N35°W,* which indicates that an outcrop trends from 35° west of magnetic north to 35° east of magnetic south. Dip is the angle of inclination of a stratum relative to the same imaginary horizontal plane. The dip direction is always perpendicular (90°) to the strike direction.

Field measurements of the strikes and dips of outcropping strata are often recorded directly on geologic maps with **T**-shaped symbols (Table 6–1). The long cross bar of the **T** is always oriented parallel to the strike of the stratum, and the shorter vertical bar points in the dip direction. The dip angle is usually noted along with this symbol.

Strike and dip can also be used to define the attitudes of planar and layered nonsedimentary rock bodies such as intrusive dikes and sills, extrusive lava flows, and foliated metamorphic rocks. In addition, they can be used to define the attitudes of faults, joints, mineral cleavage, and other planar rock structures and fabrics. The various symbols for these measurements are shown in Table 6–1. Strike and dip cannot be measured on massive granites, nonfoliated metamorphics, and other rocks which lack any bedding, foliation, or structure.

GEOLOGIC CROSS SECTIONS AND BLOCK DIAGRAMS

Geologic maps show only the surface outcrop pattern and structure of rock formations, but they can also be used to interpret the arrangement and structures of rock formations beneath the surface. This interpretation is usually presented in the forms of **geologic cross sections** and **block diagrams**.

Geologic cross sections are illustrations of vertical slices through the upper crust. They are constructed by extending or **extrapolating** field measurements on the attitudes of stratified rocks and geologic structures down into the subsurface. Block diagrams are three-dimensional models of the arrangement and structure of rock formations in the subsurface and their outcrop patterns on the surface. They consist of a geologic map of the surface, a geologic cross section oriented parallel to the strike of the major geologic structures, and a second cross section oriented perpendicular to the strike.

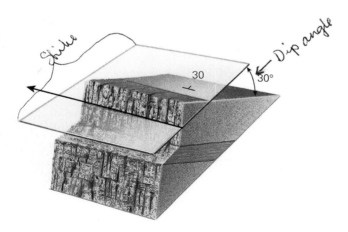

Figure 6–1 Graphic representation of the strike and dip of an outcropping rock layer. Strike is the compass bearing of a line formed by the intersection of the rock layer with an imaginary horizontal plane; the dip angle is the angle between the rock layer and the horizontal plane; and the dip direction is perpendicular to the strike. The attitude of the rock layer is recorded with a standard strike-and-dip symbol. The long cross bar of the **T** is oriented in the strike direction; the dip direction and angle are also indicated.

EXERCISE 6–1   Strike-and-Dip Symbols

1. The left-hand table contains several strike-and-dip symbols. Use your protractor to determine the strike direction of each symbol (assume that the top of the page is north), and complete the table. The first symbol has been interpreted as an example.

2. The right-hand table contains several written strike and dip measurements. Use a protractor to illustrate the strike-and-dip symbol of each pair of measurements.

SYMBOL	STRIKE	DIP
65	N45°E	65°SE
12	n45°w S45°E	12°nE
42	n90°E n90°w s90°E s90°w	42°S
36	n0°E n0°w	

ATTITUDE	SYMBOL
Strike: N45°W Dip: 20°SW	20
Strike: N10°E Dip: 45°NW	10°
Strike: N60° E Dip: 30°SE	
Strike: N75°W Dip: 60°NE	

EXERCISE 6–2   Block Diagrams

1. Complete the block diagrams in Figure 6–2 by drawing strike-and-dip symbols to indicate the attitudes of the rock strata. The first block diagram is completed as an example.

Strike + Dip

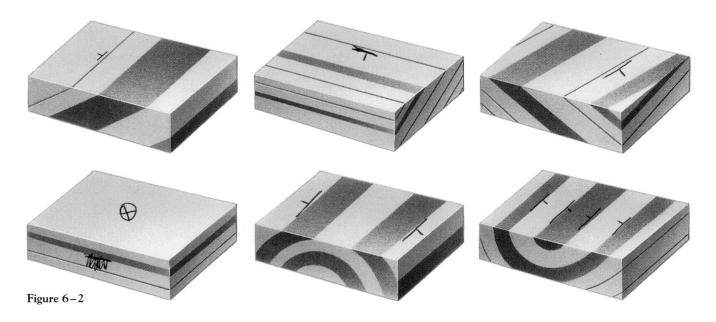

Figure 6–2

2. Complete the geologic cross sections on the block diagrams in Figure 6–3. The strikes and dips of the rock strata are given. Use your protractor to project the contacts between them into the subsurface on the forward faces of the block diagrams, and then complete their side faces. The first block diagram is completed as an example.

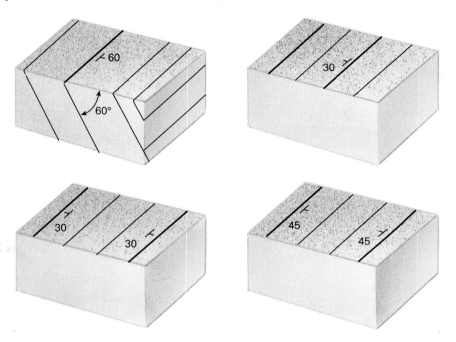

Figure 6–3

3. Mark the oldest and youngest rock layers in the block diagrams of this and the preceding exercises with the letters O and Y. Examine the ages of the outcrop belts on the upper face. Describe the downdip change in their ages—that is, do they get younger or older in the dip direction? _____

4. On the basis of the preceding answer, complete the block diagrams in Figure 6–4. Assume that all rock layers in these figures dip at an angle of 30°. Indicate the strike and dip directions on the upper faces of the block diagrams.

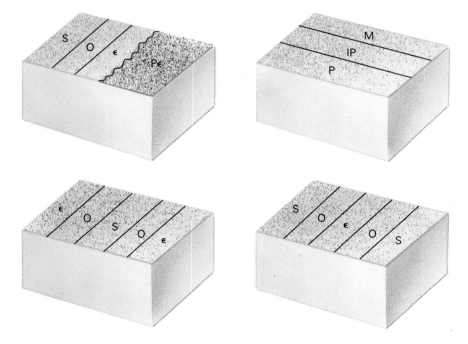

Figure 6–4

There are three classes of geologic structures in the rocks of the Earth's crust: **folds**, **faults** and **joints**, and **unconformities**.

FOLDS

Folds are upward and downward bends in rocks which are typically formed by crustal compression. They occur in all sorts of rocks, but they are most obvious in the bent and tilted layers of stratified rock bodies. Folds range in size from small bends in rock layers, just a few centimeters wide, to large crustal **upwarps** and **downwarps** which are tens to hundreds of kilometers in diameter.

A typical fold consists of two **limbs** separated by an **axis** (Fig. 6–5). The axis is the line of maximum curvature, and runs through the crest or trough of the fold. It can be oriented parallel to the Earth's surface (i.e., horizontal), or it may **plunge** downward into the Earth's crust at any angle. A fold can also be divided into two halves by an imaginary **axial plane**, which connects the axes of all of the rock layers in the fold. The axial plane can be either vertical, horizontal, or inclined, and its **trace** (the line which marks its intersection with the surface) can be oriented in any compass direction.

There are four important types of folds: **anticlines**, **synclines**, **domes**, and **basins**.

Anticlines are upward-arching elongate folds with limbs which dip away from their axes; synclines are downward-arching elongate folds with limbs which dip toward their axes. Several types of anticlinal and synclinal folds can be distinguished on the basis of the attitudes of their limbs, fold axes, and axial planes (Fig. 6–6):

- A **symmetric fold** has limbs which dip at nearly equal angles and an axial plane which is essentially vertical.
- An **asymmetric fold** has limbs which dip at different angles and an axial plane which is inclined.
- An **overturned fold** has limbs which dip in the same direction and an axial plane which is inclined. The proper stratigraphic order of the rock layers is reversed in an overturned limb: The youngest rocks underlie the oldest rocks.
- A **recumbent fold** is an overturned fold with limbs and an axial plane, which are all essentially horizontal.
- An **isoclinal fold** has limbs which dip in the same direction *and* at the same angle, and an axial plane which is either vertical or inclined.

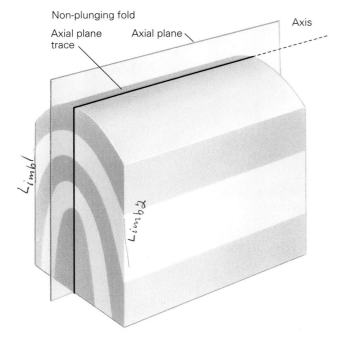

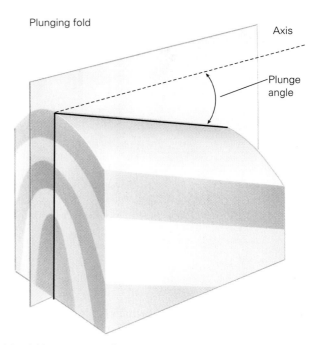

Figure 6–5 Components of nonplunging and plunging folds. The limbs of the folds are separated by the axial planes, which connect the axes of all the rock layers in the folds. The vertical and surface traces of the axial plane are shown in the figure. The plunge angle is the angle between the surface trace and the axis of a plunging fold.

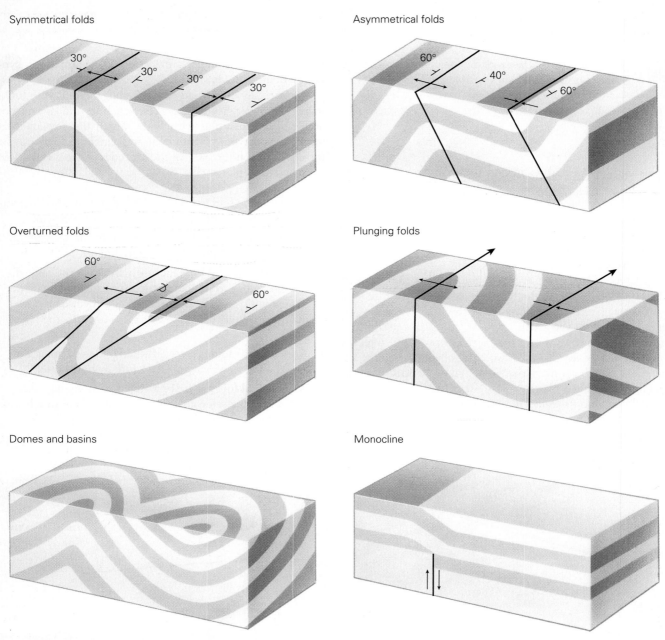

Figure 6–6 Common types of folds. The geometry of the folds and the attitudes of their limbs are indicated on the upper surfaces of the block diagrams with the appropriate map symbols.

- A **plunging fold** has an axis which is tilted downward rather than parallel to the Earth's surface. Most folds plunge, although this is not always apparent in the field.
- A **doubly plunging fold** has an axis which is itself bent and warped, so that it plunges into the ground in two directions.

The attitudes of anticlines and synclines can be defined by the compass bearings or **trends** of their axial plane traces and the **plunge angles** of their axes (Fig. 6–5). The axial plane traces of folds are often recorded directly on geologic maps with symbols which indicate their trends, the fold types, and the directions and angles of plunge (Table 6–1).

Domes and basins are circular to elliptical, upward- and downward-arching folds (Fig. 6–6). They are common in the stable interiors of continents, where they form gentle **arches** and shallow **cratonic basins** which are tens to hundreds of kilometers wide. Such large structures appear to be formed by the large-scale warping and fracturing of the crystalline basement of the continents.

Outcrop Patterns of Folded Strata

Examine the block diagrams in Figure 6–6. Note the following characteristics of the outcrop patterns of folded strata:

- The outcrop patterns of nonplunging anticlines and synclines consist of parallel belts which are elongate in the direction of strike.
- The outcrop pattern of a plunging anticline or syncline consists of a V-shaped belt which converges on the axial trace.
- The V-shaped outcrop belts of plunging anticlines point in the direction of plunge. The outcrop belts of plunging synclines point in the opposite direction.
- The V-shaped outcrop belts of adjacent plunging anticlines and synclines form zig-zagging outcrop patterns.
- The outcrop patterns of domes and basins consist of concentric belts around their centers.
- The width of the outcrop belt of a rock stratum (seen from directly above) depends on its thickness *and* dip angle: The outcrop width is great when the rock is essentially horizontal, it decreases with increasing dip, and it equals the thickness of the rock layer when it is vertical.
- The youngest rocks in the stratigraphic section are exposed in the centers of eroded synclines and basins. The oldest rocks are exposed in the centers of eroded anticlines and domes.
- Inclined strata always dip toward the outcrop of the youngest rock in the stratigraphic section unless they are overturned.

FAULTS AND JOINTS

Faults are fracture surfaces along which there has been some movement or **displacement** of rocks. This motion can occur in any direction, and it can vary in distance from a fraction of a meter to several hundred kilometers. It can also be a smooth gradual movement (called **fault creep**), or a sudden lurch which generates an earthquake.

Three types of faults can be defined on the basis of the principal direction of rock displacement along them: **dip-slip**, **strike-slip**, and **oblique faults**.

→ Goes along w/ dip

Dip-slip faults are characterized by rock displacement which is chiefly in the vertical direction—that is, up or down the fault planes. A dip-slip fault separates two rock bodies: the **hanging wall**, which lies above the fault, and the **footwall**, which lies below it (Fig. 6–7). The relative movements of these two rock bodies define three kinds of dip-slip faults:

- *footwall higher than hanging*
  Normal faults are characterized by the downward movement of the hanging wall relative to the footwall. Normal faults are typically formed by **extension**, which pulls the rocks of the crust apart and eventually fractures them. The hanging-wall rocks are then displaced downward along these fractures by the pull of gravity.
- *hanging higher than footwall*
  Reverse faults are characterized by the upward movement of the hanging wall relative to the footwall. They are typically formed by **compression**, which squeezes the rocks of the crust and eventually fractures them. The hanging-wall rocks are then displaced upward along these fractures, and the crust is shortened in the direction of the compressional stress.
- **Thrust faults** are reverse faults which are inclined at a low angle, generally less than 45° and sometimes horizontal. The hanging walls of such faults form slabs of bedrock called **thrust sheets**, which can slide laterally for considerable distances (sometimes hundreds of kilometers) over their footwalls. This style of displacement often reverses the proper stratigraphic order of rock formations by laying thrust sheets of older rocks over younger bedrock.

→ Goes along strike

Strike-slip faults are characterized by rock displacement which is chiefly in the horizontal direction (Fig. 6–7)—that is, parallel to the fault plane. Strike-slip faults are often found clustered together in narrow **fault zones** along transform margins, where the crust is being **sheared** by the movement of two plates in parallel but opposite directions. The San Andreas of California is the most famous of several strike-slip faults along the transform boundary between the North American and Pacific plates.

Strike-slip faults can be further described as either **left lateral** or **right lateral**, depending on the relative direction of displacement along the fault. Figure 6–7 shows the displacement of inclined rock layers along a strike-slip fault. Follow the outcrop of any rock layer from the bottom edge of the block diagram to the fault. Notice that the layer is displaced to the left on the opposite side of the fault. It is a left-lateral fault.

Oblique faults are characterized by significant rock displacement in *both* the strike and dip directions. Such faults are fairly common, because there is often some dip movement along strike-slip faults and some strike movement along dip-slip faults.

Joints are fractures along which there has been no appreciable movement of rocks. They can be formed by several

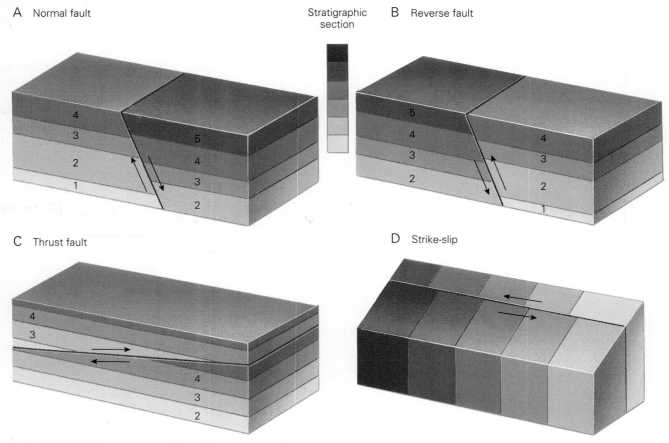

A Normal fault

Stratigraphic
section

B Reverse fault

C Thrust fault

D Strike-slip

Figure 6–7 Common types of faults. The direction of displacement along each fault is indicated by arrows.

means, including the extension and compression of the crust, the cooling and contraction of basaltic lavas and rocks, and the exfoliation of unroofed plutons. Crustal extension tends to produce a single **set** of roughly parallel joints which are oriented perpendicular to the direction of tensional stress, whereas crustal compression tends to produce two mutually cross-cutting joint sets. Basaltic rocks have a distinctive **columnar** jointing pattern which dissects them into hexagonal columns.

Outcrop Pattern of Faulted Rocks

Examine the block diagrams in Figure 6–7. Note the following characteristics of the outcrop patterns of faulted rocks:

- The surface traces of faults appear as abrupt lines which separate rocks of different lithology, age, and structure.
- The rocks on the downthrown blocks of normal faults are younger than the rocks of the footwall. The rocks on the upthrown blocks of reverse faults are older than the rocks of the footwall.

- Reverse faults can reverse the proper stratigraphic sequences of stratified rocks; that is, they can lay older rocks over younger ones.
- Stratigraphic sequences are thickened or **expanded** by reverse faults due to the repetition of rock layers. They are shortened or **condensed** by normal faults due to the displacement of rock layers.
- Outcrop belts, fault traces, streams, roads, and other natural and cultural surficial features are laterally offset by strike-slip faults.

UNCONFORMITIES —represents missing Time

Unconformities are planar and wavy contacts between rocks which represent gaps in the geologic record. Three types of unconformities can be defined on the basis of the lithology and attitude of the rocks above and below them: **disconformities**, **angular unconformities**, and **nonconformities** (Fig. 6–8).

A Disconformity

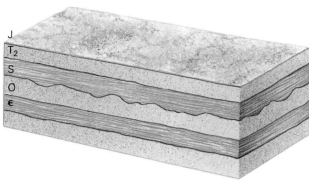

B Angular unconformity

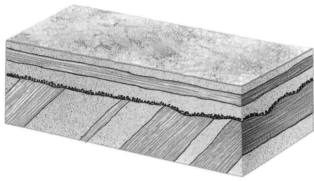

C Nonconformity

Figure 6–8 Common types of unconformities.

Disconformities are unconformable surfaces which separate parallel sequences of sedimentary rocks. They are a record of a period of time during which there was a temporary pause (or **hiatus**) in sedimentation, and probably some erosion of the underlying rock layers. Disconformities are often difficult to detect, because they separate rocks which are parallel and sometimes similar in appearance. However, they can be identified by fossil studies and any other age-dating techniques which can reveal the significant age difference between the rocks above and below them.

Angular unconformities are unconformable erosion surfaces which separate nonparallel sequences of sedimentary rocks. You may recall that the law of original horizontality states that sedimentary rocks are deposited parallel to the Earth's surface. Consequently, angular unconformities represent periods of time during which the underlying rocks were first deformed and tilted by mechanical stresses, then exposed and eroded, and finally buried beneath the overlying sedimentary rocks.

Nonconformities are unconformable erosion surfaces which separate sedimentary and nonsedimentary rocks. Nonconformities are formed by the uplift and erosion of igneous and metamorphic rock bodies and their subsequent burial beneath sedimentary rocks.

Sequences of sedimentary rocks are described as **conformable** when they were deposited without any interruption by periods of erosion or deformation.

Outcrop Patterns of Unconformities

Unconformities (like faults) appear as abrupt lines which separate rocks of different lithology, age, and/or structure. Angular unconformities and nonconformities have the most obvious outcrop patterns. The first type juxtaposes rocks of very different attitudes and ages; the second type juxtaposes sedimentary and crystalline rocks.

Unconformity: erosion took place
don't know missing time.

EXERCISE 6–3	Geometry of Folds

Complete the geologic cross sections on the block diagrams in Figure 6–9. The first block diagram is completed as an example.

1. Draw the axial plane traces on the front and top faces of each block. Describe the geometry of each fold.

2. Use the appropriate symbol from Table 6–1 to indicate the fold type (anticline, syncline, plunging or not) on the surface (top) trace of the axial plane.

3. Draw arrows to indicate the direction of the stresses which created each fold.

Symmetrical anticline _____   _____

Asymmetrical _____   _____

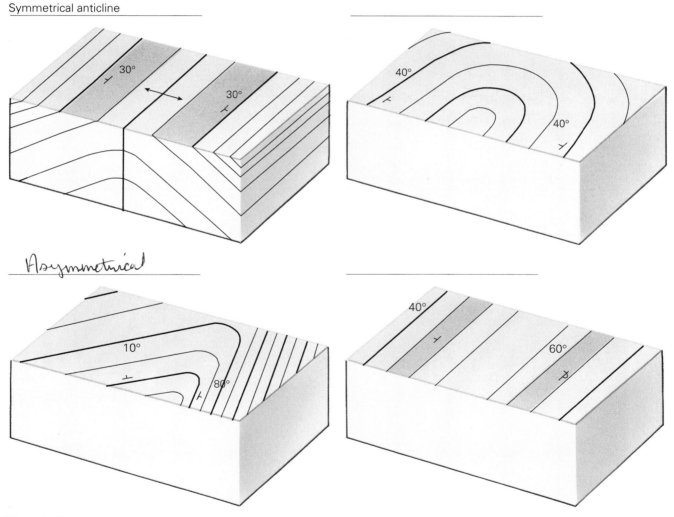

Figure 6–9

Complete the block diagrams of the dip-slip faults (A and B) in Figure 6–10:

1.   Identify the hanging walls and footwalls of each fault.

2.   Indicate the directions of displacement on the forward face of each block.

3.   Indicate whether each fault is normal or reverse in the spaces provided.

4.   Use the appropriate symbols from Table 6–1 to indicate the dip directions of each fault and its upthrown and downthrown blocks on the upper faces of each block.

5.   Measure the dip of each fault and record this datum on the fault trace.

6.   Indicate the directions of the stresses which created each fault.

Complete block diagrams C and D:

1.   Mark the hanging wall and footwall of each fault.

2.   Show the direction of displacement on the forward face of each block.

3.   Describe each fault type in the space provided.

Complete the block diagrams of strike-slip faults (E and F):

1.   Indicate the directions of displacement on each fault.

2.   Note whether each fault is left lateral or right lateral in the space provided.

3.   Use arrows to indicate the direction of the stresses which created each fault.

E X E R C I S E   6 – 5   Faults and Stratigraphic Sequences

This exercise illustrates the condensation and expansion of stratigraphic sections by faults. Examine the stratigraphic section which accompanies the block diagrams in Figure 6–7. This section consists of six rock layers, numbered 1 (the oldest) through 6 (the youngest).

Refer to the forward faces of the block diagrams of normal, reverse, and thrust faults. Draw a vertical line through the center of each face, and examine the sequence of rocks which it dissects. Draw each sequence in the space below, and describe the effect of the fault on the stratigraphic section.

*A _Dip slip Normal Fault_

*B _Dip-slip Reverse fault_

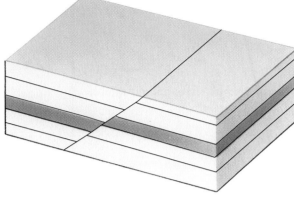

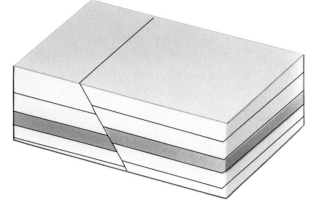

C

D

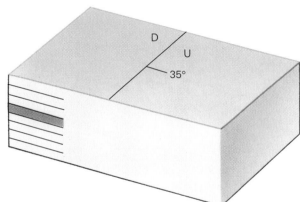

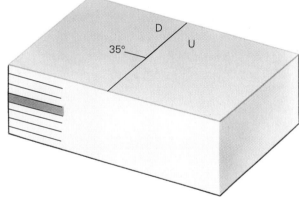

*E

*F

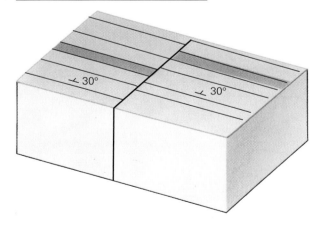

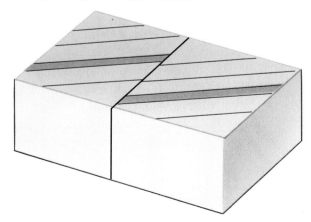

Figure 6–10

Map 6–1 is a geologic map of part of the stable cratonic interior of the North American continent. Please see the folded map at the back of this lab manual. The North American craton is composed of Precambrian igneous and metamorphic basement rocks overlain by a thin veneer of largely Paleozoic sedimentary rocks. The stratigraphic sequence for this region is given with the geologic map. The deformation of the crystalline basement beneath this region has created several large-scale geologic structures, which will be examined in this exercise.

1. Examine the circular structure centered in northern Kentucky.
    a. What is the age of the rocks in the center of the structure? _____
    b. What is the age of the rocks along its edges? _____
    c. Indicate the general attitudes of the Paleozoic strata in this structure with strike-and-dip symbols.
    d. The name of this structure is the Cincinnati _____ (fill in with either "Arch" or "Basin").

2. Examine the circular structure centered in Illinois.
    a. What is the age of the rocks in the center of this structure? _____
    b. What is the age of the rocks along its edges? _____
    c. Indicate the general attitudes of the Paleozoic strata in this structure with strike-and-dip symbols.
    d. The name of this structural feature is the Illinois ("Arch" or "Basin"). _____

3. The Paleozoic strata of the stable interior are buried along the southern margin of the United States by a seaward-inclined wedge of **coastal plain** strata. The contact between the Paleozoic strata and these younger strata can be seen in the **Mississippi Embayment**, which is located in the southwest corner of the map. The embayment is a crustal depression which has been slowly filling with fluvial, coastal, and shallow marine sediments since the late Mesozoic era.
    a. Locate the embayment, and trace the contact between the Paleozoic and coastal plain strata with a dark colored pencil.
    b. What is the age of the rock directly above the contact in the eastern part of the embayment? _____ the western part? _____
    c. What is the age of the rock directly below the contact in the eastern part of the embayment? _____ the western part? _____
    d. What kind of contact is this? _____
    e. What is the source of the alluvium in the western part of the embayment? _____

4. Construct a geologic cross section along line A–A′. Follow these instructions, and refer to the examples in the space below.

Step 1. Place the edge of a piece of paper along line A–A′. Mark the locations of the end points. Mark off the locations of formation contacts on the paper, and label the outcrop belt of each formation with its symbol (such as Cpv or Cmu).

Step 2. Transfer these data to the space below. Place the edge of the paper along the line, mark off the locations of the contacts, and label the outcrop belt of each formation.

Step 3. Inclined strata always dip toward the outcrop belt of the youngest rock layer in the stratigraphic section unless they are overturned. Given this rule, indicate the dip direction of each formation on the cross section line with an arrow.

Step 4. Extend the strata into the subsurface by drawing the formation contacts with the appropriate dips. Use solid lines for conformable contacts and wavy lines for nonconformable ones. The dip directions of the strata are now established, but what are the appropriate dip angles to use? (They are not recorded on this map.) The coastal plain strata dip at very low angles toward the sea, so we suggest that you use dip angles of approximately 10° for them. The Paleozoic strata can be drawn with slightly higher dips (because they are folded), and we have exaggerated their dip angles in the example so that the structure is easier to draw. We have also assumed that the Paleozoic strata have constant thicknesses in the map area. Given this assumption, the dip angles of strata are indicated by outcrop widths: Dip angles are high wherever the outcrop belt is narrow, and low wherever the outcrop belt is wide.

Step 5. Interpret your data, and complete the structural cross section. Sketch the vertical trace of the axial plane of the structure. Describe the attitudes of the limbs and the axial plane, and the overall symmetry of the structure.

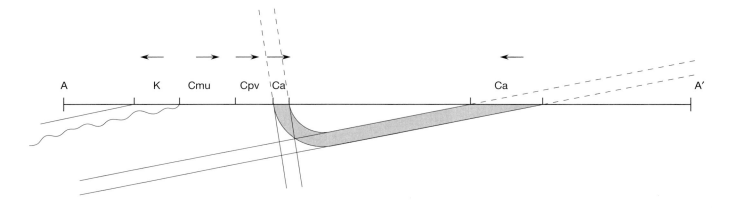

5. Locate and label the Michigan Basin, Wisconsin Dome, Nashville Dome, and Ozark Uplift. How does the Ozark Uplift differ structurally and lithologically from the first three folds?

6. Locate the town of Port Clinton, Ohio, on the southern shore of Lake Erie. The town is situated in the middle of an elongate northeast-trending structure.

    a. What is the age of the rocks in the center of this structure? _____

    b. What is the age of the rocks along its margins? _____

    c. Indicate the general attitudes of the Paleozoic strata in this structure with strike-and-dip symbols.

    d. Construct a geologic cross section along line B–B′ in the space below. Note: The formations labeled Dml and Dm are rocks of the same age.

    e. What type of structure is this? _____ Draw the surface trace of its axial plane on the map, and use the appropriate symbols from Table 6–1 to indicate the fold's type and attitude.

7. Notice the elongate structure south of Chicago, Illinois. What type of structure is this? _____ Draw the surface trace of its axial plane on the map, using the appropriate symbols from Table 6–1 to indicate the fold's type and attitude.

8. Examine the fault which runs across the Michigan Basin near Lansing. This fault dips to the southwest. What kind of fault is it? _____ Indicate the fault type and the upthrown and downthrown blocks on the map with the appropriate symbols from Table 6–1.

EXERCISE 6–7   Valley and Ridge, Pennsylvania

Map 6–2 is a geologic map of the Valley and Ridge of Pennsylvania. The Valley and Ridge is the western province of the Appalachian Mountains, and the picturesque mountains and valleys of central Pennsylvania. Its bedrock is composed largely of sandstones, shales, and limestones which were deposited along the eastern margin of North America in the early Paleozoic and then compressed, folded, and thrust-faulted during the Allegheny orogeny of the middle Paleozoic era. The Valley and Ridge is bordered to the northwest by the Allegheny Plateau (shown on this map), which is underlain sedimentary rocks shed from the Appalachians after the Allegheny orogeny. It is bordered to the east by the Blue Ridge Mountains and Piedmont Plateau, which are not shown on this map.

1. Examine the fold in the southeast corner of the map (around the town of Ardenheim):
    a. What is the age of the rocks in the center of the fold? _____
    b. What is the age of the rocks along its edges? _____

    c. Indicate the general attitudes of the strata in this fold with strike-and-dip symbols.

    d. Compare the outcrop widths of the formations in the two limbs of the structure. What difference do you see?

    e. Assume that the formations maintain a constant thickness in the area of the fold. What does the difference in outcrop width indicate about the relative dips of its two limbs?

    f. Construct a geologic cross section across this fold. Draw the vertical trace of the axial plane of the fold on the cross section.

    g. What type of fold is this:
        • anticline or syncline? _____
        • symmetric or asymmetric? _____
        • plunging or nonplunging? _____

    h. Draw the surface trace of the axial plane of the fold on the map, using the appropriate symbols from Table 6–1 to indicate the fold type and its attitude. Measure the general trend of the fold with your protractor. Record that information here:

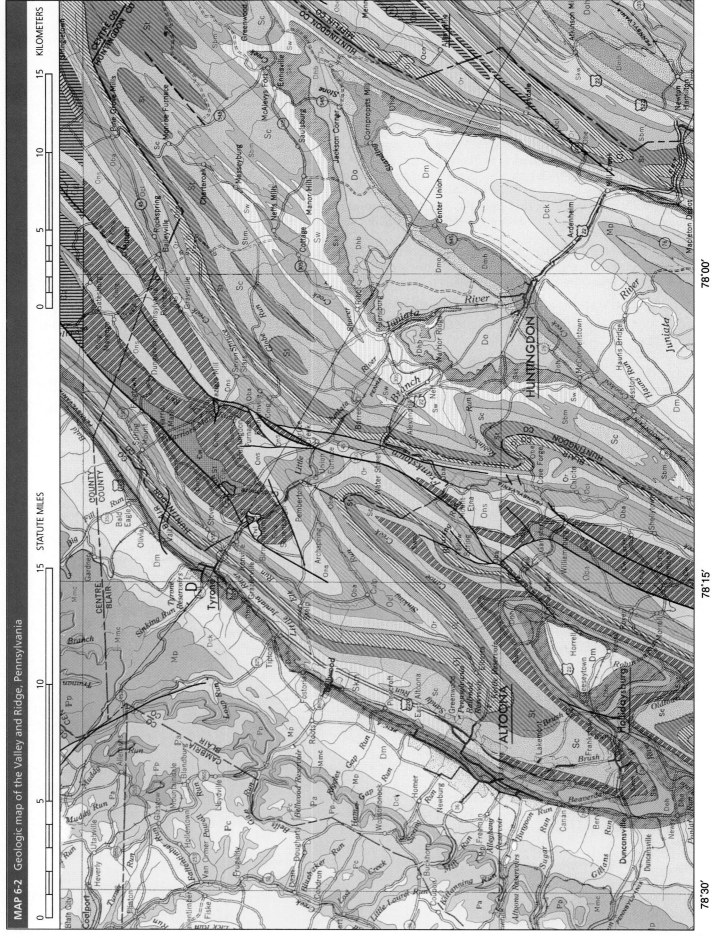

MAP 6-2 Geologic map of the Valley and Ridge, Pennsylvania

KEY

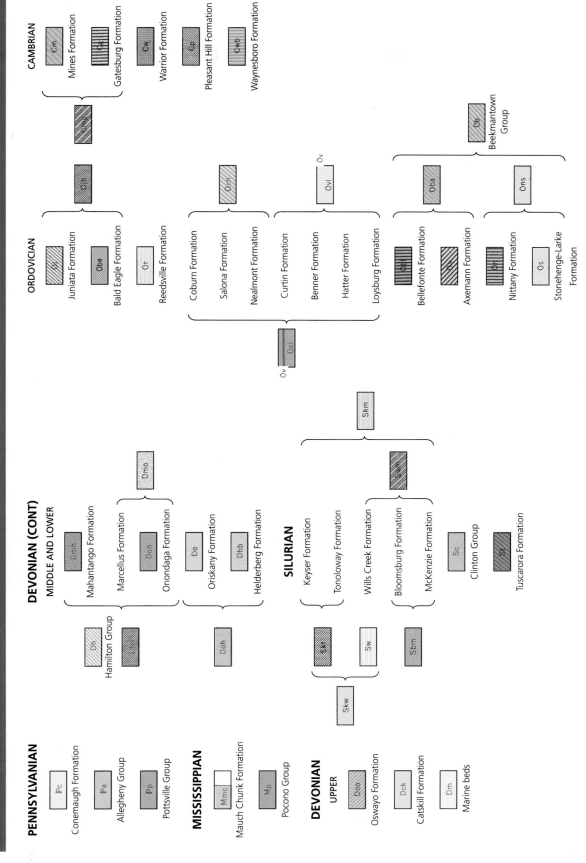

PENNSYLVANIAN

Pc — Conemaugh Formation
Pa — Allegheny Group
Pp — Pottsville Group

MISSISSIPPIAN

Mmc — Mauch Chunk Formation
Mp — Pocono Group

DEVONIAN

UPPER

Doo — Oswayo Formation
Dck — Catskill Formation
Dm — Marine beds

DEVONIAN (CONT)

MIDDLE AND LOWER

Dh — Hamilton Group
Dmh — Mahantango Formation
Dm — Marcellus Formation
Dmo — Doh — Onondaga Formation
Doh — Do — Oriskany Formation
Dhb — Helderberg Formation

SILURIAN

Skw — Skt — Keyser Formation
Sw — Tonoloway Formation
Skm — Wills Creek Formation
Swm — Bloomsburg Formation
Sbm — McKenzie Formation
Sc — Clinton Group
St — Tuscarora Formation

ORDOVICIAN

Cmu — Oju — Oj — Juniata Formation
Obe — Bald Eagle Formation
Or — Reedsville Formation
Osn — Coburn Formation
Salona Formation
Nealmont Formation
Ocl — Ov — Ovl — Curtin Formation
Benner Formation
Hatter Formation
Loysburg Formation
Oba — Obl — Bellefonte Formation
Ov — Axemann Formation
Ons — Ob — Beekmantown Group
Ohl — Nittany Formation
Os — Stonehenge-Larke Formation

CAMBRIAN

Cm — Mines Formation
Cg — Gatesburg Formation
Cw — Warrior Formation
Cp — Pleasant Hill Formation
Cwb — Waynesboro Formation

117

2. Examine the fold in the southwest corner of the map (around the town of Altoona).
   a. What is the age of the rocks in the center of the fold? _____
   b. What is the age of the rocks along its edges? _____
   c. Indicate the general attitudes of the strata in this fold with strike-and-dip symbols.

   d. Compare the outcrop widths of the formations in the two limbs of the structure. What difference do you see?
   e. Assume again that all the formations maintain a constant thickness in the area of the fold. What does the out-crop width indicate about the respective dips of its two limbs?

   f. Construct a geologic cross section through the fold. Draw the vertical trace of the axial plane of the fold on the cross section.

   g. What type of fold is this:
      • anticline or syncline? _____

      • symmetric or asymmetric? _____

      • plunging or nonplunging? _____

   h. Draw the surface trace of the axial plane of the fold on the map, using the appropriate symbols from Table 6–1 to indicate the fold type and its attitude. Measure the general trend of the fold with your protractor. Record that information here: _____

3. Locate other examples of folds in the map area, and draw the surface traces of their axial planes on the map with the appropriate symbols.

4. Measure the trends of three additional folds with your protractor. Calculate the average trend of these three folds and the Ardenheim and Altoona structures. (Sum the trends of the five folds, and then divide by 5.) Record that in-formation here: _____ What does this value indicate about the general direction of the tectonic stresses which deformed this part of the Valley and Ridge?

5. The western part of the Valley and Ridge contains several long faults which strike north-northeast to south-south-west and dip to the east at low angles. Note the ages of the strata on either side of these faults. What kind of faults are they? Explain your answer. Mark the upthrown and downthrown blocks of two of these faults with the appropriate symbols from Table 6–1.

EXERCISE 6–8	Glen Creek, Montana

Map 6–3 is a geologic map of the Rocky Mountains of northwest Montana. The Rockies are a fold-and-thrust belt which was formed by the tectonic compression of the western margin of North America during the Laramide orogeny. They mark the eastern edge of the Cordillera, the great expanse of mountain ranges, plateaus, and basins which lies between the cratonic interior and the Pacific coast.

The Rocky Mountains of northwest Montana consist of overlapping thrust sheets separated by thrust faults which dip toward the west. The thrust sheets are themselves composed of sedimentary and metasedimentary rocks, and they are sometimes internally deformed by small folds. The thrust sheets and thrust faults are also dissected by high-angle dip-slip faults, such as Glen fault, which are also inclined toward the west.

1. Construct a east-west geologic cross section across the center of the map in the space provided. Follow these steps:

    Step 1.   Locate and draw the thrust faults on the cross section with a dip angle of 30° to 45°, and the dip-slip fault with a higher dip. Remember: Glen fault is younger than the thrust faults. Indicate the direction of rock displacement along the faults with arrows. What kind of fault is Glen fault: normal or reverse? _____

    Step 2.   Examine the structures of the thrust sheets. The formations in all but one of the thrust sheets dip in the same direction (indicated by strike-and-dip symbols). The single exception contains folded strata. Locate this thrust sheet. What is the dip direction of the western limb of the fold? _____ the eastern limb? _____ What kind of fold is it? _____

    Step 3.   Locate and draw the formation contacts on the cross section, using the strike and dip data on the map. Show the vertical trace of the axial plane of the folded thrust sheet.

2. What effect has thrust faulting had on the thickness and outcrop of the Kootenai Formation (Kk) in the central part of the map?

MAP 6-3 Geologic map of Glen Creek, Montana

CONTOUR INTERVAL = 40 FEET

KEY

QUATERNARY

Pleistocene

Surficial deposits

CRETACEOUS

LOWER PLATE

Upper Cretaceous

Montana Group

- Ktm — Two Medicine Formation
- Kv — Virgelle Sandstone
- Kt — Telegraph Creek Formation
- Ki — Sills
- Kmk / Kmsf / Kmf / Kmcf — Marias River Shale

Colorado Group

UNCONFORMITY

- Kbv / Kbt / Kbf — Blackleaf Formation

Lower Cretaceous

UNCONFORMITY(?)

- Kk — Kootenai Formation

JURASSIC

UNCONFORMITY(?)

- Jm — Morrison Formation

Upper Jurassic

Ellis Group

- Js — Swift Formation

UNCONFORMITY

- Jr — Rierdon Formation

Middle Jurassic

- Jsa — Sawtooth Formation

CARBONIFEROUS

MISSISSIPPIAN

UNCONFORMITY

Madison Group

Upper Mississippian

- Mc — Castle Reef Dolomite

Lower Mississippian

- Ma — Allan Mountain Limestone

CAMBRIAN

UPPER PLATE

Upper Cambrian

- €s — Switchback Shale
- €st — Steamboat Limestone
- €p — Pagoda Limestone
- €d — Dearborn Limestone

Middle Cambrian

- €da — Damnation Limestone
- €g — Gordon Shale
- €f — Flathead Sandstone

PRECAMBRIAN

UNCONFORMITY

- p€i — Sills
- p€au / p€al — Ahorn Sandstone*

Belt Series

- p€hd / p€hc / p€hb / p€ha — Hoadley Formation*
- p€he — Helena Dolomite
- p€e — Empire and Spokane Formations

Geologic maps and geologic cross sections are very useful for interpreting the geologic history of the crust—that is, the sequence in which the crust was created by igneous, sedimentary, and metamorphic rock-forming processes deformed by tectonic processes, and eroded and buried by surficial processes. Generally, the relative ages of rocks and the relative timing of geologic events can be determined by applying three of the fundamental principles of stratigraphy: **original horizontality, superposition, and cross-cutting relationships**.

The principles of original horizontality and superposition were defined in 1669 by **Nicholas Steno**, one of the early pioneers in the field of stratigraphy. The first principle states that sedimentary rocks were originally deposited in horizontal layers or strata of sediment. (There are some exceptions to this principle, such as the deposition of cross-bedded sand on dunes, but they can usually be identified by a geologist.) The second principle states that in a vertical sequence of undisturbed sedimentary rocks, the oldest stratum lies at the base of the sequence and the youngest stratum lies at the top. The principle of cross-cutting relationships, which was derived after Steno's time, states that any rock or structure which cuts across a second rock or structure is younger than that second rock or structure.

Together, these three principles have several major implications regarding the relative ages of rocks and the relative timing of geologic events (Table 6–2). The application of these principles can be demonstrated with a consideration of the development of Hutton's Unconformity.

Figure 6–11 is a photograph of an outcrop on the coast of Scotland near Edinburgh, which was described by James Hutton in his *Theory of the Earth* (1795). The outcrop exposes two formations separated by an angular unconformity. The lower formation consists of lower Silurian graywackes and shales, and the upper formation is the upper Devonian arkoses of the Old Red Sandstone Formation. The angular unconformity is overlain by a poorly sorted layer of sediment containing gravel clasts of the Silurian rocks.

Figure 6–11 The exposure of Hutton's Unconformity at Siccar Point on the coast of Scotland.

TABLE 6-2   Relative Ages of Rocks and Events

A nonhorizontal sedimentary rock layer must have been displaced and tilted by folding or faulting.

A stratigraphic sequence which becomes older from base to top must have been overturned by folding or reversed by thrust faulting.

A fold is younger than the youngest rock formation in it.

Faults and joints are younger than the youngest rock formation or structure which they dissect.

An unconformity is older than the youngest rock formation beneath it but younger than the oldest rock formation above it.

An intrusive rock must be younger than the country rock it intrudes.

A metamorphic rock is older than the tectonic event or geologic agent which metamorphosed them.

An extrusive rock layer, like a sedimentary one, must be younger than the rock formation beneath it but older than the rock formation above it.

Igneous and sedimentary rocks which contain inclusions or pebbles of another rock are younger than that other rock.

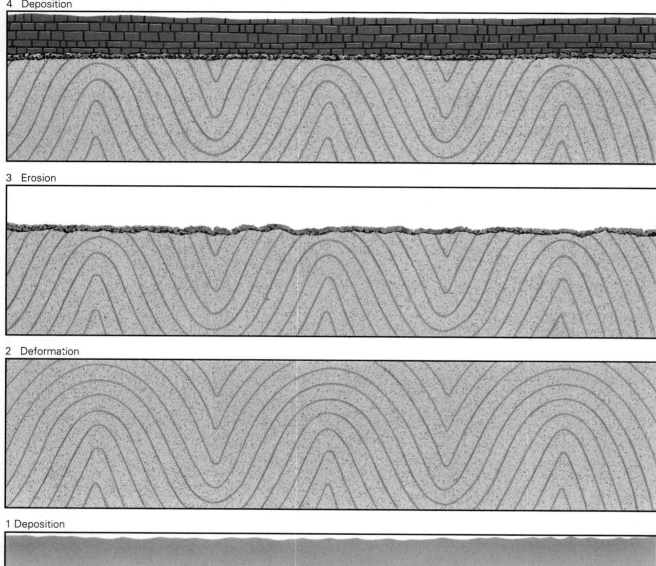

4 Deposition

3 Erosion

2 Deformation

1 Deposition

Figure 6–12 The four stages in the geologic history of the outcrops at Siccar Point.

The application of the three principles of stratigraphy reveals the following history for this sequence. This chronology is traditionally given in stratigraphic order—that is, with the oldest geologic event at the bottom of the list and the most recent event at the top (Fig. 6–12).

4. Weathering and erosion of the Old Red Sandstone after the Devonian period

3. Deposition of the Old Red Sandstone during the late Devonian period

2. Deformation, tilting, and erosion of the shales and graywackes after the early Silurian period but before the late Devonian period

1. Deposition of shales and graywackes during the Ordovician and early Silurian periods

EXERCISE 6-9   Relative Dating

Figure 6–13 shows a hypothetical geologic cross section through the Earth's crust. Apply the principles of original horizontality, superposition, and cross-cutting relationships to determine the relative timing of geologic events in this cross section. List these events in their proper stratigraphic order.

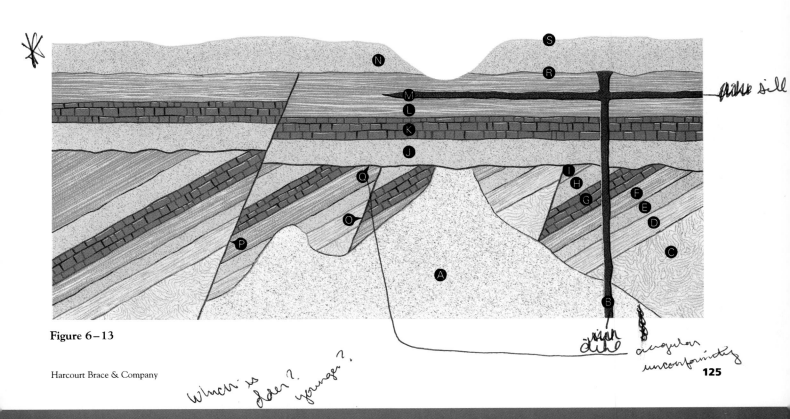

Figure 6-13

1. Return to Exercise 6–7 and Map 6–2.
    a. What is the age of the oldest strata in the Valley and Ridge? _____

    b. What is the age of the youngest strata in the Valley and Ridge? _____

    c. What does the preceding answer indicate about the timing of the Allegheny orogeny and the beginning of deposition in the Allegheny Plateau? Explain your answer.

    d. Given this information, locate and trace the boundary between the Valley and Ridge and the Allegheny Plateau. What is the age of the oldest bedrock in the plateau? _____

2. Return to Exercise 6–8 and Map 6–3.
    a. What is the age of the oldest rocks in the thrust sheets? _____

    b. What is the age of the youngest rocks in the thrust sheets? _____

    c. What does the preceding answer indicate about the timing of the Laramide orogeny? Explain your answer.

    d. There are several places along the Rocky Mountain cross section where thrust faulting has produced vertical sequences which violate the law of superposition. Identify such sequences with the letter V.

STRUCTURAL AND VOLCANIC LANDFORMS

KEY WORDS

Differential weathering and stream erosion . . . resistant and nonresistant rocks . . . domes and strike ridges and valleys . . . dip and scarp slopes . . . strike, dip, and scarp streams . . . drainage density and drainage pattern . . . parallel, trellis, annular, rectangular, radial, and dendritic drainages . . . plateaus, mesas, and buttes . . . fault scarps, grabens, horsts, and pull-apart basins . . . flood basalts, lava plateaus, and inverted valleys . . . volcanic necks and radial dikes . . . craters and calderas

CINDER CONES OF WUPATKI SUNSET CRATER NATIONAL MONUMENT, ARIZONA

(Wupatki National Monument)

Materials Needed *Colored pens or pencils, lead pencils with erasers, ruler, protractor, calculator, stereoscope*

The previous chapter discussed the deformation of the crust and the representation of the arrangement and structure of rock bodies on geologic maps and cross sections. This chapter continues on this subject with an examination of the influence of the structure and lithology of bedrock on the topography of its outcrops.

WEATHERING, EROSION, AND TOPOGRAPHY

It was mentioned previously that geologic mapping requires many long hours of field work, describing outcrops, measuring the strikes and dips of rock bodies and geologic structures, and surveying and mapping the outcrop patterns and contacts of formations. The study of landforms is also an important part of the preparation of geologic maps, because the general shape of the land—its topographic highs and lows, the slope of its surface, and the paths of streams—typically reflects the structure and lithology of the underlying bedrock. This is due to the combined effects of two geologic processes: **differential weathering** and **stream erosion**.

DIFFERENTIAL WEATHERING

Differential weathering describes the tendency of rocks in the same area to weather and erode at different rates. This variation produces differences in the topography of the exposed bedrock.

Granites, gneisses, conglomerates, and quartzose sandstones are usually weathered and eroded relatively slowly because they are resistant to breakage, chemical alteration, and dissolution. Such **resistant rocks** tend to form prominent **domes, ridges,** and other sorts of **topographic highs.** On the other hand, shales and evaporites are weathered fairly rapidly because they are soft and soluble. Such **nonresistant rocks** tend to be eroded more rapidly into valleys and other sorts of **topographic lows.** In addition, resistant rocks are commonly cliff formers, and nonresistant rocks usually have gentler, talus-covered slopes.

The topographies of limestone and basalt outcrops are dependent on the weathering climate. Generally, such rocks form topographic lows and gentle slopes in humid climates, where there is abundant moisture and fairly rapid rates of chemical weathering. They form topographic highs and steep cliffs in arid climates, where chemical weathering is slow (Fig. 7–1).

STREAM EROSION

Streams tend to erode their valleys along the fastest and least resistant paths to the sea, and these paths are strongly influenced by the structure and lithology of the bedrock. The fastest paths to the sea, for instance, are often directly down the slopes of inclined strata, domes, and volcanoes. The least resistant paths are typically through the outcrop belts of nonresistant rocks and along the surficial traces of faults and joints, which are usually lined by weathered and fragmented rocks.

This habit of **preferential erosion** by streams creates several distinctive landforms which reflect the structure and lithology of the bedrock. For example, consider the topographic expression of inclined sandstone and shale strata (Fig. 7–2), and note the following features:

- The resistant sandstones form the topographic highs, whereas the nonresistant shales form the lows.
- The uplifted edges of the sandstone strata form long strike-parallel **strike ridges** or **hogbacks.**
- The shale outcrops between the sandstone ridges are strike-parallel **strike valleys** eroded by **strike streams.**

Figure 7–1 El Capitan, Texas. Reefal limestone forms the cliffs at the peak of the mountain in this arid desert setting, whereas nonresistant siltstones and mudstones form its lower, gentler slopes.

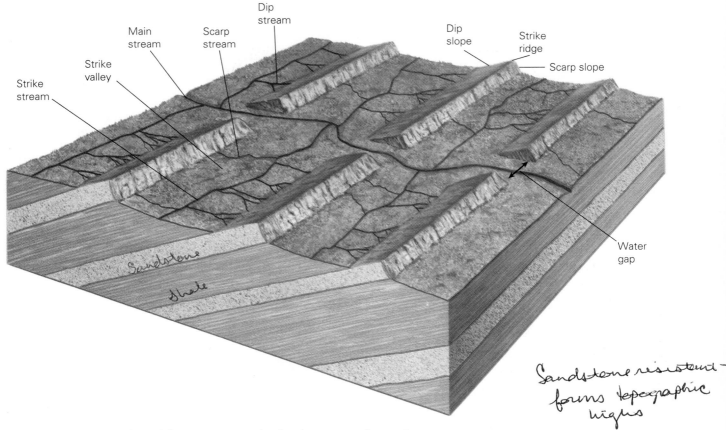

Figure 7–2 The topography and drainage patterns of inclined resistant and nonresistant strata.

- The sandstone ridges are asymmetrical in profile. This profile consists of a **dip slope** which parallels the upper bedding surfaces of the sandstone strata, and a generally steeper **scarp slope** which is inclined opposite to the dip.
- The sandstone ridges are eroded on either side by **consequent** or **dip streams** and shorter **scarp streams**, both of which drain into the strike streams.

Two characteristics of streams are particularly indicative of the structure and lithology of the bedrock: their **drainage density** and **drainage pattern**.

Drainage Density

Drainage density describes the number of streams in a given area. The drainage density of any land surface is obviously influenced to a great degree by its climate: The more precipitation (and less evaporation), the greater the density of streams needed to drain the surficial runoff. However, it is also influenced by the **infiltration capacity** of the bedrock, which is its ability to absorb water from the surface and thereby rob streams of their runoff.

Rocks which contain interconnected (or **permeable**) networks of pores and fractures tend to have high infiltration capacities and can absorb large volumes of rainwater from the surface before it runs into streams. The outcrops of such rocks, which include many sandstones, limestones, and basalts, are characterized by low drainage densities. On the other hand, nonpermeable rocks tend to have lower infiltration capacities and allow large volumes of rainwater to escape into streams or ponds on the surface. The outcrops of such rocks, which include many shales, mudstones, and massive rocks, are characterized by higher drainage densities, as well as by the presence of bogs, swamps, and lakes.

Drainage Pattern

Drainage pattern describes the areal arrangement of the streams in a given area. There are several common types of stream drainage patterns, including **parallel**, **trellis**, **annular**, **rectangular**, **radial**, and **dendritic** (Fig. 7–3), which are highly indicative of the bedrock structure.

129

Dendritic pattern   Trellis pattern   Parallel pattern

Rectangular pattern   Radial pattern

Figure 7–3 Drainage patterns and the structure of the bedrock.

Parallel drainage patterns consist entirely of parallel dip streams. They are generally found on the dip slopes of inclined strata and exposed fault planes. **Trellis** and **annular** drainage patterns are common to folded strata. They are composed of three types of streams (in order of increasing size): dip and scarp streams which flow off the strike ridges, strike streams which flow through the outcrop belts of non-resistant strata, and a single **main stream** which flows through **water gaps** in the strike ridges. Trellis patterns are typical of anticlines and synclines, and annular patterns are typical of domes and basins.

Rectangular drainage patterns consist of strike streams which follow the surficial traces of faults and joints. They are common to all rock types, including stratified rocks and massive crystalline rocks such as granites. Wherever there are two or more intersecting sets of fractures, the streams can consist of several long, straight stretches separated by sharp bends.

Radial drainage patterns consist of dip streams which radiate away from the centers of plutonic domes and volcanic cones.

A **dendritic** drainage pattern consists of a main stream with tributaries that branch off in many different directions in a treelike pattern. This drainage pattern develops wherever there is no lithologic or structural control over the paths of streams, so they can assume a fairly random arrangement. Dendritic patterns are common to horizontal strata and massive crystalline rocks with no prominent faults and joints.

EXERCISE 7 – 1   **Drainage Patterns in the United States**

1.   Map 7–1 contains four maps of different sedimentary-rock terranes in the United States. Sketch and describe the drainage pattern of the streams in each map area.

2.   Locate and label examples of dip and strike streams on the maps. Use these streams to determine the general strike and dip directions of the strata in each map area. Record these directions on the map with strike-and-dip symbols.

3.   Describe the structure of the strata in each area.

4.   Return to the geologic map of the Stable Interior of the United States (Map 6–1) in the previous chapter. Trace and describe the drainage pattern around the Ozark Uplift.

5.   Return to the geologic map of the Valley and Ridge of Pennsylvania (Map 6–2) in the previous chapter. Locate and list the names of the major strike streams in the map area. What is the lithology of the bedrock beneath each stream?

EXERCISE 7 – 2   **Front Range, Colorado**

Figure 7–4 is a stereophotograph of the Front Range of the Rocky Mountains near Loveland, Colorado. The bedrock is composed of inclined beds of sandstones, shales, siltstones, and carbonates. The sandstones and carbonates are the more resistant strata, and the shales and siltstones are less resistant. The resistant strata form two long hogbacks separated by Thompson Creek in the western part of the area.

1.   Locate the two hogbacks, and draw their upper and lower contacts. Label their dip and scarp slopes.

2.   Use colored pencils to illustrate the outcrop patterns of the resistant and nonresistant strata in the photographs.

3.   Examine the hogbacks carefully. Locate and label examples of the dip (D) and scarp (S) streams which flow down them. Examine the topographic lows around the hogbacks, and locate and label examples of strike (St) streams. Trace and describe the drainage patterns which are formed by all these streams. What kind of stream is Thompson Creek?

4.   Use the strike-and-dip symbol to indicate the attitudes of the hogbacks on the stereophotograph.

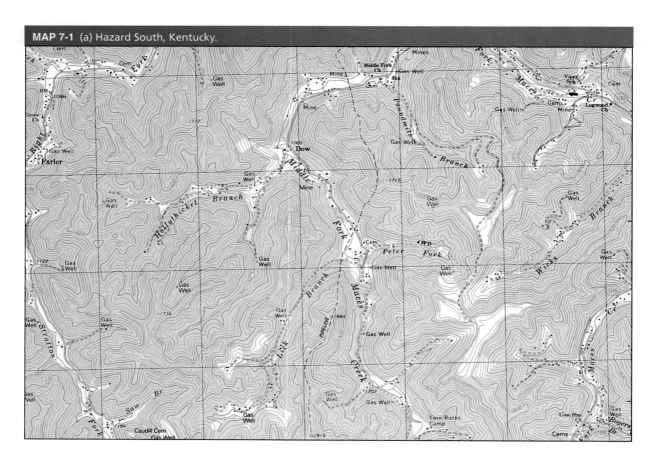

MAP 7-1 (a) Hazard South, Kentucky.

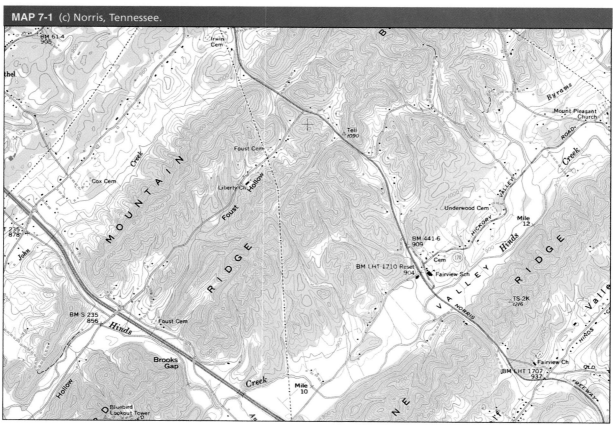

MAP 7-1 (c) Norris, Tennessee.

MAP 7-1 (b) Sinclair, Wyoming.

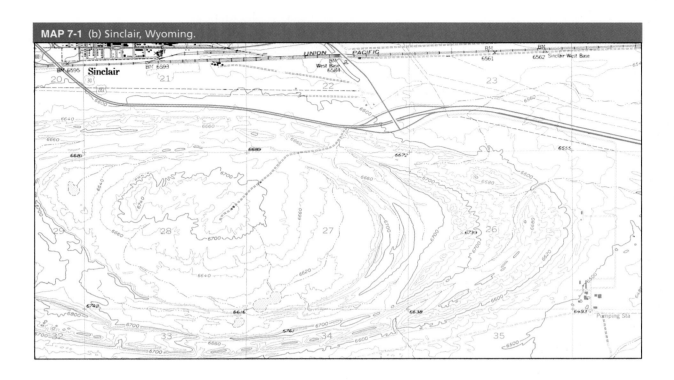

MAP 7-1 (d) Cascade Springs, South Dakota

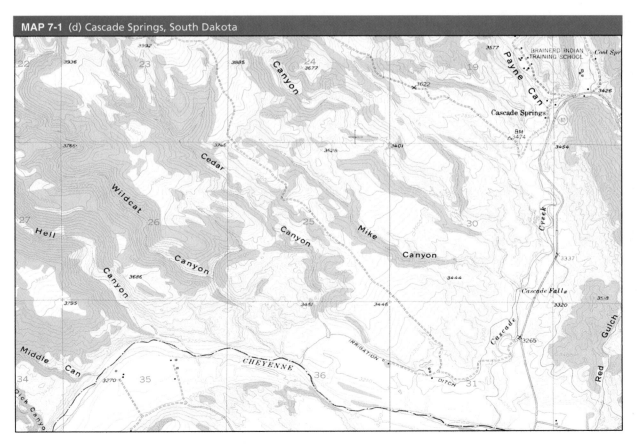

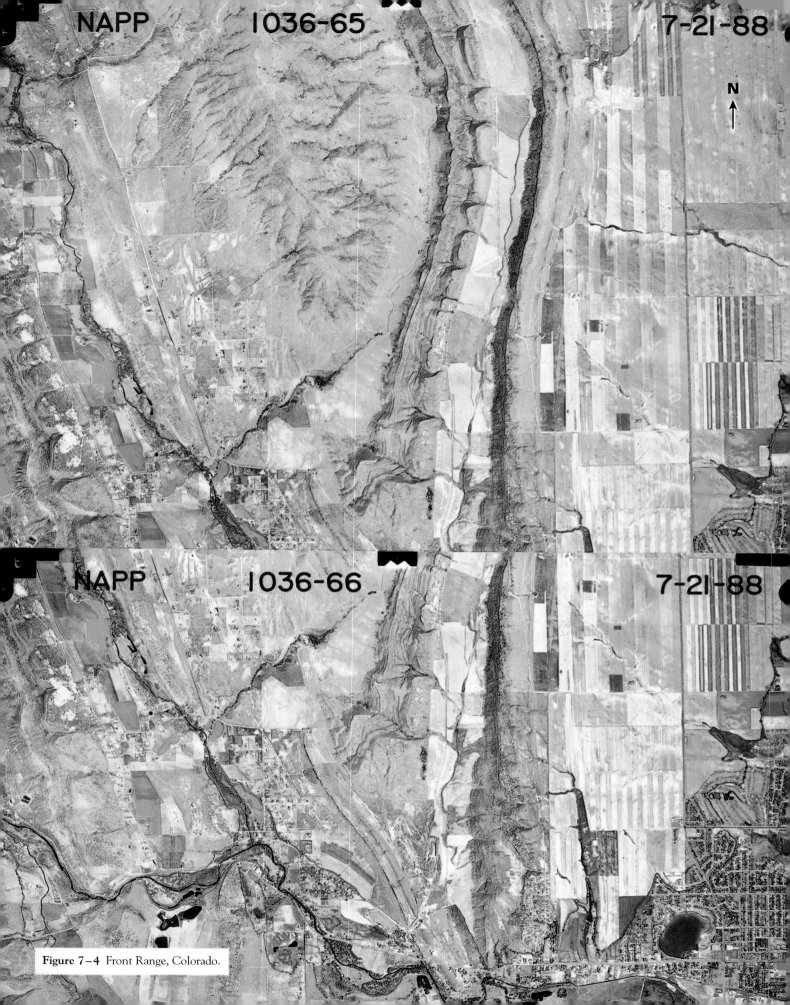

NAPP   1036-65   7-21-88

N

NAPP   1036-66   7-21-88

Figure 7–4 Front Range, Colorado.

| EXERCISE 7–3 | Hollidaysburg, Pennsylvania |

The Valley and Ridge province of the Appalachian Mountains offers excellent examples of the relationship between the structure and lithology of bedrock and its topographic expression. Map 7–2 is a combined geologic-topographic map of the Valley and Ridge near Hollidaysburg, Pennsylvania, where the bedrock consists of lower Paleozoic limestones, dolomites, sandstones, and shales.

1. Examine the fold in the northwest part of the map (centered around the town of Dry Gap).
   a. What is the age of the rocks in the center of the fold? _____
   b. What is the age of the rocks along its edges? _____
   c. What kind of fold is this: anticline or syncline? _____
   d. Does the fold plunge? If so, in what direction? _____
   e. What is the general dip direction of the eastern limb of the fold? _____ of the western limb of the fold (note the strike-and-dip symbols around McKee Gap)? _____ What does this indicate about the attitude of the axial plane of the fold? _____
   f. Sketch the surface trace of the axial plane on the map. Indicate the fold type and attitude with the appropriate symbols from Table 6–1 in the previous chapter.
   g. Measure the trend of the axial plane trace with a protractor. Record it here: _____

2. Examine the fold in the north-central part of the map (east of the Dry Gap fold). Sketch the surface trace of its axial plane on the map, and indicate the fold type and its attitude with the appropriate symbols.

3. Construct a combined topographic-geologic cross section along line A-A'. First construct the topographic profile, and then overlay the geology on it. Draw the vertical traces of the axial planes of the folds on the cross section, and label the dip and scarp slopes of the strata with the letters D and S.

    a. What are the age and lithology of the rock formations which form the topographic highs along this transect?

    b. Calculate the dip slope of the outcrops of these rock formations on the eastern side of Loop Mountain.

    c. What are the age and lithology of the rock formations which form the topographic lows along this transect?

    d. Calculate the dip slope of the outcrops of these strata.

    e. The upper Ordovician strata consist of interbedded sandstones and shales. Examine the cross section, and indicate the outcrops of the sandstone beds with the letters SS. Explain your answer.

4. Examine the drainage pattern and density in and around the two northern folds.
   a. Locate the major strike streams in the map area, and trace them with a yellow or red marker. Record their names in the following space, and indicate the lithology of the rock formations beneath them.

   b. What is the origin of McKee's Gap? _____
   c. Examine the outcrop belt of the upper Ordovician strata between East Sharpsburg and Dry Gap. Trace the streams which drain this belt. What kind of streams are they: dip, scarp, or strike streams? Explain your answer.

   d. Examine the outcrop belt of the lower and middle Silurian strata on the eastern slope of Loop Mountain. Trace the streams which drain this belt. What kind of streams are they: dip, scarp, or strike streams? Explain your answer.

   e. Compare and contrast the drainage densities of the upper Ordovician and lower to middle Silurian strata. Which strata have the greater drainage density? Why?

   f. Describe the overall drainage pattern of the northern part of the map area. _____

KEY TO HOLLIDAYSBURG MAP

Symbol	Age	Lithology
uD	**UPPER DEVONIAN**	Shales and sandstones
mD	**MIDDLE DEVONIAN**	Shales
lD	**LOWER DEVONIAN**	Sandstones and limestones
uS	**UPPER SILURIAN**	Limestones, sandstones, and shales
mS	**MIDDLE SILURIAN**	Shales and sandstones
lS	**LOWER SILURIAN**	Quartzose sandstones
uO	**UPPER ORDOVICIAN**	Sandstones and shales
mO	**MIDDLE ORDOVICIAN**	Limestones
lO	**LOWER ORDOVICIAN**	Dolomites and limestones
u€	**UPPER CAMBRIAN**	Dolomites and limestones
l-m€	**LOWER-MIDDLE CAMBRIAN**	Limestones and sandstones

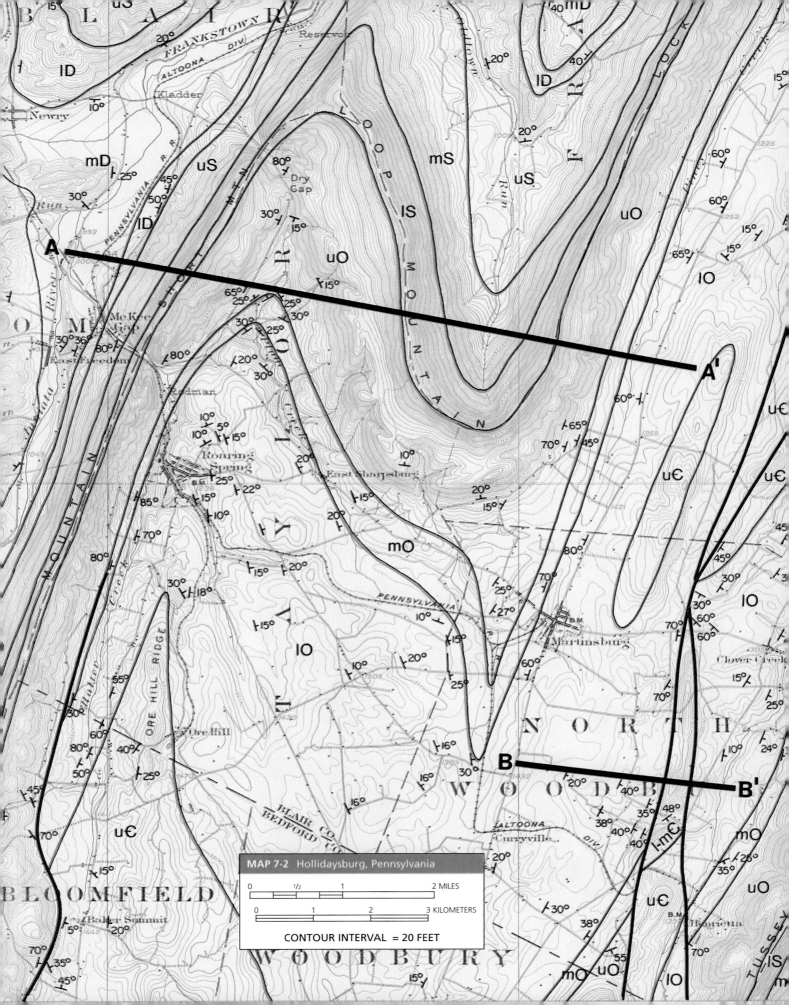

MAP 7-2 Hollidaysburg, Pennsylvania

0   1/2   1   2 MILES

0   1   2   3 KILOMETERS

CONTOUR INTERVAL = 20 FEET

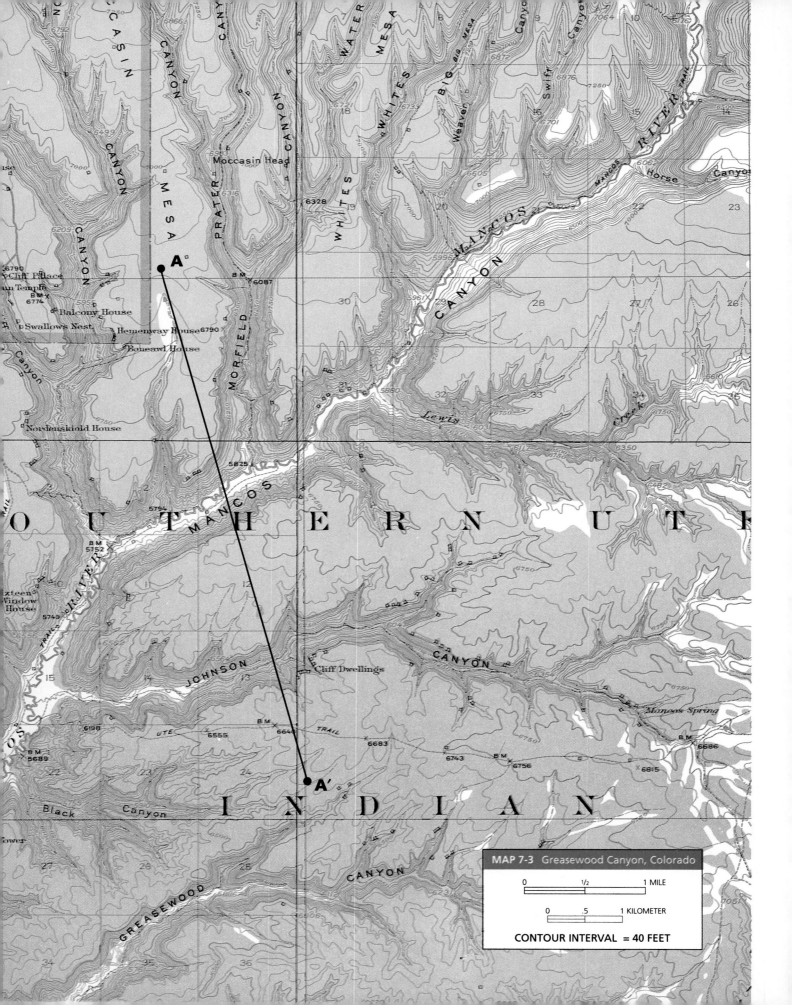

A

A'

MAP 7-3 Greasewood Canyon, Colorado

| 0 | ½ | 1 MILE |

| 0 | .5 | 1 KILOMETER |

CONTOUR INTERVAL = 40 FEET

MAP 7-5 Dinosaur National Monument, Utah

CONTOUR INTERVAL = 80 FEET

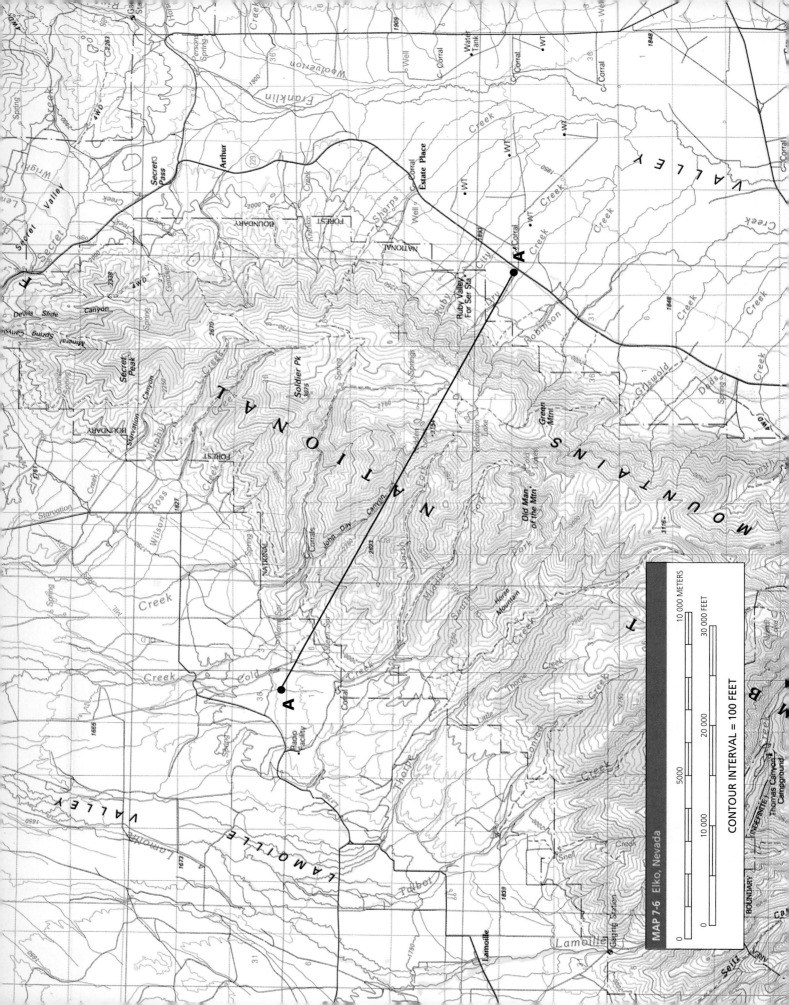

MAP 7-6 Elko, Nevada

CONTOUR INTERVAL = 100 FEET

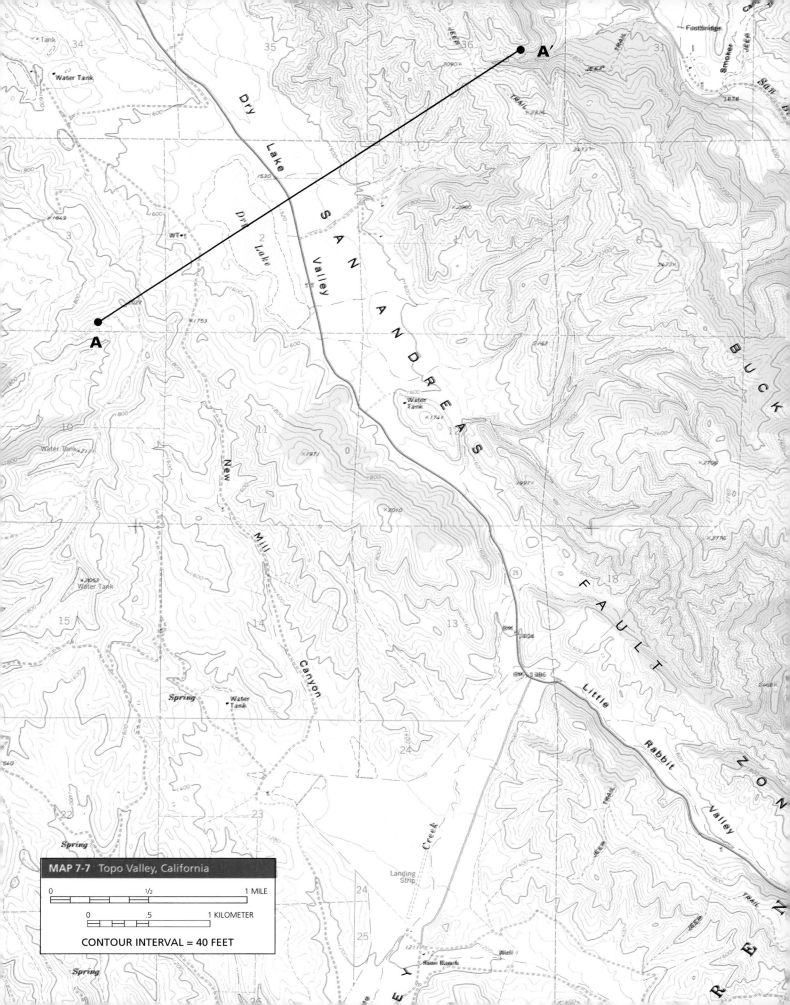

A

A'

Dry Lake

Dry Lake

SAN ANDREAS

Valley

New Mill Canyon

BUCK

FAULT

Little Rabbit Valley

ZON

Spring

Spring

Spring

Tank

Water Tank

Water Tank

Water Tank

Water Tank

Water Tank

WT

Footbridge

JEEP

TRAIL

JEEP

TRAIL

Smoker Can.

Sun Pr.

Landing Strip

Creek

Sans Ranch

Well

34 35 36 31

3 2

10 11 12

15 14 13 18

22 23 24

MAP 7-7 Topo Valley, California

0   1/2   1 MILE

0   .5   1 KILOMETER

CONTOUR INTERVAL = 40 FEET

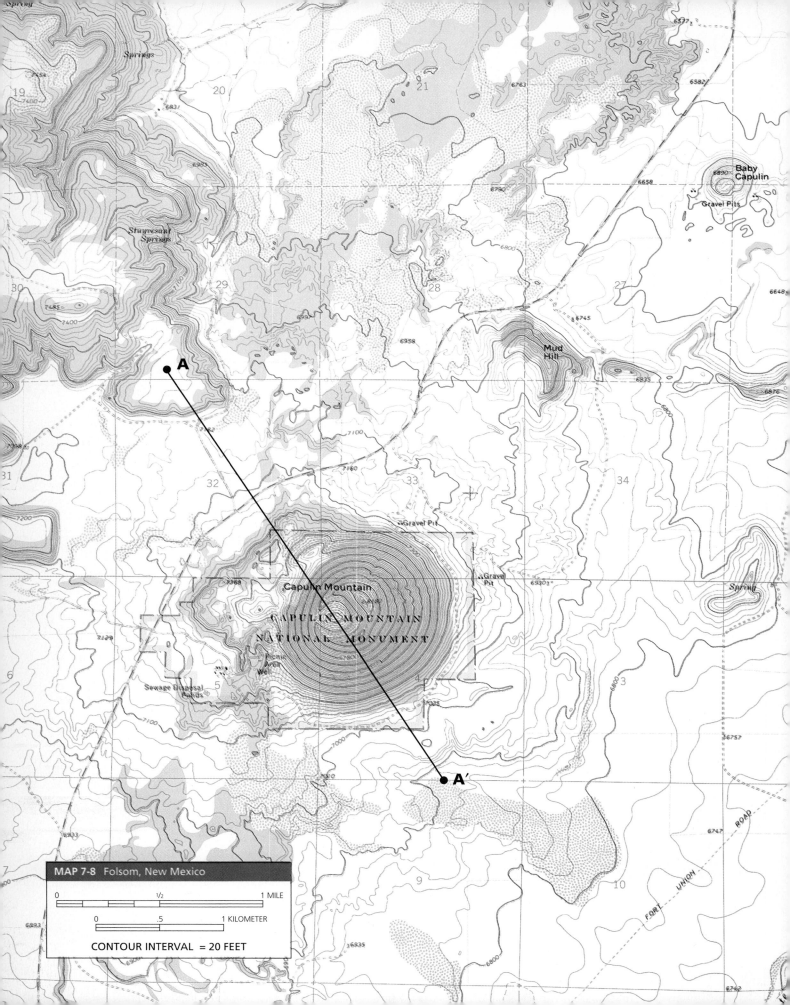

A

A′

Baby
Capulin

Gravel Pits

Mud
Hill

Gravel Pit

Gravel Pit

Capulin Mountain

CAPULIN MOUNTAIN
NATIONAL MONUMENT

Spring

Spring

Picnic
Area

Well

Sewage Disposal
Ponds

Springs

Stuyvesant
Springs

Spring

FORT UNION ROAD

MAP 7-8 Folsom, New Mexico

0	1/2	1 MILE

0	.5	1 KILOMETER

CONTOUR INTERVAL = 20 FEET

MAP 7-9 Crater Lake National Park, Oregon

CONTOUR INTERVAL = 50 FEET

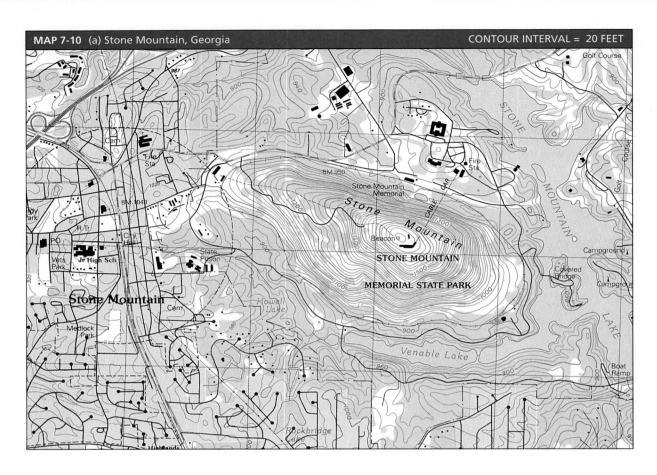

MAP 7-10 (a) Stone Mountain, Georgia — CONTOUR INTERVAL = 20 FEET

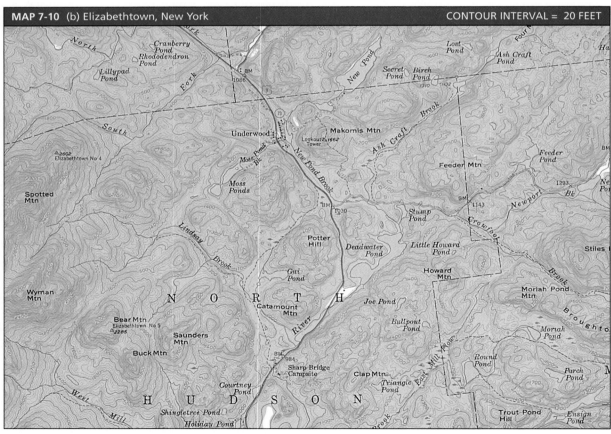

MAP 7-10 (b) Elizabethtown, New York — CONTOUR INTERVAL = 20 FEET

MAP 7-10 (c) Rocky Bar, Idaho

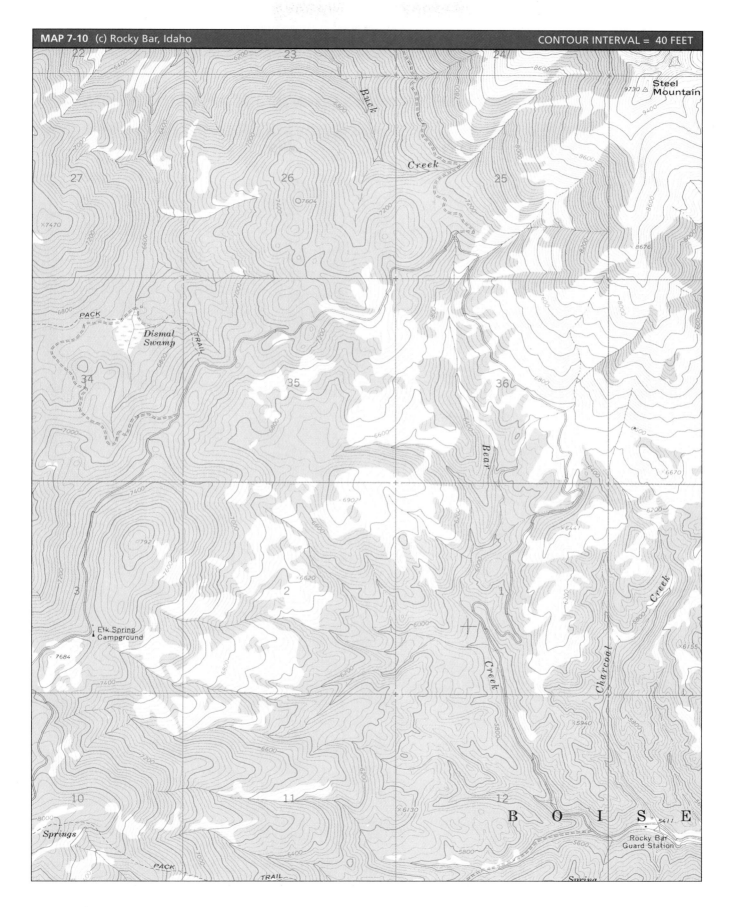

This section examines the topographic expression of stratified and massive rock bodies and associated geologic structures. The term *stratified* or *layered* is used to describe sedimentary rocks as well as volcanic and foliated metamorphic rocks, whereas the term *massive* describes plutonic rocks as well as nonfoliated metamorphic rocks and nonstratified sedimentary rocks (such as reef limestones). The purpose of this section is to show that the topography and drainage patterns of the land indicate the structure and lithology of the underlying bedrock; that is, the geometry of topographic highs and lows reflects the outcrop patterns of resistant and nonresistant rock bodies, and the drainage patterns of streams define their attitudes and structures.

TOPOGRAPHY OF HORIZONTAL STRATA

Horizontal strata are typically uplifted into broad, low-relief **plateaus**, which are dissected by deep, steep-walled **canyons** (Fig. 7–5). The oldest strata are exposed in the deepest parts of the canyons, and successively younger strata outcrop at higher elevations. The erosion of these plateaus creates a **badlands topography** consisting of small, flat-topped plateau

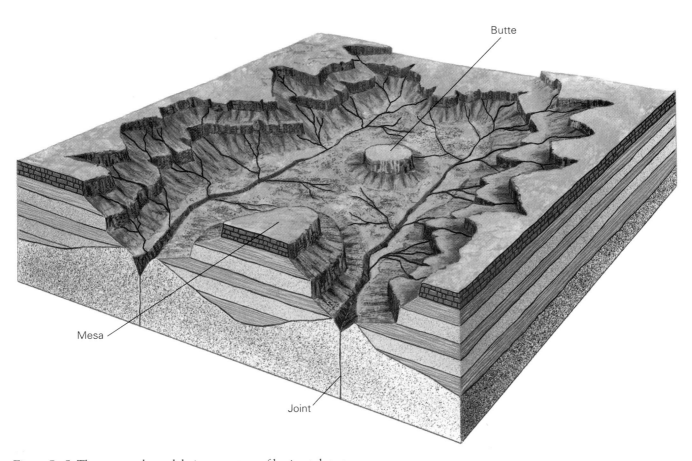

Figure 7–5 The topography and drainage pattern of horizontal strata.

remnants (**mesas**) and isolated **buttes**. The outcrops of the resistant strata form steep cliffs which rim the walls of the canyons and encircle the buttes and mesas at constant elevations, whereas the outcrops of the nonresistant strata form gentle, talus-covered slopes.

The drainage patterns of horizontal strata depend on their structure. The streams assume a random or dendritic pattern wherever there are no structural controls on their paths, and a rectangular pattern wherever the bedrock is faulted or jointed.

TOPOGRAPHY OF FOLDED STRATA

The topography of folded strata varies with the geometry of the folds. The topography of nonplunging anticlines and synclines is similar to the topography of inclined strata (Fig. 7–2)—that is, the resistant strata form long, straight strike ridges which run parallel to the axial planes of the folds, and the nonresistant strata are eroded into straight, parallel strike valleys. The strike ridges of plunging folds converge to

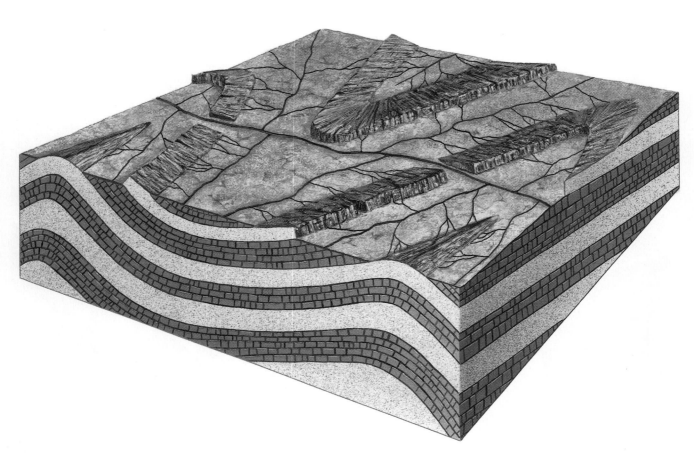

Figure 7–6 The topography and drainage pattern of plunging folds.

form V-shaped patterns (Fig. 7–6). The Vs of plunging anticlines point in the direction of plunge, whereas the Vs of plunging synclines open in the direction of plunge. The strike ridges of domes and basins are circular or elliptical in shape and concentric around the centers of these structures (Fig. 7–7).

The strike directions of folded strata are apparent from the orientation of the strike streams, ridges, and valleys. The dip directions are apparent from the attitudes of the dip slopes and their dip streams: The dip slopes are inclined and the streams flow outward from the axes of domes and anticlines and inward toward the axes of basins and synclines. Together, dip and strike streams form **trellis** drainage patterns over anticlines and synclines and **annular** patterns over domes and basins.

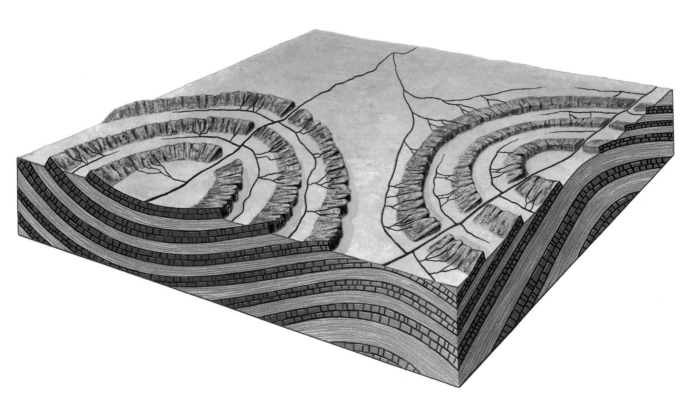

Figure 7–7 The topography and drainage pattern of domes and basins.

TOPOGRAPHY OF FAULTS AND JOINTS

The preferential erosion of the traces of faults and joints creates long, straight stream valleys and other linear topographic depressions. Wherever the bedrock is dissected by two or more intersecting sets of faults or joints, the streams assume a rectangular drainage pattern consisting of straight stretches separated by sharp bends.

The traces of high-angle normal faults are often exposed as steep, planar **fault scarps** (Fig. 7–8A). The downthrown blocks between two normal faults appear as topographic depressions called **grabens**, whereas the upthrown blocks between two normal faults appear as topographic highs called **horsts**. On the other hand, small, lake-filled **pull-apart basins** and sliver-shaped ridges are typically formed by the shearing motion associated with strike-slip faults (Fig. 7–8B). Stream channels, ridges and valleys, roads and fences, and other natural and human-made features are also laterally offset, sometimes for great distances, by strike-slip faults.

TOPOGRAPHY OF PLUTONIC ROCKS

The outcrops of plutonic rocks are generally highly resistant to weathering and erosion, and they form prominent topographic highs with steep slopes (Fig. 7–9). The outcrops of batholiths and stocks form rugged, barren mountain ridges and circular and elliptical **domes** that cut discordantly through the surrounding country rocks; the outcrops of dikes form narrow tabular ridges, which also cut discordantly through the country rocks; and the outcrops of sills and laccoliths form steep buttes, mesas, and strike ridges, which are concordant with the country rocks. On the other hand, the outcrops of the country rocks tend to form topographic lows with gentler slopes, because such rocks are usually structurally and thermally deformed and nonresistant to erosion.

The drainage patterns of plutonic rocks depend on their structure. The streams assume a random or dendritic pattern wherever there are no structural controls on their paths, a rectangular pattern wherever the bedrock is faulted or jointed, and a radial pattern on massive granite domes.

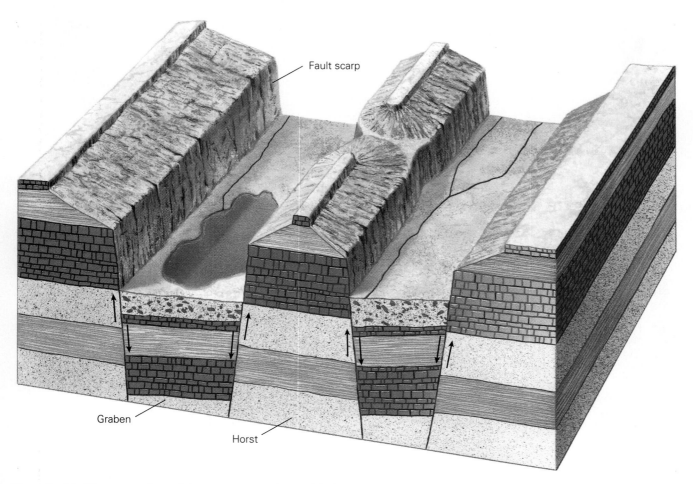

Figure 7–8A The topography and drainage pattern of normal faults.

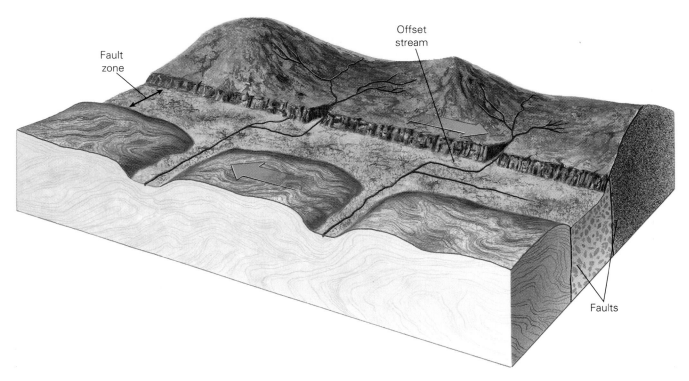

Figure 7–8B The topography and drainage pattern of strike-slip faults.

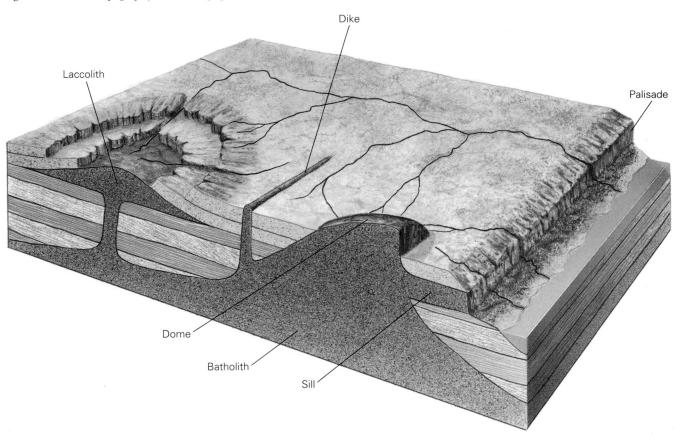

Figure 7–9 The topography and drainage pattern of plutonic rock bodies.

TOPOGRAPHY OF VOLCANIC ROCKS

Fissure eruptions of low-viscosity **flood basalts** create broad and low-relief **lava plateaus**. Flood basalts are fairly resistant to weathering and erosion (particularly in arid climates) and thus often form steep **mesas, buttes**, and **inverted valleys** (Fig. 7–10). Similarly, active shield volcanoes have broad, dome-shaped profiles and gentle slopes which are composed of multiple layers of overlapping lava flows.

Active cinder cones are typically small (less than 400 meters high) and steep. However, they are also eroded fairly rapidly because they are composed of poorly consolidated pyroclastic debris. The common remnants of the erosion of cinder cones are **volcanic necks** and **radial dikes**, which consist of igneous rocks that crystallized in their vents and feeder dikes.

Active stratovolcanoes have concave profiles with gentle, lower slopes and steep summits. They form large and imposing mountains, often with beautiful snowcapped peaks. However, they are also prone to violent eruptions which pulverize their summits and leave gaping holes in their slopes.

Circular **craters** form around the central vent at the summit of a volcano and around smaller vents along its flanks. Violent volcanic eruptions often cause the walls and floors of these craters to collapse into deep, wide **calderas**.

Volcanic cones are characterized by radial drainage patterns consisting of dip streams which flow outward in all directions from their summits. In addition, they may have a second radial drainage system which flows into their craters and calderas.

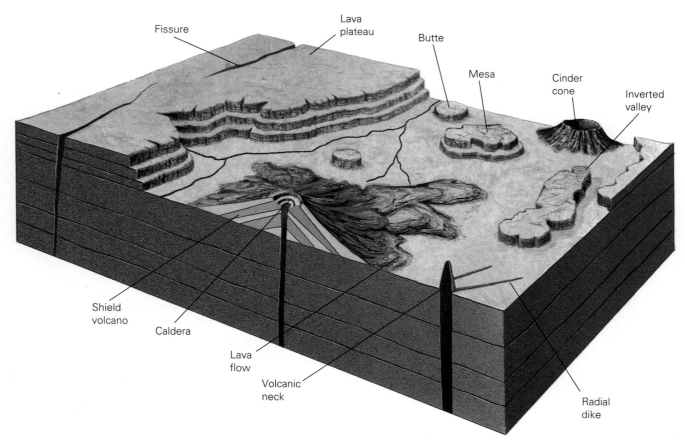

Figure 7–10 The topography and drainage pattern of volcanic rock bodies.

EXERCISE 7-4   Greasewood Canyon, Colorado

Examine the topographic map of the canyons which have been eroded into the late Cretaceous sedimentary rocks of the Colorado Plateau by the Mancos River (Map 7–3).

1. Trace the drainage pattern of the Mancos and its tributaries. What is the drainage pattern?

2. The major cliff formers in the Canyon are the fine quartzose sandstones of the Mesa Verde Group; the major slope former is the Mancos Shale, which underlies this group. Use colored pencils to shade the outcrops of the Mesa Verde Group and Mancos Shale along the walls of the canyons. How are their outcrops distinguished on the topographic map?

3. Describe the outcrop pattern of the Mesa Verde Group, and compare it to the stream drainage pattern. What do the outcrop and drainage patterns suggest about the general attitude of the strata?

4. Construct a topographic profile along line A-A′ on the accompanying sheet of graph paper. Locate and mark the topographic elevation of the contact between the Mesa Verde Group and Mancos Shale along the northern and southern walls of the canyons. Connect these points in your profile.
   a. What is the attitude of the contact between the Mesa Verde Group and Mancos Shale (and the regional structure of the formations)?

   b. What is the maximum thickness of the Mesa Verde Group in this region? _____

5. Use the vertical and horizontal scales of the map to determine the slope of the contact between the Mesa Verde Group and Mancos Shale.

EXERCISE 7-5   Grand Canyon of the Colorado River, Arizona

This exercise examines the geology, structure, and topography of Grand Canyon of the Colorado River. Grand Canyon is the largest and most spectacular of the canyons which have been cut into the Colorado Plateau by the Colorado River and its tributaries. It is 277 miles (443 km) long and 9 to 18 miles (14 to 28 km) wide from its northern to its southern rim, and more than 1 mile (1.6 km) deep at its Inner Gorge.

Grand Canyon is only a few million years old, but it exposes a sequence of igneous, sedimentary, and metamorphic rocks which encompasses 2 billion years of geologic history (Fig. 7-11). The oldest exposed rocks are the Vishnu and Brahma Schists (early Precambrian), which were deformed, metamorphosed, and intruded by the Zoroaster Granite during a period of mountain building about 1.8 billion years ago. They are overlain unconformably by the Grand Canyon Supergroup, which comprises shales, sandstones, limestones, and lava flows of late Precambrian age (0.8 to 1.2 billion years). The youngest formations in the canyon consist of several hundred meters of largely Paleozoic sedimentary rocks. This section overlies a prominent erosion surface, called the Great Unconformity, which resulted from the deformation and faulting of the Colorado Plateau at the end of the Precambrian.

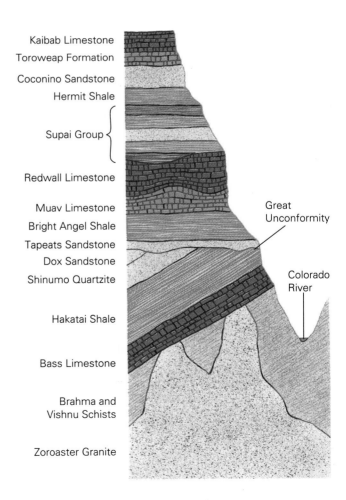

Kaibab Limestone
Toroweap Formation
Coconino Sandstone
Hermit Shale
Supai Group
Redwall Limestone
Muav Limestone
Bright Angel Shale
Tapeats Sandstone
Dox Sandstone
Shinumo Quartzite
Hakatai Shale
Bass Limestone
Brahma and Vishnu Schists
Zoroaster Granite

Great Unconformity

Colorado River

Figure 7-11 The stratigraphy of Grand Canyon of the Colorado River, Arizona.

1. Examine the satellite photograph of Grand Canyon (Fig. 7–12). Grand Canyon separates the Kaibab Plateau (to the north) from the Coconino Plateau (to the south). Label these two plateaus.

2. The Colorado Plateau was gently uplifted and slightly deformed during a long period of structural deformation which lasted from the early to middle Tertiary (about 65 to 20 million years ago). This part of the Colorado Plateau was dissected by two major sets of fractures which intersect each other at a high angle. Examine the topography of the plateaus. Locate and trace with a dark marker the traces of these two fracture sets. Measure the strikes of their traces with your protractor (assume that the top of the photograph is north). Record your data here:

3. Describe the regional drainage pattern of the Colorado River and its tributaries in Grand Canyon. What is the origin of this pattern, and what is its relationship to the fractures on the plateaus?

4. Examine the geologic-topographic map of the Grand Canyon (Map 7–4). You will find a large foldout copy of this map in the back of this lab manual.

    a. Locate the Inner Gorge of the Grand Canyon. What is its elevation in the center of the map? _____ What are the ages of the rocks exposed in the gorge? _____ What are their lithologies? _____

    b. Locate the Kaibab and Coconino plateaus in the northern and southern parts of the map, respectively. What are the ages of the rocks exposed along the plateau edges? _____ What is the elevation of the northern canyon rim (that is, the edge of the Kaibab plateau)? _____ of the southern canyon rim? _____ What does this indicate about the attitude of this part of the Colorado Plateau?

    c. What other evidence of this attitude do you see in the satellite photograph of the Grand Canyon (Figure 7-12)? Highlight this evidence with a colored pencil or marker.

    d. What is the approximate depth of the Grand Canyon in the center of the map area in feet? _____ in miles? _____

    e. What kind of geomorphic features are the various "temples" in the Grand Canyon? _____

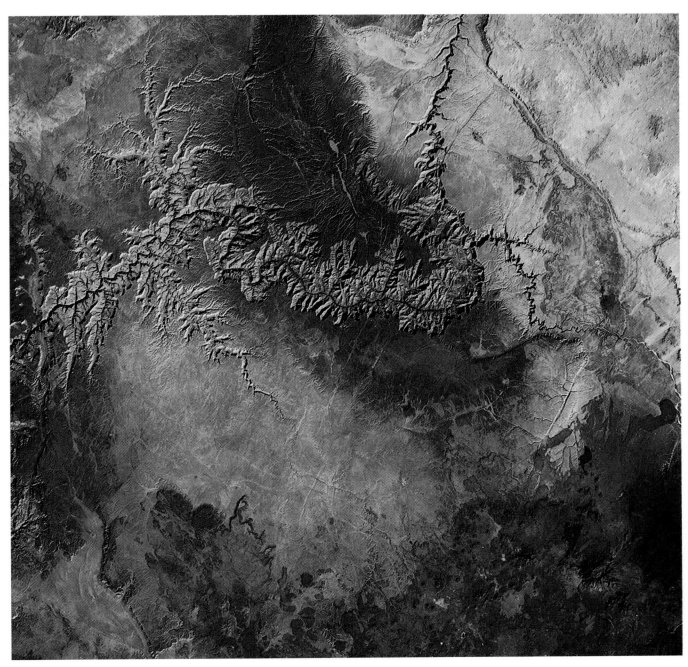

Figure 7–12 Satellite photograph of Grand Canyon of the Colorado River, Arizona.

5. Examine the structure of the Grand Canyon on the geologic-topographic map that is folded and inserted at the back of this manual (Map 7–4).

a. The canyon is dissected by several sets of cross-cutting faults. There are two principal strike directions for these faults. Measure these directions with your protractor.

b. What is the effect of this fault pattern on the general drainage pattern of the Colorado River and its tributary streams in the canyon?

c. Examine Bright Angel fault. What is the age of the oldest rock which is broken by this fault? _____ the youngest rock? _____ What is the relative age of this structure? Explain your answer.

d. Directly to the east of Bright Angel Canyon, there is a fault which runs parallel to it between Deva Temple and the Inner Gorge. What is the age of the oldest rock broken by this fault? _____ the youngest rock? _____ What is the relative age of this structure? _____

e. Determine the ages of the northwest-southeast trending faults. Are they older or younger than Bright Angel fault? _____ Are they all the same age? _____ Indicate the ages of some of these faults on the map with the appropriate symbols from Table 6–1.

f. Monoclines are structures which are characterized by the local steepening or step-like folding of strata. There are two sets of monoclines in the Grand Canyon. One set runs parallel to Bright Angel fault. Locate an example of these monoclines. What is the age of the youngest rock deformed by them? _____ Speculate on the relationship between these monoclines and Bright Angel fault.

g. Locate the second set of monoclines. Measure the general trend of their axes with your protractor. _____ What is the age of these monoclines relative to Bright Angel fault? Explain your answer.

6. Locate the outcrops of the Precambrian formations in Bright Angel Canyon. Trace the Great Unconformity (which separates these rocks from the Paleozoic strata) with a solid dark line through the length of the canyon.

    a. What formation lies directly above this surface? _____

    b. What is the approximate elevation of the Great Unconformity in the northern part of Bright Angel Canyon? _____ the southern part of the canyon? _____ in the western part of Inner Gorge? _____ the eastern part of Inner Gorge? _____ What does this indicate about the topography of the unconformity?

    c. What formation lies beneath the Great Unconformity in the northern part of Bright Angel Canyon? _____ the southern part of the canyon? _____ in the western part of Inner Gorge? _____ the eastern part of Inner Gorge? _____

    d. Trace (with a dashed line) the unconformity which separates the early Precambrian and late Precambrian rocks in Bright Angel Canyon. What happens to this unconformity in the southern part of the canyon? What is its relationship to the Great Unconformity?

    e. What do the answers to the two preceding questions indicate about the thickness and geometry of the late Precambrian rocks in Bright Angel Canyon?

    f. Describe the attitudes of the late Precambrian rocks exposed in Bright Angel Canyon.

    g. Examine the Paleozoic strata which overlie the late Precambrian rocks on the western walls of Bright Angel Canyon. Trace the drainage patterns of two streams which erode into them. What kind of drainage pattern do they have? _____

    h. Examine the outcrop patterns of the Paleozoic strata on the western side of Bright Angel Canyon. Describe the relationship between the relative ages of the strata and the elevations of their outcrops.

    i. What do the answers to the two preceding questions indicate about the attitudes of the Paleozoic strata here?

    j. What kind of unconformity is the Great Unconformity? _____

7. Examine the Paleozoic strata on the western walls of Bright Angel Canyon. Locate and trace the contact between the Muav Formation and Redwall Limestone.

    a. What is the nature of this contact? _____

    b. What is the approximate elevation of this contact on the western walls of Bright Angel Canyon? _____ along the southwest margin of the Kaibab Plateau? _____ along the north margin of the Coconino Plateau? _____ Describe the morphology and attitude of this contact in the map area.

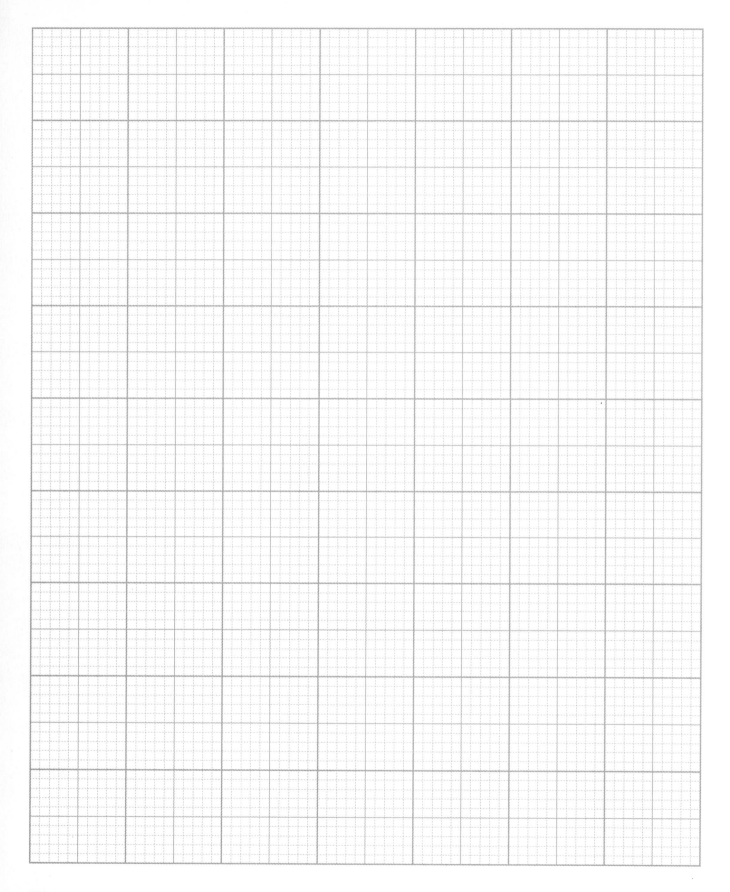

8. Summarize in proper stratigraphic order the history of the structural deformation of Grand Canyon.

9. Construct a combined geologic-topographic cross section along a line between Buddha and Brahma temples. Draw the topography first and then overlay the geology on it. Use the information from the preceding questions on the attitudes of the rock formations to correlate them through the cross section. Draw the unconformities with a wavy line.

    The line passes through Bright Angel fault, which dips to the southeast at a high angle. Show the fault on your cross section, and use arrows to show the direction of displacement along it. What kind of fault is it: normal or reverse? _____

    The line also passes through a normal fault east of Bright Angel Canyon. The downthrown block of this fault is labeled with the letter D. Indicate this fault on the cross section as well.

    What is the approximate thickness of the Paleozoic section in your profile? _____ the Precambrian _____.

10. List the names and lithologies of some of the cliff-forming and slope-forming Paleozoic strata in Grand Canyon.

11. Calculate the slopes of one cliff-forming formation and one slope-forming formation on your topographic profile.

12. Why do you find cliffs on limestone outcrops here?

Map 7–5 is a topographic map of Split Mountain in Dinosaur National Monument, Utah. The mountain exposes folded sandstones, siltstones, and shales of Mesozoic age, including the Morrison Formation, which has been a rich source of dinosaur fossils.

1. Construct a topographic profile along line A-A'. Identify the dip and scarp slopes along this profile. Locate the major cliff-forming strata in the formation, and mark their lower and upper contacts with the nonresistant strata.

2. Construct a geologic cross section which shows the general structure of the strata along your topographic profile. What kind of structure is Split Mountain? _____

3. Examine the topographic map. Identify and trace examples of dip and scarp streams, and indicate their flow directions.
    a. What kind of streams are Cottonwood and Red Wash? _____
    b. Trace and describe the drainage pattern around Split Mountain. _____

4. Locate the contacts between resistant and nonresistant strata on the topographic map, and trace them as far as possible through the map area. Color the outcrop belts of the different strata, and describe the general outcrop pattern. Indicate the oldest and youngest exposed strata in the map area with the letters O and Y.

5. Complete your geologic map by recording the strike-and-dip directions of the strata and marking the axial trace of the fold with the appropriate symbols (see previous chapter, Table 6–1). In addition, indicate the fold type (anticline or syncline) and its plunge angle and direction (if any) on your cross section.

| EXERCISE 7–7 | Folded Strata, Utah and Wyoming |

1. Examine the stereophotographs of folded strata in Figure 7–13. Trace and describe the drainage pattern in each area.

2. Trace the crests of the strike ridges in each stereophotograph, and identify their dip and scarp slopes. Indicate the strike-and-dip directions, the fold types, and the attitudes of the folds on the dip slopes with the appropriate symbols. Describe the structure of each area.

3. Locate the upper and lower contacts of the resistant strata and trace them as far as possible through each photo area. Shade the outcrop belts of these strata with a blue pencil and the outcrop belts of the nonresistant strata between them with a yellow pencil. Describe the outcrop pattern in each photo area. Indicate the oldest and youngest exposed strata in the map area with the letters O and Y.

4. The nonresistant strata in these two areas consist of thick beds of shales as well as interbedded shales and siltstones. Examine the stereophotographs carefully, and locate the outcrop belts of these two sets of strata. Compare and contrast the topography of their outcrops below. Mark their contacts and trace them as far as possible through each photo area.

5. The fold in stereophotograph B is broken by a large fault along its northern limb. Examine this limb carefully, and trace the crests of some of the strike ridges to locate the fault. Locate other faults along the northern and southern limbs of the fold. Draw the fault trace on the map. What other surficial feature(s) marks the fault trace?

A

Figure 7–13 Folded strata, Utah and Wyoming.

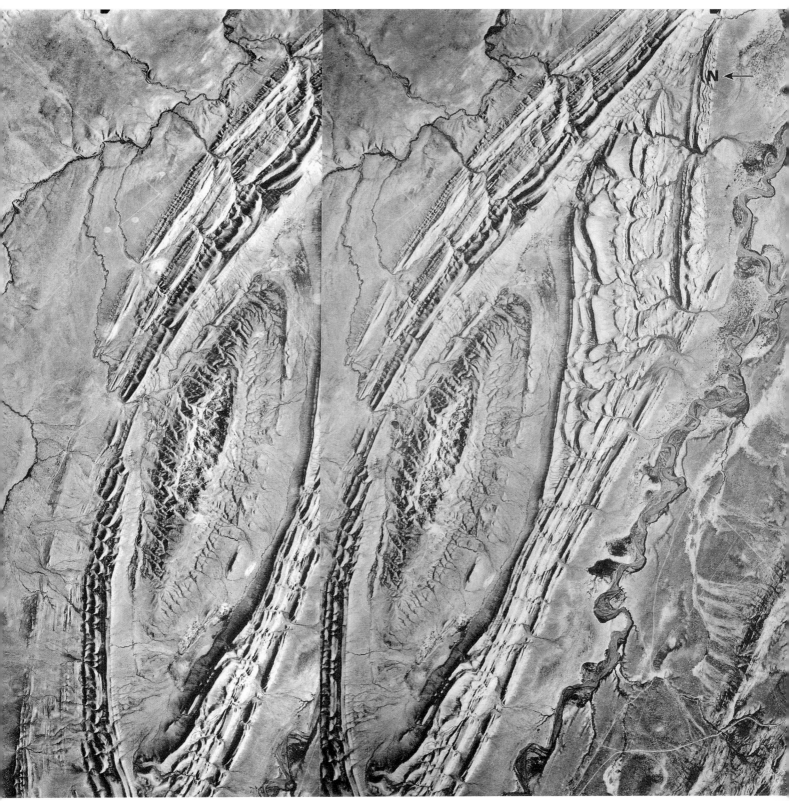

N ←

B

EXERCISE 7–8   Hollidaysburg, Pennsylvania

Refer back to the geologic-topographic map of Hollidaysburg, Pennsylvania (Map 7–2).

1.   Locate the Baker Summit fault, which runs through Bloomfield County in the southwest corner of the map, and the East and West Henrietta faults, which run through North Woodbury County in the southeast corner of the map. These three faults dip toward the east at low angles.
   a. What kind of faults are they? _____
   b. What is the evidence for your answer?

   c. Mark the upthrown and downthrown blocks of these faults.
   d. Construct a combined geologic-topographic cross section along line B-B′. Label the upthrown and down-thrown blocks, and indicate the direction of rock displacement with arrows.
   e. Describe the topographic expression of the faults intersected by line B-B′.

	KEY TO HOLLIDAYSBURG MAP	
Symbol	*Age*	*Lithology*
uD	**UPPER DEVONIAN**	Shales and sandstones
mD	**MIDDLE DEVONIAN**	Shales
lD	**LOWER DEVONIAN**	Sandstones and limestones
uS	**UPPER SILURIAN**	Limestones, sandstones, and shales
mS	**MIDDLE SILURIAN**	Shales and sandstones
lS	**LOWER SILURIAN**	Quartzose sandstones
uO	**UPPER ORDOVICIAN**	Sandstones and shales
mO	**MIDDLE ORDOVICIAN**	Limestones
lO	**LOWER ORDOVICIAN**	Dolomites and limestones
u€	**UPPER CAMBRIAN**	Dolomites and limestones
l-m€	**LOWER-MIDDLE CAMBRIAN**	Limestones and sandstones

EXERCISE 7–9   Elko, Nevada

The Basin and Range province was created by the tectonic deformation of the Cordilleran region of the United States during the late Tertiary and Quaternary. The topography consists of fault-block mountains of Paleozoic and Mesozoic rocks which alternate with basins filled with Tertiary and Quaternary sediments. Map 7–6 is a topographic map of the Ruby Mountains in northern Nevada that illustrates the typical landscape of the Basin and Range province.

1. Construct a topographic profile along line A-A'. The sharp change in topography between the mountain and the adjacent valleys marks the locations of high-angle dip-slip faults. Indicate these faults on the topographic map and topographic profile.
   a. What kind of faults are they? _____
   b. What kind of topographic feature is the mountain range? _____ the Ruby Valley? _____

2. Examine the topography of the Ruby Mountains closely.
   a. Trace and describe its drainage pattern _____
   b. What do the topography and drainage pattern indicate about the attitude and lithology of the bedrock in the Ruby Mountains? Indicate the attitude of the bedrock on the topographic map with a strike-and-dip symbol.

Map 7–7 shows the trace of the San Andreas fault zone in central California. The San Andreas is one of several strike-slip fault zones which form the boundary between the North American and Pacific plates.

1. Construct a topographic profile across the fault zone along line A-A′. Indicate the boundaries of the zone with vertical lines. What is its width? _____ the local relief in the fault zone? _____ the relief between the fault zone and the mountains? _____

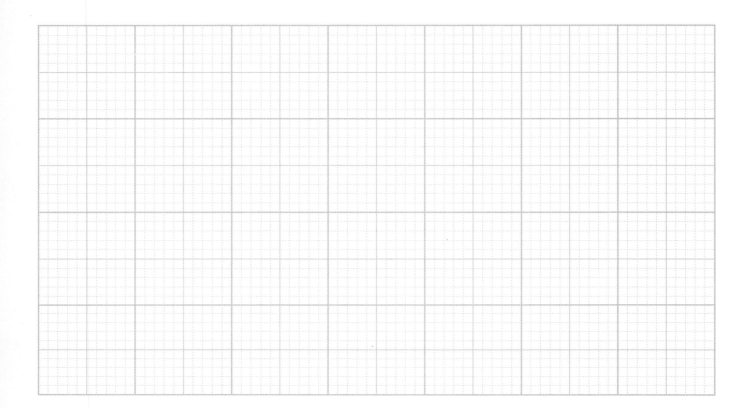

2. Trace the paths of some of the streams emerging from Buck Ridge. What happens to them when they reach the fault zone? What does this reveal about the relative directions of movement along the fault? Use arrows to indicate your answer on the map.

3. Why does the fault zone form a topographic low?

EXERCISE 7–11   Mount Capulin, New Mexico

Mount Capulin is a recently extinct volcano in northeastern New Mexico (Map 7–8). It is one of several volcanoes which formed on the Colorado Plateau during the Tertiary and Quaternary.

1. Draw a topographic profile through its cone along A-A'.
   a. What are the height and slope of the cone? _____
   b. What kind of volcano is Mount Capulin (shield, cinder, or strato)? _____
   c. What is the depression at its summit? _____

2. Mount Capulin is surrounded by erosional remnants of basaltic lava flows. Identify such remnants on the map and profile. Describe their geometry and topography.

3. Which is older: Mount Capulin or the lava flows? _____ How do you know? What does this suggest about the last phase of volcanic activity at this location?

Examine the stereophotograph of Shiprock (Fig. 7–14).

1. Describe its topography. What kind of topographic feature is Shiprock?

2. What is the origin of the steep vertical faces and profile of Shiprock?

3. Describe the topography of the surrounding desert floor. What is the origin of the linear ridges which extend away from Shiprock?

4. How will the topography of the photo area change in the future?

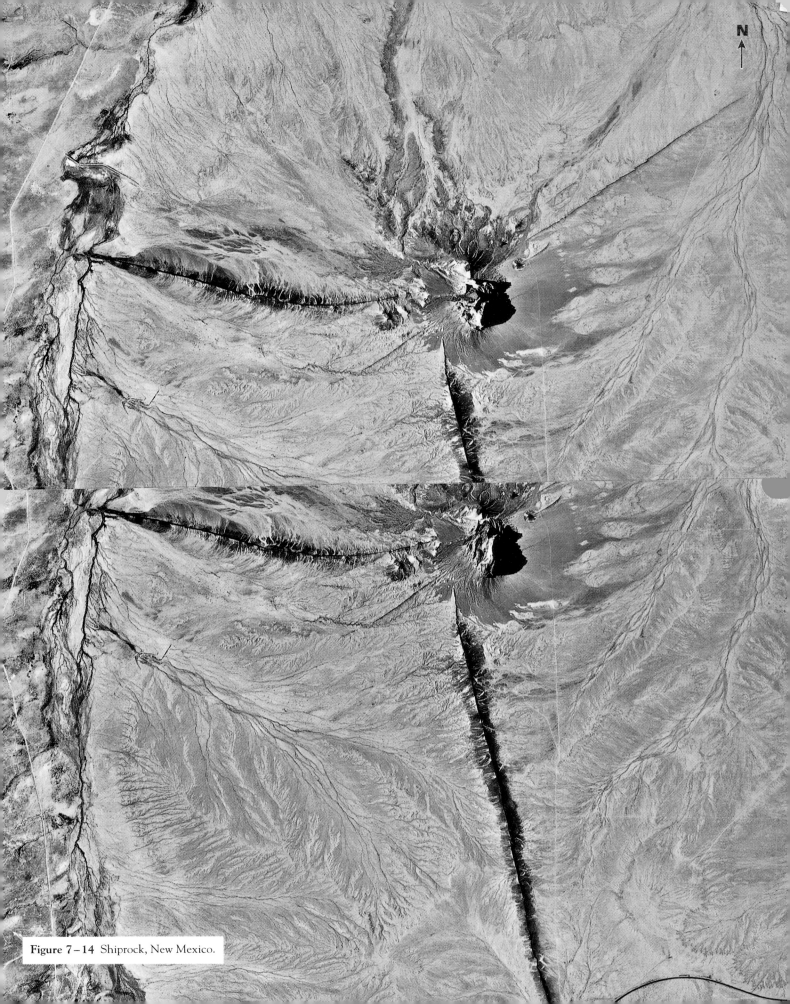

Figure 7–14 Shiprock, New Mexico.

EXERCISE 7–13   Crater Lake National Park, Oregon

Mount Mazama is one of the string of majestic stratovolcanoes which comprise the Cascade Range of the northwestern United States (Map 7–9). This volcano once stood some 12,000 feet high; it was reduced to its present height during a violent eruption about 6900 years ago. The eruption pulverized the summit, threw tons of rocks into the atmosphere, and caused the lower slopes of the volcano to collapse into a deep caldera. This caldera is now filled by Crater Lake, which is the deepest lake in the United States. Since the eruption, several smaller eruptions have constructed a small cone, Wizard Island, within the caldera.

1. Construct a topographic profile through the center of the caldera. The cross section should run through Mazama Rock on the northeast slope and Rim Village Campground on the southwest rim, and it should include the submerged part of the caldera. Use a vertical scale of 1″ = 2000′, and extend the vertical axis of the graph up to 12,000 feet.

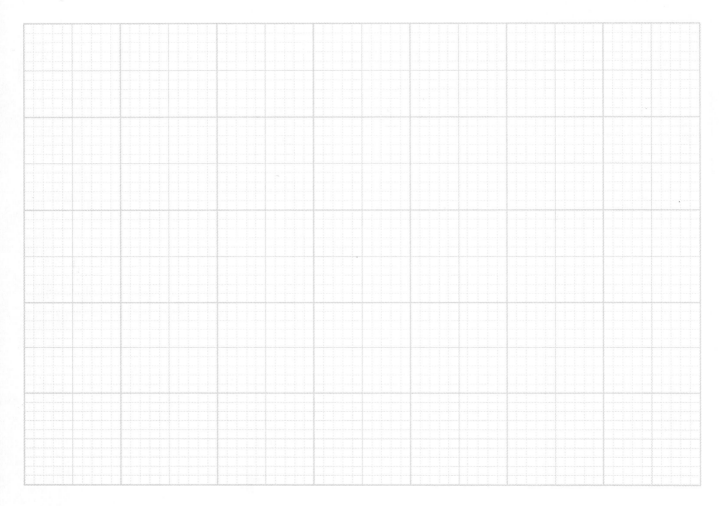

2. Refer to Table 5–3 (Chapter 5) in order to answer these questions:
   a. What is the diameter of the caldera in miles? _____ in kilometers? _____
   b. What is the radius of the caldera in miles? _____ in kilometers? _____
   c. What is its maximum depth in feet? _____ in meters? _____
   d. What is the elevation of its rim in feet? _____ in meters? _____

3. Estimate the volume of summit rock ejected into the atmosphere during the eruption. To do this, you can assume that the pulverized summit of the volcano was shaped roughly like a cone, with a round base at the caldera rim and an apex at 12,000 feet.

    a. The cross section of a cone-shaped summit has a roughly triangular profile. Sketch this profile on the cross section with dotted lines.

    b. What is the height of the cone (the lost summit between the present rim and the former peak) in feet? in meters? _____ in kilometers? _____

    c. The volume of a cone is calculated with the formula

        Volume = (Square of the radius (r^2) of the base $\times$ pi (3.14) $\times$ height of the cone) $\div$ 3

    Calculate the volume of the lost summit in cubic kilometers.

    d. A cubic meter of andesite weighs approximately 2.5 tons. Calculate the weight of rock which was ejected into the atmosphere when the summit exploded using these steps:

        1. Calculate the number of cubic meters in one cubic kilometer. (Hint: A cubic kilometer has sides which are 1000 meters long.)

        2. Calculate the weight of a cubic kilometer of andesite.

        3. Calculate the weight of the pulverized summit.

    4. Trace and describe the drainage pattern on Mount Mazama. _____

    5. Examine the topographic map of the caldera rim carefully. The caldera rim is riddled with radial dikes composed of andesite, which is fairly resistant to weathering. What topographic evidence do you see for the dikes? Indicate the approximate trends of these dikes on the map, and measure them with your protractor.

    6. Describe the topography, relief, and slope of Wizard Island. What kind of volcanic cone is it?

Figure 7–15 is a stereophotograph of Green Mountain, a domal structure which has been formed by the emplacement of a shallow laccolith. The igneous core of the laccolith is not exposed but lies beneath the folded beds of the Minnelusa Formation (which appears dark in the stereophotographs because it is vegetated densely with pines). The resistant sandstones and limestones in the Minnelusa form prominent hogbacks.

Figure 7–15 Green Mountain, Wyoming.

1. Examine the stereophotographs, locate and trace the crests of the Minnelusa hogbacks, and indicate their attitudes with strike-and-dip symbols.

2. The Minnekahta Limestone, which is younger than the Minnelusa, also forms a set of hogbacks around Green Mountain. Locate and trace these ridges.

3. Trace and describe the drainage pattern around Green Mountain. _____

4. In the space provided, sketch a rough geologic cross section through the dome.

EXERCISE 7–15   Plutonic Rock Bodies

Map 7–10A (Stone Mountain) is a topographic map of a granite stock which has intruded through the igneous and metamorphic rocks of the Georgia Piedmont. Map 7–10B (Elizabethtown, New York) is a topographic map from part of the Adirondack Mountains, where the crystalline basement of the North American continent is exposed. Map 7–10C (Rocky Bar, Idaho) is a topographic map of an exposed batholith.

1. Sketch a rough north–south topographic profile through Stone Mountain. Trace and describe its stream drainage pattern. _____ This drainage pattern and the smooth domal topography, which is the result of exfoliation weathering, are common to massive crystalline rocks. Locate some examples of similar topographic features from the other two map areas.

2. Trace and describe the regional drainage patterns in the Elizabethtown and Rocky Bar map areas. What is the origin of their drainage patterns?

8

SEISMOLOGY

EARTHQUAKE CRACKS. VOLCANO NATIONAL PARK, HAWAII. *(Tony Stone Worldwide, Ltd./Nancy Simmerman)*

Materials Needed *Colored pens or pencils, lead pencils with erasers, ruler, calculator, compass*

Seismology is the study (in Greek, *logos*) of the shaking (*seismos*) of the Earth. The Earth is often jolted and quaked by large, sudden releases of energy. These **earthquakes** originate in the rocks of the lithosphere due to the movement of rocks along faults, and they set off vibrations called **seismic waves,** which radiate outward in all directions from their source (Fig. 8–1). The directions and velocities of seismic waves often change as they travel through the Earth's interior, for they are dependent on the physical properties of the

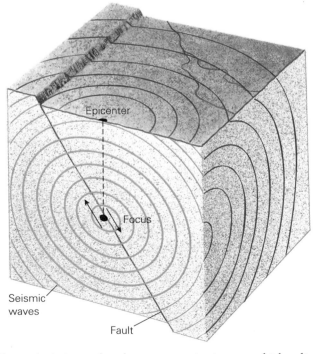

Figure 8–1 An earthquake generates seismic waves which radiate outward in all directions from its focus. The epicenter is the point on the surface which lies above the focus.

179

matter through which they move. For instance, they travel more rapidly through the dense rocks of the deep mantle than through the less dense rocks of the crust, and more rapidly through the dense mafic rocks of the oceanic crust than through the lighter sialic rocks of the continental crust.

Consequently, the study of seismic waves is the major—and, in some cases, the only—source of information about the structure, composition, and physical properties of the Earth's interior. **Seismologists** use sensitive instruments called **geophones** and **seismographs** to detect and record seismic waves, and they use powerful computers to analyze their directions and travel times to produce a dynamic picture of the whole Earth which could not be acquired by any other means. **Seismic analysis** enables us to pinpoint the source of an earthquake and measure its intensity; it locates the tectonically active **seismic zones** in the lithosphere and monitors the movement and interaction of the plates; and it reveals the nature of the Earth's deep interior and produces detailed, X-ray-like images of its crust called **seismic profiles**.

This chapter discusses the principles of seismology and its application to the study of the structure and composition of the Earth's interior. The first section deals with the causes of earthquakes, the location of their epicenters, and the measurement of their magnitudes. The second section deals with the field of seismic stratigraphy, in which the principles of seismology are applied to the detailed study of the structure and composition of the Earth's crust.

EARTHQUAKES

CAUSES OF EARTHQUAKES

Earthquakes are generated by movement and deformation of rock and magma in the Earth's lithosphere. They occur most frequently in the Benioff zones of subduction margins, where plates are being dragged beneath other plates, massive batholiths are rising upward into the crust, and volcanoes are erupting on the surface. They are also common in collision zones, where rocks are deformed and displaced along faults by tremendous compressional forces; at transform margins, where the edges of plates rub and grind against each other; and in rift zones, where there is active volcanism and the movement of rocks along transform faults. Earthquakes are comparatively rare (but by no means unknown) in the interiors of plates.

The mechanics of earthquake generation are illustrated in Figure 8–2 by a model for the San Andreas fault of California. The San Andreas is one of several major strike-slip faults between the North American and Pacific plates, which are sliding past each other at a net rate of about 3.5 cm per year. This movement is smooth and steady along some stretches of the fault (Fig. 8–2A), but there are other stretches where it occurs in quick, sudden lurches which interrupt long periods of no movement.

This lurching movement is due to friction between the rocks on the two sides of the San Andreas fault. The friction locks these rocks together and prevents movement along the fault for years or even decades (Fig. 8–2B). During such times, the rocks along the fault absorb energy from the movement of the plates, and they are strained and **elastically**

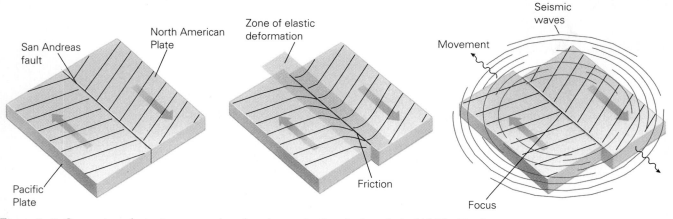

Figure 8–2 Generation of seismic waves and earthquakes on the San Andreas fault. (A) The North American and Pacific plates move past each other at a relative speed of 3.5 c per year. (B) At some places, friction between the edges of the two plates locks them together. The continued movement of the plates is absorbed by their edges, which become elastically deformed. (C) Eventually the rocks along the edges of the plates fracture and displacement occurs in a series of short jerky motions along the fault.

deformed. The friction is eventually overcome when the rocks reach their **elastic limit** and fracture. The break releases the accumulated elastic energy in the rocks, and they are displaced along the fault, causing an earthquake (Fig. 8–2C). This displacement occurs in a series of short movements which begins at the **focus** (the point of initial fracture) and travels down the length of the fault, and it generates **seismic waves** of energy which radiate outward from the focus and cause the ground to shake and tremble.

SEISMIC WAVES

Two classes of seismic waves are propagated by earthquakes: **body** and **surface**. They are distinguished on the basis of their sources, travel paths, and travel velocities.

Body Waves

Body waves radiate outward from the initial fracture point (focus) of an earthquake and travel through the Earth's interior (Fig. 8–1). There are several types of body waves, but the two most important in the study of earthquakes are **primary** and **secondary waves**, or more simply **P waves** and **S waves**.

P waves are also known as push-pull or compressional waves because they alternately compress and expand the rocks through which they travel in the direction of propagation. They are called primary waves because they travel faster (5 to 7 km/sec in the crust) than S waves and therefore arrive first at seismograph stations. P waves are able to travel through all forms of matter; that is, they can pass through the Earth's solid crust and mantle, its liquid outer core, and its solid inner core.

S waves are also known as side-to-side or shear waves because they move at right angles to the direction of propagation. They travel more slowly (3 to 4 km/sec in the crust) and arrive later at seismograph stations than P waves. They cannot travel through liquids, and thus they cannot penetrate the outer core.

Surface Waves

Surface waves radiate outward from the **epicenter** of an earthquake (the point on the surface directly above the focus) and travel across the Earth's surface (Fig. 8–1). They are also called long or L waves because they travel very slowly (2 to 3 km/sec) and thus arrive last at seismic stations. Surface waves have two distinct components of movement: an up-and-down motion (which appears as a wave crest that sweeps across the land) and a side-to-side motion caused by shearing. The movement of surface waves, and particularly their shearing motion, is the principal cause of earthquake damage to structures such as buildings and bridges.

EARTHQUAKE DETECTION

Seismic waves are detected by **seismographs**, which produce continuous records of earthquake activity called **seismograms** (Fig. 8–3).

The seismogram of an earthquake consists of three sets of wavy traces which record the successive arrivals of P waves, S waves, and L waves. The **arrival times** of these waves, and the **time lag** between them, are measured in minutes and seconds on the horizontal scale of the seismogram. The time lag between the P waves and S waves is directly proportional to the distance between a seismograph station and the epicenter of an earthquake: If the station is fairly close to the epicenter, the S waves arrive shortly after the P waves, but as distance from the epicenter increases, the time lag between the waves also increases. This relationship, which is summarized in **travel-time curves** (Fig. 8–4), is the key to locating the epicenter of an earthquake.

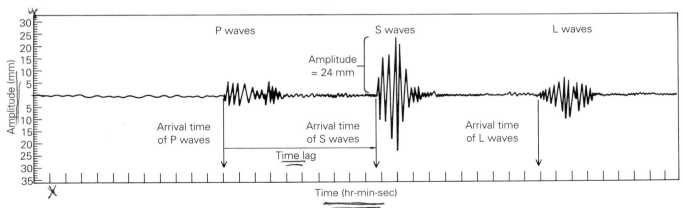

Figure 8–3 A seismogram of an earthquake consists of three sets of wave traces. The traces record the arrival times and amplitudes of P waves, S waves, and L waves.

The **amplitudes** of seismic waves (i.e., the heights of their peaks) are measured in millimeters with the vertical scale of the seismogram. The recorded amplitudes of seismic waves can be used to calculate the absolute **magnitude**, or size, of an earthquake, because they are directly proportional to the amount of elastic energy released by it. However, this calculation must also take into account the distance between the seismograph station and the epicenter of a quake, because seismic waves are dampened and weakened as they travel through the Earth's interior.

Once the distance to an earthquake's epicenter has been determined from the arrival times of the P and S waves and the amplitudes of its seismic waves have been measured from the seismogram, the magnitude of the earthquake can be calculated with the aid of a graphical device called a **nomogram**, shown in Figure 8–5. This nomogram contains a logarithmic scale of earthquake magnitude which bears the name of its inventor, **Charles Richter**.

However, the magnitude of an earthquake is not always as important as its **intensity**, which describes its effect on surficial features, including buildings, roads, and other structures. For example, a magnitude 5.0 earthquake with an epicenter in the Mojave Desert would have less effect on the city of San Francisco than a magnitude 5.0 quake with an epicenter in San Francisco Bay, because the seismic waves from the Mojave quake would be dampened as they traveled the long distance between the desert and the city.

The intensity of an earthquake is not measured with seismograms. Rather, it is assessed on the basis of structural damage, human reactions, and other similar criteria. The **Modified Mercalli scale** defines 12 levels of earthquake intensity (I through XII) on the basis of these criteria (Table 8–1).

The subjectivity and potential for error of the Mercalli scale should be evident from criteria such as "many frightened" and "damage negligible in buildings of good construction." However, Mercalli-scale measurements are useful for predicting earthquake damage and gauging the sizes of earthquakes which occurred before the invention of the seismograph.

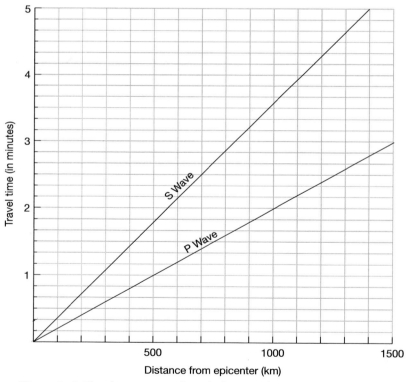

Figure 8–4 Travel-time curves through the crust for P waves and S waves.

TABLE 8–1   Modified Mercalli Intensity Scale of 1931

I.   Not felt except by a very few under especially favorable circumstances.

II. Felt only by a few persons at rest, especially on upper floors of buildings

III. Felt quite noticeably indoors, especially on upper floors, but many people do not recognize it as an earthquake. Vibration like the passing of a truck.

IV. During the day felt indoors by many, outdoors by few. At night some awakened. Dishes, windows, doors disturbed; walls make cracking sound. Sensation like heavy truck striking building.

V.   Felt by nearly everyone; many awakened. Some dishes, windows, etc., broken; a few instances of cracked plaster; unstable objects overturned. Disturbance of trees, poles, and other tall objects sometimes noticed. Pendulum clocks may stop.

VI. Felt by all; many frightened and run outdoors. Some furniture moved; a few instances of damaged chimneys. Damage slight.

VII. Everybody runs outdoors. Damage *negligible* in buildings of good con-struction; *slight to moderate* in well-built ordinary structures; *consider-able* in poorly built or badly designed structures.

VIII. Damage *slight* in specially designed structures; *considerable* in ordinary substantial buildings; *great* in poorly built structures. Fall of chimneys, fac-tory stacks, columns, monuments, walls.

IX. Damage *considerable* in specially designed structures; well-designed frame structures thrown out of plumb; *great* in substantial buildings, with partial collapse. Ground cracked conspicuously. Underground pipes broken.

X.   Some well-built wooden structures destroyed; most masonry and frame structures destroyed; ground badly cracked. Considerable landslides from river banks and steep slopes.

XI. Few, if any, masonry structures remain standing. Bridges destroyed. Broad fissures in ground. Underground pipelines completely out of service. Earth slumps and land slips in soft ground.

XII. Damage total. Waves seen on ground surface. Lines of sight and level dis-torted. Objects thrown upward into the air.

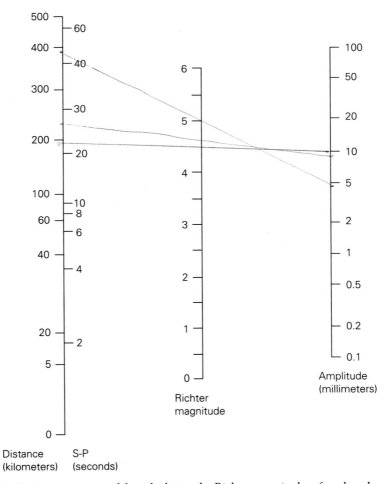

Figure 8–5 A nomogram, used for calculating the Richter magnitudes of earthquakes.

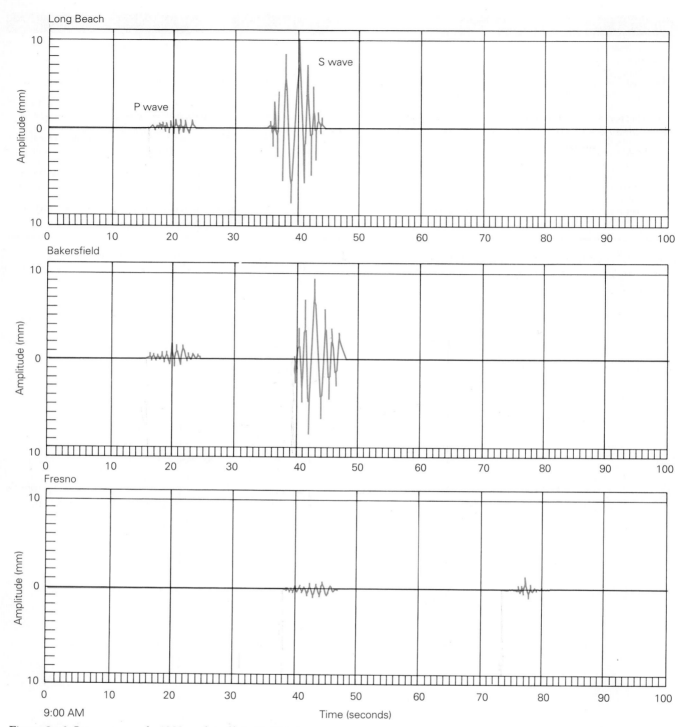

Figure 8–6 Seismograms of a 1989 earthquake in California.

Name _____ Section _____ Date _____

EXERCISE 8 – 1   Velocity of Seismic Waves

An earthquake occurs 1000 km from a seismograph station at exactly 9:00:00 A.M. This earthquake generates P waves which travel at 7 km/sec, S waves which travel at 4 km/sec, and L waves which travel at 3 km/sec. Calculate the arrival times at the seismograph station for each set of waves.

P- waves = 9:02:23   L- waves = 9:05:33

S- waves = 9:04:10

EXERCISE 8 – 2   Analysis of Seismic Records

This exercise demonstrates the analysis of the seismic record of an earthquake along the southern California coast in 1989. You are to locate the epicenter of the earthquake and measure its magnitude. Summarize your data on the accompanying table.

STATION	ARRIVAL TIMES		TIME LAG	DISTANCE TO EPICENTER	AMPLITUDE OF S WAVE	MAGNITUDE
	P wave	S wave				
Long Beach	16 sec.	35 sec.	19 sec.	195 km	10 mm	4.5
Bakersfield	16 sec.	39.5 sec	23 sec.	235 km	9 mm	4.6
Fresno	38 sec.	73 sec.	35 sec	390 km	4 mm	5.0

I. Locating the Epicenter

The time lag between the P waves and S waves in a seismic record measures the distance between an epicenter and a seismograph station, but it does not define the direction to that epicenter. To determine the location of an epicenter, it is necessary to analyze data from at least three stations. Such data, from three stations in California (Fig. 8–6), are shown in Figure 8–7.

Locate the epicenter of the earthquake according to the following steps:

1. Measure the arrival time lags between the P waves and S waves at each station with the horizontal scales on the seismograms.

2. Refer to the travel-time curves in Figure 8–4. For each station, place the edge of a sheet of paper along the vertical time axis, and mark off the length of the time lag between the P waves and S waves. Slide the edge of the paper vertically between the P and S curves until the distance between the marks on the paper corresponds to the distance between the two curves. The value of the horizontal axis at this point is the distance to the epicenter for that station.

3. Use a compass to draw a circle around each station. The radius of each circle is equal to the distance between each station and the epicenter. For each station, the position of the epicenter could be at any point along its circle. The actual position of the epicenter is the point of intersection between these three circles. What are its latitude and longitude?

II. Measuring the Magnitude

For one station, measure the amplitude of the largest seismic wave (the amplitude is equal to half the vertical distance between its two peaks). Plot this value and the distance to the epicenter on the nomogram in Figure 8–5, and draw a line to connect them. The magnitude of the earthquake is defined by the intersection between this line and the Richter scale in the center of the nomogram.

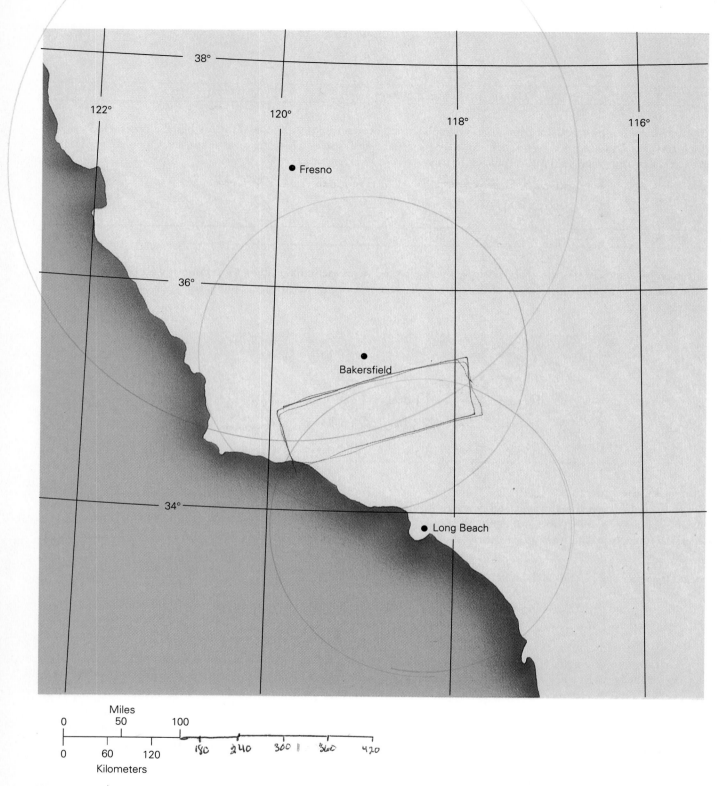

Miles
0   50   100
0   60   120   180 240 360 | 360   420
Kilometers

Figure 8–7 Locations of three seismograph stations in California.

EXERCISE 8-3 Mercalli Intensity

The San Francisco Bay Area lies in one of the most seismically active regions of the world. The city was devastated by a magnitude 8.3 earthquake and its ensuing fire in 1906, and it is periodically struck with lesser-magnitude quakes. San Francisco has strict construction codes for new buildings, roads, and other structures which are intended to avoid or minimize the effects of earthquakes, and it has invested heavily in reinforcing its existing infrastructure to bring it up to code. However, even the best engineering is no guarantee against earthquake damage, as the city learned when it was shaken for a full 30 seconds during the height of rush hour on October 17, 1989. The earthquake measured 7.1 on the Richter scale, and its epicenter was located 70 miles away in the Santa Cruz Mountains.

1. The following table contains newspaper accounts of the 1989 earthquake. Use the observations and incidents described in each account to estimate the level of intensity of the earthquake.

   Estimate the level of each incident. Record your estimate on the right side of the table. The first account has been completed as an example. Notice that this estimate is represented with a bar which extends from level V to level VIII. This indicates that it would vary depending on the quality of construction of the home, the placement of the pottery, and whether the Donaldsons felt that the destruction of their pottery was slight or considerable, among other factors.

   Complete the rest of the table in a similar fashion. Once you have done this, a pattern will emerge and you will be able to estimate the overall Mercalli level. Record that value here:

2. Note the variance in your estimates of the incidents. Discuss with your classmates the possible sources of this variance.

3. The composition of the earth foundation beneath each structure is indicated in parentheses at the end of each incident. What role does it play in earthquake intensity and damage?

INCIDENT	MERCALLI LEVEL
Rene and Tony Donaldson live near Stanford University. Their $425,000 home escaped major damage in the Pretty Big One, but the tremors did smash their collection of American Indian pottery. "Now I know why California Indians didn't have a pottery tradition," Rene says. (Alluvium)	I II III IV V VI VII VIII IX X XI XII
The shattered portion of the I-880 freeway in Oakland will have to be torn down, and the Embarcadero Freeway, a double-decker that skirts downtown San Francisco, is riddled with cracks in the support columns. (Tidal flats)	I II III IV V VI VII VIII IX X XI XII
Pier 45, the city's main fishing pier, was closed because inspectors found deep fissures running the length of the pier floor. (Landfill)	I II III IV V VI VII VIII IX X XI XII
Cracks appeared in the paving of the main runway at Oakland International Airport. (Landfill)	I II III IV V VI VII VIII IX X XI XII
In the South Bay town of Watsonville, 125,000 crates of strawberries spoiled when an electrical failure knocked out refrigeration equipment. (Landfill)	I II III IV V VI VII VIII IX X XI XII
Six people were crushed to death in their cars when part of an old four-story brick building toppled onto the vehicles on Bluxome Street in San Francisco, said Police Lt. Jerry Kilroy. (Alluvium)	I II III IV V VI VII VIII IX X XI XII
The World Series was postponed because of a power loss and a crack in the stadium's upper tier. (Solid bedrock)	I II III IV V VI VII VIII IX X XI XII
"Then I realized, No, I was the one who was moving, that the whole press box was moving. It moved three times, then it began to sway back and forth. There was a good 15 seconds of bouncing and swaying." (Solid bedrock)	I II III IV V VI VII VIII IX X XI XII
A section in the upper stands in right field separated by about 6 inches, several cracks ran down the concrete aisle, and pieces broke off. Fans were carrying them home as souvenirs. (Solid bedrock)	I II III IV V VI VII VIII IX X XI XII
Mayor Art Agnos' press secretary, Eileen Mahoney, said as many as 20 people had been injured at a fire in the Marina section. Another fire was blazing near downtown Berkeley. (Landfill)	I II III IV V VI VII VIII IX X XI XII
In downtown San Francisco, plaster fell off buildings and windows popped off high-rises. (Alluvium)	I II III IV V VI VII VIII IX X XI XII
Some people were hit with falling plaster at San Francisco Airport. (Landfill)	I II III IV V VI VII VIII IX X XI XII
Several motorists were reportedly tossed from the Bay Bridge into the water, more than 100 feet below, when the 50-foot, upper-level span collapsed onto a lower level. (Tidal flats)	I II III IV V VI VII VIII IX X XI XII
Texas Treasurer Ann Richards, a Democratic candidate for governor, was in San Francisco Tuesday for a political fund-raiser when the earthquake hit. "I was walking on the street near Union Square when suddenly a large granite building began swaying," she told her press secretary. (Solid bedrock)	I II III IV V VI VII VIII IX X XI XII
All the windows shattered in the Saks Fifth Avenue store on Union Square. (Solid bedrock)	I II III IV V VI VII VIII IX X XI XII
An apartment building covering an entire block caught fire in San Francisco's Marina District. (Landfill)	I II III IV V VI VII VIII IX X XI XII
San Francisco City Hall and the state building housing the California Supreme Court were evacuated. Both buildings had severe internal damage. (Alluvium)	I II III IV V VI VII VIII IX X XI XII
Near the Civic Center, a large section of a four-story building came down, and witnesses said at least two bodies of women were removed. (Alluvium)	I II III IV V VI VII VIII IX X XI XII
In the Financial District, huge cracks formed in the streets. Large windows at street level shattered. Witnesses said older buildings collapsed, and the dust was so thick that people could not see. Newer buildings were swaying. (Landfill)	I II III IV V VI VII VIII IX X XI XII
Officials of the Bay Area Rapid Transit system said initial reports indicated there was no damage to the tube under the bay. (Tidal flats)	I II III IV V VI VII VIII IX X XI XII
Several major roadways sustained heavy damage. A portion of Highway 101, one of the main arteries to the south, buckled in several places near San Francisco International Airport. (Landfill)	I II III IV V VI VII VIII IX X XI XII
Catastrophic failure of Interstate 880 freeway, as an upper deck collapses onto a lower deck.	I II III IV V VI VII VIII IX X XI XII
There are ruptured water lines in the Marina District. (Tidal flats)	I II III IV V VI VII VIII IX X XI XII
There are cracks in the street, resulting from compaction of underlying soft bay muds, in the Marina District. (Landfill)	I II III IV V VI VII VIII IX X XI XII
Buildings have collapsed in the Marina District. (Landfill)	I II III IV V VI VII VIII IX X XI XII

Seismic stratigraphy is the application of the principles of seismology to the study of the structure and composition of the Earth's interior.

It was previously pointed out that seismic waves change velocity and direction as they travel through the Earth's interior, for they are dependent on the density of the matter through which they pass. For example, they travel at 8 km/sec through the mantle, which consists of highly pressured ultramafic rocks with densities of 3300 to 5500 kg/m^3, but they decelerate to 6 km/sec when they pass into continental crust, which consists of sialic rocks with densities of only 2500 to 2700 kg/m^3. This characteristic enables geologists to map out the structure of the Earth's interior, such as the dimensions of its crust, mantle, and core, as well as to determine its overall composition and physical properties.

This same principle can also be applied to a more detailed study of the structure and composition of the Earth's crust, because seismic waves also change direction and velocity as they travel through the crust's many different types of rocks. For example, they travel at different velocities in rocks of different lithologies; they are often reflected when they cross the boundary between a **low-velocity zone** (such as a shale layer) into a **high-velocity zone** (such as a limestone layer); and they are scattered by massive and structureless rocks such as reefs, salt domes, and granite batholiths.

The analysis of the travel of shallow seismic waves produces detailed cross-sectional images of the crust called **seismic profiles**. Seismic profiles record the behavior of waves which are generated on the Earth's surface by the detonation of explosives, the pounding of a large trunk-mounted hammer or thumper against the ground, the explosion of an underwater air gun, or some other means. These waves travel downward into the crust until they are reflected back to the surface by a **reflecting horizon**, such as a layer of quartz-cemented sandstone or limestone. The reflected waves are then detected at the surface by an array of listening devices called **geophones**, which feed the arrival times and strengths of the returning wave signals into a computer. The computer records these data and then analyzes them to produce a seismic profile (Fig. 8–8).

A seismic profile shows the locations of the major reflecting horizons in the crust, as well as the travel velocities of waves as they pass through the layers of rocks between these horizons.

Reflecting horizons mark the depths at which abrupt changes occur in the seismic velocity. A reflecting horizon can have many possible origins. For example, it may mark the boundary between a low-velocity shale formation and a high-velocity limestone formation; it may be a cherty layer within a limestone formation; or it may mark an erosional unconformity between two formations. Reflecting horizons are useful for dividing the sediment fill of basins into smaller intervals called **sequences** and **parasequences**, which consist of rocks of different lithologies, ages, and origins. The shapes and lateral continuities of reflecting horizons are also useful for identifying geologic structures such as folds and faults in a seismic profile.

A seismic profile appears to show an exact cross-sectional representation of the Earth's crust, but this is not entirely true. The horizontal scale of a seismic profile is truly proportional to the horizontal distance across a profiled area. However, the image is distorted in the vertical direction, for the vertical axis does not measure the depths or thicknesses of rock layers; rather, it measures the **two-way travel time**, which is the time required for waves to travel to a reflecting horizon and then back to the surface.

Generally, it is not possible to determine either the origins of reflecting horizons or the thicknesses of the rock

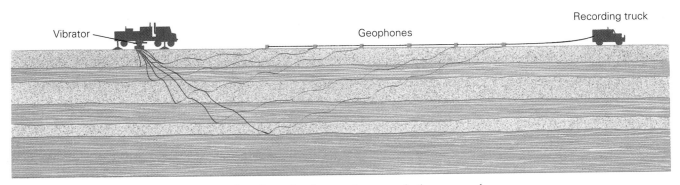

Figure 8–8 Shallow seismic waves, generated on the surface by a truck-mounted vibrator, travel downward into the crust until they reach a reflecting horizon. They are then bounced back to the surface, where their arrival times and amplitudes are recorded by geophones.

layers between them with a seismic profile alone. Rather, a seismic profile must usually be accompanied by a nearby **control well**, where the rock layers have been drilled and cored. This allows geologists to examine the rocks of the reflecting horizons and thus determine their origins and significance, as well as to describe the lithology and measure the physical properties (particularly the densities and travel times) of the different rock layers. Once the travel times of the rock layers have been determined in this fashion, it is possible to calculate their actual thicknesses.

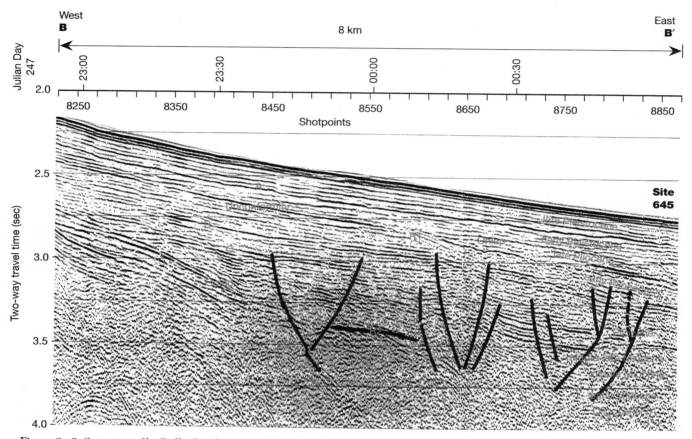

Figure 8–9 Seismic profile, Baffin Bay (courtesy of Ocean Drilling Program, Texas A&M University). The R_2 reflector is outlined by the subhorizontal black line in the center of the profile.

EXERCISE 8-4   Seismic Profile, Baffin Bay

The *JOIDES Resolution,* the drill ship for the Ocean Drilling Program, drilled and cored a 1147-meter-thick sequence of sedimentary rocks at Site 645 in Baffin Bay (situated between Greenland and Baffin Island, Canada) in August of 1985. Figure 8–9 shows an interpreted seismic profile for the area around this site, which was surveyed prior to the drilling expedition.

The seismic profile shows that Site 645 is underlain by many layers of sedimentary rock, which the core shows are composed of sandstones, siltstones, and claystones of Oligocene to late Pleistocene age. The rock layers can be divided into two major sequences by a prominent reflecting horizon, R1, which the core shows to be an erosional unconformity. The upper sequence consists of late Pliocene through Pleistocene rocks, and it can be further subdivided into two units (I and II) by another reflector, which is marked A on the seismic profile. The lower sequence (Unit III) consists of Oligocene through early Pliocene rocks, which are extensively faulted. It can also be subdivided into two units (IIIa and IIIb) by a reflecting horizon, which is marked R2. Unit III also contains many minor reflectors, including one marked B and a pair of closely spaced horizons marked C.

1.   Use different-colored pencils to trace the five principal reflecting horizons (A, B, C, R1, and R2) across the profile. Describe the effects of the faults in Unit III on the vertical positions of reflecting horizons B and C.

   a. What kind of faults are they? Indicate the relative direction of displacement along each fault with an arrow.

   b. Have Units I and II been affected by these faults? What is your evidence?

2.   The average travel velocities of the three principal units (I, II, and IIIA) at Site 645—that is, the velocities at which seismic waves travel through them—were determined by geophysical analysis of samples from each unit. They are as follows:

> Unit I: Velocity = 1825 m/sec
> Unit II: Velocity = 1875 m/sec
> Unit IIIA: Velocity = 2100 m/sec

Determine the thickness of each of the principal units at Shotpoint 8550 by the following steps:

   a.   Measure the two-way travel time (in seconds) for each unit with the vertical scale.

   b.   Calculate the one-way travel time (in seconds) by dividing this value by 2.

   c.   Multiply the one-way travel time by the average travel velocity for that unit.

3.   Units I, II, and III are all composed of the same types of sedimentary rocks, and yet Unit III has a faster travel time than Units I and II. Speculate on the possible reason(s) for this difference.

Two-way travel time (sec)

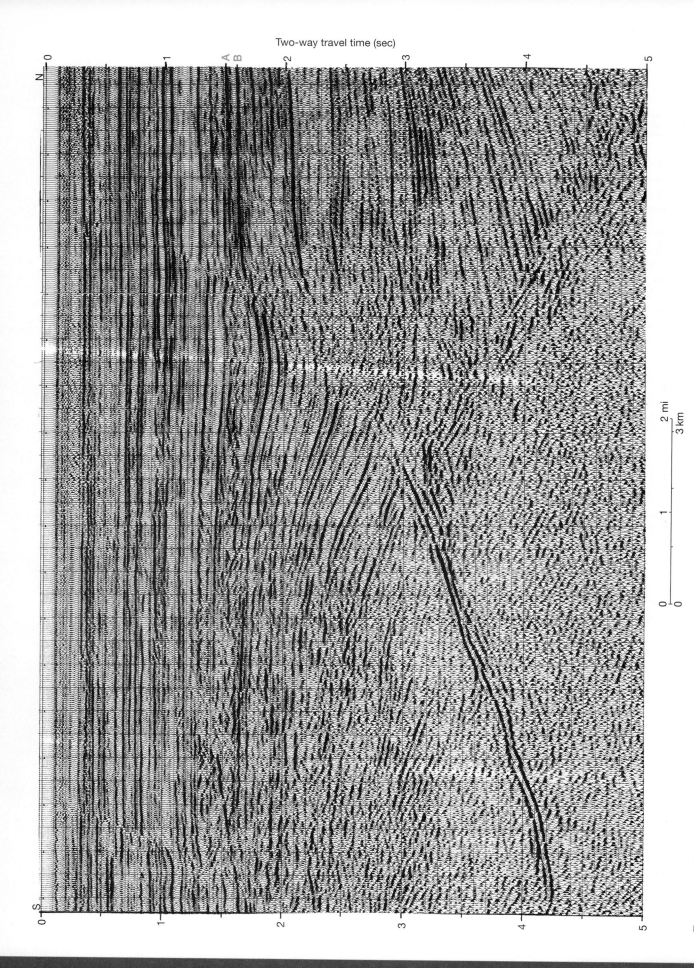

Figure 8–10 Seismic profile, Louisiana continental shelf (courtesy of Dr. Robert Berg, Texas A&M University).

| EXERCISE 8–5 | Seismic Profile, Louisiana Continental Shelf |

Figure 8–10 is an uninterpreted seismic profile across the continental shelf of Louisiana. The profile shows that the shelf is largely underlain by layered sedimentary rocks, which control wells indicate are sandy shales and shaly sandstones of Pliocene and Pleistocene age. However, the lower left corner of the seismic profile shows that these rock layers have been intruded from below by a salt dome, which can be recognized by its massive and unlayered seismic signature.

1. Draw the outline of the salt dome on the seismic profile. Note that the upward-moving salt has deformed the layers of rock through which it intrudes. Locate and describe the deformed layers.

2. The layered sedimentary rocks and the salt dome in this seismic profile are broken by a fault. Locate this fault and determine its extent by tracing several reflecting horizons across the seismic profile and noting their lateral continuity and vertical displacement. Draw the fault on the seismic profile and mark its hanging wall and footwall. What kind of fault (normal or reverse) is this? _____

3. Trace reflecting horizons A and B across the seismic profile. Notice that they are fairly close to each other in the footwall section but are farther separated in the hanging wall. Measure the two-way travel time of the rock layer between these two horizons on both sides of the fault, and then calculate the one-way travel times. If the average travel velocity for seismic waves through sandy shale and shaly sandstone is 15,000 ft/sec, what is the approximate thickness of this rock layer on either side of the fault?

4. The difference in thickness of the rock layer across the fault is due to the fact that the fault was active during its deposition. The slow, gradual movement along the fault created a small basin on the hanging wall side of the fault, thus leading to the accumulation of a thicker section of sediment there than on the footwall side. This type of fault is called a **growth fault.** If we assume that the rock layer was deposited in 100,000 years, what is the average rate of vertical movement along this fault?

Harcourt Brace & Company

9

PLATE TECTONICS

KEY WORDS

Alfred Wegener, Gondwanaland, and continental drift . . . mid-ocean ridges, rift valleys, and fractures . . . abyssal plains, rises, and hills . . . seamounts, guyots, and atolls . . . trenches and magmatic arcs . . . continental margins, shelves, slopes, and rises . . . aseismic ridges and submarine canyons . . . Harry Hess and sea-floor spreading . . . magnetic anomalies and reversals . . . divergent, convergent, and transform boundaries . . . collision and subduction zones . . . mantle plumes

THE ANDES MOUNTAINS, CONDORIRI, BOLIVIA.

(Karl Mueller)

Materials Needed *Colored pens or pencils, lead pencils with erasers, ruler, calculator, scissors*

The theory of plate tectonics states that the cold, rigid lithosphere of the Earth—that is, the upper 100 km or so of its crust and upper mantle—is broken into 14 large **lithospheric plates** (and several small **microblocks**) which are moving in different directions over the plastic rocks of the asthenosphere. The formation and deformation of rocks, the origin and evolution of the continents and oceans, and the causes of volcanism and earthquakes (among other things) can be explained by the movement and interaction of these

plates. For this reason, the theory of plate tectonics is considered the central unifying theory of geology.

Given the importance of the theory of plate tectonics, it is remarkable to learn that it has come to be understood and accepted only within the final three decades of the twentieth century. For most of the history of geology, it was thought that the continents and oceans were stable and ancient features of the Earth's surface. But slowly, evidence gathered that the crust of the Earth is dynamic and constantly changing, and that this change is the driving force for its formation and evolution. Some of this evidence came from the study of continental rock outcrops during the late 1800s and early 1900s, but the most convincing proof of the theory of plate tectonics came during an explosive period of ocean exploration which began after the Second World War.

This chapter begins with an examination of the three principal lines of evidence which support the theory of plate tectonics: the shared geology and physiography of the continents, the bathymetry of the ocean floors, and the age and magnetism of the rocks of the ocean crust. It concludes with a consideration of plate boundaries and the types of rock-forming and rock-deforming processes which characterize them.

There are three principal lines of evidence which support the theory of plate tectonics:

sim. fossils

- The geology and physiography of the continents
- The bathymetry of the sea floor *(topography)*
- The age and magnetism of oceanic crust

GEOLOGY AND PHYSIOGRAPHY OF THE CONTINENTS

The possibility that the continents are not permanently fixed on the surface of the Earth became apparent as soon as their shores were mapped by the explorers and traders of the 1400s and 1500s. The remarkable fit of the coasts of South America and Africa led several scientists to speculate about an ancient connection between them and the possibility that the continents are drifting across the Earth's surface.

By 1915, there was sufficient knowledge of the geology, paleontology, and structure of the continents for a German meteorologist named Alfred Wegener to advance the hypothesis of continental drift in a book titled *The Origin of Continents and Oceans*. Wegener presented a wealth of data on the structure of the continents, the global distribution of plant and animal fossils, and the paleoclimate of ancient sedimentary environments, which indicated that the continents were once united in a single landmass which we now call **Gondwanaland**. According to Wegener, this **supercontinent** broke up during the early Mesozoic Era, and since that time, the continents have drifted apart slowly and the oceans have gradually opened and widened between them.

Wegener was essentially correct about the existence of Gondwanaland and the timing of its breakup, but he could not satisfactorily explain the mechanism which moved the continents across the surface of the Earth. This was considered a fatal flaw in the hypothesis of continental drift, and it was initially dismissed by the scientific community.

BATHYMETRY OF THE SEA FLOOR

Little was known about the morphology of the sea floor until the 1950s, when scientists began to survey the oceans systematically with **echo sounders, seismic profilers**, and other instruments which had been developed for the purpose of naval (and particularly submarine) warfare during the Second World War. This surveying led to an amazing discovery:

that the sea floor has a bathymetry and structure which rivals the topography of the continents in its complexity and diversity.

There are four principal bathymetric zones in the world oceans: the **mid-ocean ridge**, the **abyssal floor**, the **oceanic trenches**, and the **continental margins** (Fig. 9–1).

Mid-Ocean Ridge

The **mid-ocean ridge system** is a long chain of submarine volcanoes which circles the Earth. It is the largest mountain chain in the world: It is 80,000 km long and averages 1500 km wide and 3 km high, and it covers nearly 20 percent of the Earth's surface.

The principal sites of volcanism on the mid-ocean ridge are the **rift** or **medial valleys**, which run along its crest. These rift valleys are 1 to 2 km deep and 2 to 50 km wide, and their floors and walls are extensively fractured and faulted. Basaltic magma rises up through these fractures to the Earth's surface, where it erupts from submarine volcanic **seamounts** and volcanic islands and crystallizes to form new oceanic crust. The flanks of the mid-ocean ridge dip gently away from the crest and eventually disappear beneath the sedimentary deposits of the abyssal plain. The ridge flanks are comprised of many smaller seamounts and block-faulted basins that are filled largely with volcaniclastic sediments.

The course of the mid-ocean ridge system is broken and offset into several segments by cross-cutting transform faults or **fracture zones**. The fractures often extend several hundreds of kilometers across the abyssal floor to either side of the ridge, and they form deep linear gouges and steep scarps on the sea floor. Volcanic seamounts and volcanic islands are also common along the lengths of these fractures.

The mid-ocean ridge is fairly active seismically: Approximately 1 in 20 earthquakes occurs along the ridge and the associated fracture zones. This seismicity is largely in the form of shallow-focus earthquakes, which are generated by the movement and emplacement of lava and the fracturing and faulting of the oceanic crust at spreading centers.

Abyssal Floor

The abyssal floor forms the broad, low-relief bottom of the ocean basins. The deepest parts of the abyssal floor are the **abyssal plains**, where the rough and rocky topography of the volcanic rocks of the oceanic crust is completely buried by a flat and featureless blanket of marine sediments; and the **abyssal rises**, broad areas where the abyssal floor has

① land bridge
② Continental drift
③ Plate tectonics

diff. is that thought cont. just floated around

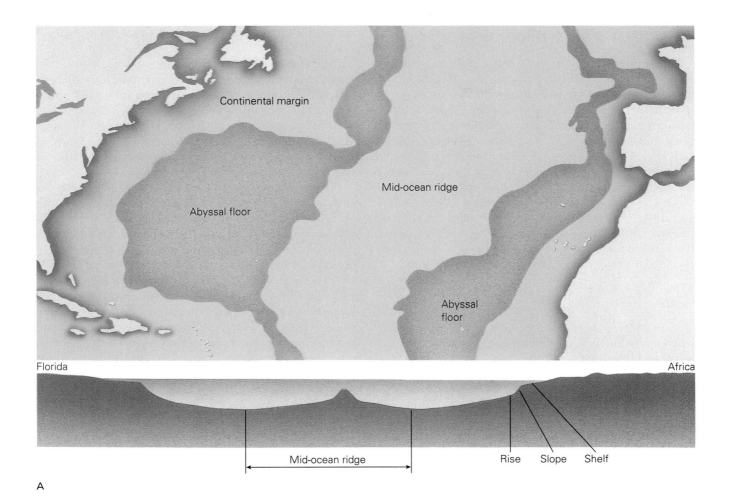

Florida

Africa

Mid-ocean ridge

Rise   Slope   Shelf

A

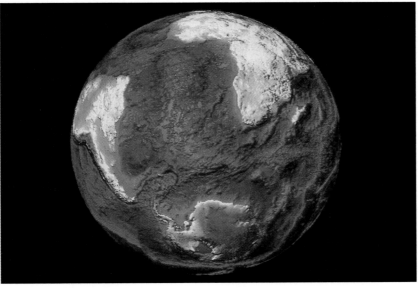

B

Figure 9–1 (A) The general bathymetry of the Atlantic Ocean. (B) Digital satellite photograph of the southern Atlantic Ocean.

(National Oceanic and Atmospheric Administration)

plates consist of oceanic & continental crust.

been elevated several hundred meters above the abyssal plains.

The generally low relief of the abyssal floor is commonly interrupted by volcanic seamounts, smaller (less than 1000 m) seamounts called **abyssal hills,** and other exposures of the volcanic basement rocks of the oceanic crust. Volcanic seamounts and abyssal hills can occur in clusters (called **abyssal hill provinces**) and long chains (**aseismic ridges**) which stretch across the ocean floor. (The term *aseismic* is used to distinguish these seamounts from the seamounts of the mid-ocean ridge, which is more seismically active.)

The summits of the largest seamounts of the abyssal floor break the sea surface to form islands, although they are often quickly eroded by waves and gravity into flat-topped **guyots.** Seamounts can also be capped by coral reefs, which form circular islands called **atolls** around their summits.

Oceanic Trenches

Trenches are deep depressions in the sea floor. They stretch for hundreds of kilometers across the sea floor, and they are always associated with equally long chains of volcanic mountains and volcanic island arcs which are collectively called **magmatic arcs.** Some trenches are greater than 8000 m deep, and form the lowest elevations on the face of the Earth.

The trenches are the most seismically active regions of the world: Nearly nine out of ten earthquakes occur in their vicinity. The foci of these earthquakes are fairly shallow near the trenches but become progressively deeper in the direction of the magmatic arcs.

Continental Margins

Continental margins are transition zones between continents and the deep ocean. They can be divided into three provinces based on their bathymetry: **shelves, slopes,** and **rises.**

The continental shelves are seaward-sloping, low-relief plains which extend from the shorelines of the continents to water depths of approximately 200 m. They are essentially the seaward extensions of the coastal plains of the continents, and (like the coastal plains) they are underlain by sedimentary rocks and the sialic crystalline basement of the continents.

The continental slopes are separated from the shelves by the shelf-slope breaks, which mark a sudden increase in the slope of the sea floor. The slopes plunge steeply down to water depths of approximately 1500 to 3000 m, where they level off into the gently sloping continental rise. The slope and rise directly overlie the transition zone between the continental and oceanic crustal segments of the lithospheric plates.

The outer continental shelf and slope are often incised by **submarine canyons,** which serve as conduits for sediment transport from the continent and continental shelf to the deep ocean. Canyons are common offshore of rivers, reefs, and other sources of large volumes of sediment. This sediment is transported through the canyons by slumps, debris flows, and turbidity currents to the continental rise, where it is deposited in **submarine fans** or **cones.**

The basement rocks of some continental margins are overlain by a thick wedge of sedimentary rocks. This sediment accumulates in basins which are formed by the subsidence and block faulting of the edges of the continents. One example of this type of basin is the northern Gulf of Mexico between Florida and Texas, where the Precambrian and Paleozoic basement rocks of the southern United States are overlain by more than 8000 m of Mesozoic and Cenozoic sedimentary rocks. On the other hand, the basement rocks of other continental margins are overlain by a very thin and discontinuous veneer of sedimentary rocks, so that they are widely exposed throughout the continental shelf, slope, and rise.

AGE AND MAGNETISM OF THE OCEANIC CRUST

The discovery of mid-ocean ridges and trenches led the geophysicist Harry Hess to advance the concept of **sea-floor spreading** in 1962. Hess hypothesized that the mid-ocean ridges were **spreading centers** where new crustal rocks were continually being emplaced between two diverging fragments of older crustal rocks. He also postulated that, to maintain the constant surface area of the Earth, older crustal rocks were continually being consumed in the oceanic trenches. The test of his hypothesis was the determination of the age and magnetism of the oceanic crust.

Age of the Oceanic Crust

During the past few decades, the oceanic crust has been sampled and cored by the Deep Sea Drilling Program, the Ocean Drilling Program, and other scientific programs. The age of crystallization of the basaltic basement rocks of the oceanic crust has been determined by radiometric dating techniques, whereas the age of deposition of the sedimentary rocks which directly overlie the basaltic basement has been determined by paleontological techniques. Both techniques support Hess' hypothesis of sea-floor spreading, for they show that the oceanic crust becomes progressively older with increasing distance from the mid-ocean ridges (Fig. 9–2).

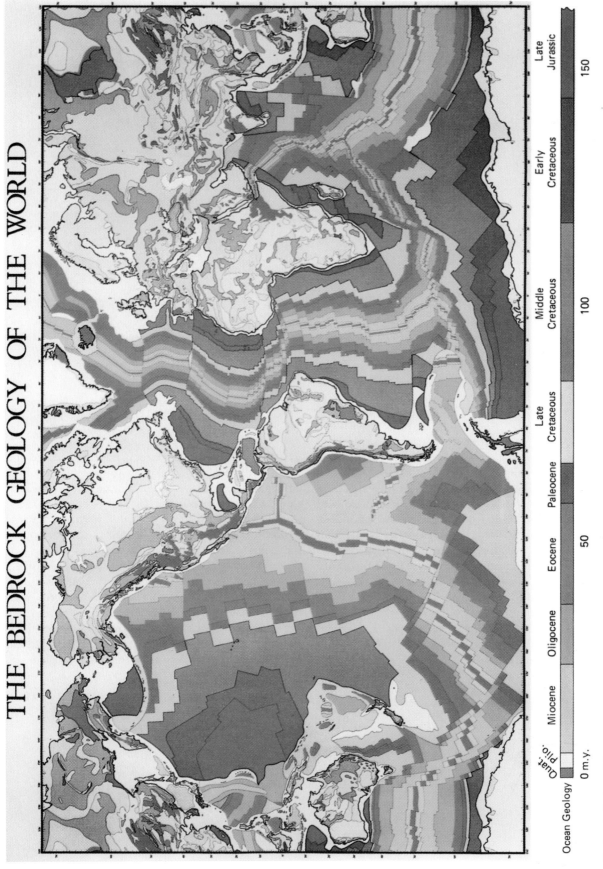

THE BEDROCK GEOLOGY OF THE WORLD

Ocean Geology

| Quat. Plio. | Miocene | Oligocene | Eocene | Paleocene | Late Cretaceous | Middle Cretaceous | Early Cretaceous | Late Jurassic |

0 m.y.　　　　50　　　　100　　　　150

Figure 9–2 Age of the ocean crust.

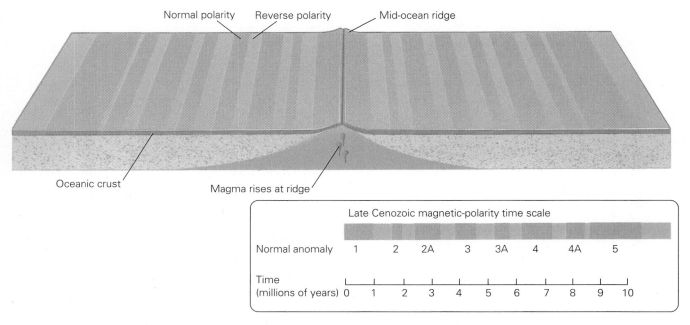

Figure 9–3 Model for the formation of new oceanic crust and symmetrical magnetic polarity stripes at a mid-ocean ridge. When new crust is created at the mid-ocean ridge, it assumes the prevailing polarity of the Earth's magnetic field. The alternation of normal and reverse polarity stripes is due to the periodic reversals in the Earth's magnetic field. The dates of these reversals, which are shown in the time scale, are determined by radiometric dating of the rocks of the crust.

Magnetism of the Oceanic Crust

This same conclusion about the age of the oceanic crust has been reached by studies of its magnetism. Since the early 1900s, it has been known that the Earth's magnetic field changes polarity periodically from its present **normal polarity** to **reverse polarity** (when magnetic north becomes south, and magnetic south becomes north). Figure 9–3 shows the history of magnetic reversals (called **polarity events**) and the major normal and reverse-polarity epochs (or **magnetic anomalies**) which have been identified for the last 10 million years of the Cenozoic Era.

The rocks of the oceanic crust are composed of several iron-bearing minerals, such as magnetite, which tend to align parallel to the direction of the Earth's magnetic field when they crystallize. Such rocks contain a record of the polarity of the magnetic field at the time of their formation. This magnetism is weak, but it can be measured by ship-towed **magnetometers**.

Hess predicted that a record of magnetic reversals during geologic time should be present in the volcanic rocks of the oceanic crust if his hypothesis was correct—that is, there should be parallel symmetrical bands of normal and reverse-polarized volcanic rocks on either side of a mid-ocean ridge (Fig. 9–3). Magnetometer surveys of the mid-ocean ridges and abyssal plains by Vine, Matthews, and Morley in the 1950s and 1960s confirmed the existence of the parallel bands of magnetic anomalies and thus provided the final proof for what is now known as the theory of plate tectonics.

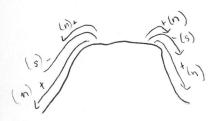

EXERCISE 9 – 1   Seismic Profiles from the Ocean Floor

Figure 9–4A is a seismic profile across part of the Juan de Fuca Ridge, which forms the western boundary of the Juan de Fuca plate. Notice that the volcanic rocks of this ridge are partially buried beneath sediments.

1. Use a blue pencil to trace the contact between the volcanics and the sediments. Compare and contrast their seismic signatures.

2. Examine the exposed part of the ridge in the eastern part of the figure (near Site 855). Locate the dark reflecting horizon near the top of this exposure, and trace it across the seismic profile. Notice that it is laterally displaced by faults in several locations. Trace these faults on the profile, and indicate the relative directions of displacement along them with arrows.
   a. What kind of faults are they? _____
   b. What kind of structural features have formed between them? _____
   c. What kind of tectonic force creates them: compression, tension, or shear? _____

3. Leg 139 of the Ocean Drilling Program drilled several cores into the volcanics and sediments of this ridge at the two sites which are indicated on the profile. Shipboard analysis of samples from these cores indicated that seismic waves travel through the sediments at an average velocity of 1500 m per second. Estimate the maximum thickness of the sediment fill in the basin in the western part of the profile.

4. Figure 9–4B is a seismic profile across part of the Iberian abyssal plain, off the western coast of Spain. Use a blue pencil to trace the contact between the oceanic crust and the sediment of the abyssal plain. Describe the structure and seismic signature of the sediment in the abyssal plain.

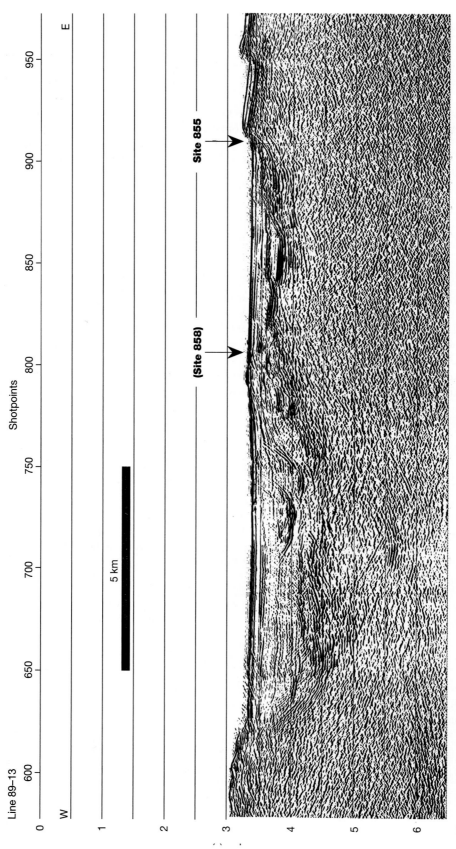

Figure 9–4A Seismic profile, Juan de Fuca Ridge, Pacific Ocean. (*Ocean Drilling Program, Texas A&M University*)

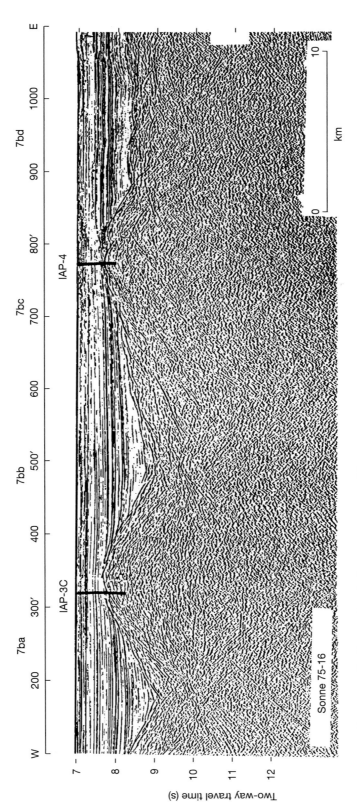

Figure 9–4B Seismic profile, Iberian abyssal plain, western coast of Spain.

| EXERCISE 9–2 | Bathymetry of the North Atlantic Ocean |

Examine the map of the floor of the northern Atlantic Ocean (Map 9–2).

  1. Locate and mark the western and eastern boundaries of the mid-ocean ridge. Locate and trace the rift valleys at its crest. What is the relief between the ridge crest and its flanks? between the ridge crest and nearby abyssal plains?

  2. Locate and trace the principal fracture zones. Indicate the relative directions of displacement of the crest of the ridge with arrows. Describe the bathymetric expression of the fracture zones.

  3. Locate the epicenters of the following earthquakes, and describe the bathymetric features which are closest to them.

DATE	EPICENTER	BATHYMETRY	DATE	EPICENTER	BATHYMETRY
August 1962	32°N, 41°W		November 1965	31°N, 42°W	
May 1963	24°N, 46°W		July 1966	38°N, 25°W	
May 1964	35°N, 36°W				

  4. Locate and outline the Sohm, Hatteras, and Nares abyssal plains. Record the maximum depth of each plain.

  5. Record the locations of some of the major seamounts in the map area. Describe their relief.

  6. Locate and identify an example of an aseismic ridge in the map area.

  7. Locate and outline the Bermuda Rise. Describe its bathymetry. What is the relief between the rise and the Hatteras abyssal plain? between the rise and the island of Bermuda?

8. Locate and trace the shelf-slope break and the borders between the continental slope, rise, and abyssal plains along the North American continental margin. Measure the widths of the shelf, slope, and rise as well as the width of the entire continental margin. Provide both maximum and minimum values, and then calculate their averages.

	MAXIMUM	MINIMUM	AVERAGE
Shelf			
Slope			
Rise			
Continental margin			

9. Notice that the continental slope of the southeastern United States levels off to form a **marginal plateau** called the Blake Plateau. Locate and outline this feature on the map. Describe its bathymetry, including its width and relief.

10. Locate and outline the Laurentian Cone and its submarine canyon. What part of North America serves as the source of sediment for this cone? What kind of sediment sources exist in this area?

11. Locate and record the names of two additional submarine canyons along this continental margin. These canyons are located offshore from large rivers which drain the Appalachian Mountains and Atlantic coastal plain. The rivers are the major source of sediment for the canyons. Identify the fluvial source of each canyon.

EXERCISE 9–3   Bathymetry of the Eastern Pacific Ocean

Examine the map of the floor of the eastern Pacific Ocean (Map 9–3).

1. The floor of the eastern Pacific Ocean is traversed by several segments of the mid-ocean ridge system, the largest of which is the East Pacific Rise. Trace the crest of the East Pacific Rise and the major fracture zones which dissect it. Indicate the relative directions of displacement along the fracture zones.

2. Locate and label examples of seamounts (S), abyssal hills (H), abyssal plains (P), and aseismic ridges (R). What is the approximate relief between the seamounts along the East O'Gorman fracture zone and the abyssal floor?

3. Construct a cross section along line T–T′. Label the ridge crest, abyssal plain, and trench floor; the continental shelf, slope, rise, and flank; and the shelf-slope break and shoreline. What is the relief between the ridge crest and the abyssal floor?

4. Examine the western continental margin of Central America. Trace the shelf-slope break as well as the boundaries between the slope, rise, trench, and abyssal plain. Measure the widths of the continental shelf, slope, rise, and trench as well as the width of the entire margin. Compare these data to the comparable data from the Atlantic continental margin. Speculate on the differences in bathymetry between these margins.

	MAXIMUM	MINIMUM	AVERAGE
Shelf			
Slope			
Rise			
Continental margin			

5. Which has the steeper gradient: the continental margin of Central America or that of North America?

Map 9—4 is a bathymetric map of part of the Mid-Atlantic Ridge between Brazil and Angola.

1. Construct a bathymetric profile through the ridge crest along line A-A′. Indicate the rift valley on the profile. What is its maximum depth and width?

2. Use a colored pencil to shade other rift valleys on the map.

3. Construct a bathymetric profile through the fracture zone along line B-B′. Describe the bathymetric expression and relief of the fracture zone. What other bathymetric features are associated with the fractures in the map area?

EXERCISE 9–5   Bathymetry of the Northern Pacific Ocean

The Hawaiian and Emperor Islands form a chain of volcanic islands and seamounts which runs from the central Pacific Ocean to the Aleutian trench. Map 9–5 shows the bathymetry of the Hawaiian aseismic ridge, which consists of the volcanic seamounts and island in the southern part of this chain.

1. Construct a bathymetric profile from Oahu to Midway through the crests of the largest seamounts. Construct a second profile along line A-A′.

2. Describe the shape and relief (above the sea floor) of the seamounts. How do they change through the map area?

3. Compare and contrast the width and relief of the Hawaiian Ridge in the areas of the Hawaiian Islands and Midway Island.

4. Locate and label examples of atolls.

1. Return to the exercise on the bathymetry of the north Atlantic Ocean (Map 9–2).
   a. The ridge crest and its flanks are both composed largely of basaltic volcanic rocks. However, the younger rocks of the ridge crest have far more relief than the older rocks of its flanks. Why?

   b. If you were operating a drill ship and wanted to core the oldest oceanic crust in this part of the Atlantic Ocean, where would you drill and why?

   c. Where is the marine sediment column thickest in the map area? Why?

2. Return to the exercise on the bathymetry of the Hawaiian Ridge (Map 9–5).
   a. Consider the change in the geometry of the seamounts. What does this imply about the relative ages of the islands and seamounts? Which is youngest? oldest?

   b. Explain the change in the geometry of the Hawaiian Ridge in the map area.

EXERCISE 9 – 7   Magnetic Profiles Across a Mid-Ocean Ridge

Figure 9–5 shows a set of magnetic profiles across a segment of the East Pacific Rise. The locations of these profiles are indicated on Map 9–3.

1. The normal magnetic anomaly which marks the crest of the mid-ocean ridge is shaded red. Shade the remaining normal anomalies red, and shade the reverse anomalies green.

2. Transfer this information to Map 9–3. Lay each magnetic profile on the map, and mark off the contacts between the normal and reverse magnetic anomalies.

3. Correlate the normal and reverse anomalies from profile to profile. Shade the normal magnetic anomaly stripes with a red pencil, and shade the reverse anomaly stripes with a green pencil.

4. There have been seven normal magnetic anomalies (numbers 1, 2, 2a, 3, 3a, 4, and 4a) during the past 10 million years (Fig. 9–3). The four magnetic profiles in Figure 9–5 traverse between five and seven of these anomalies. Identify and number the normal magnetic anomaly stripes which you have drawn in Map 9–3, beginning with 1 for the normal magnetic anomaly at the ridge crest.

5. Describe the symmetry of the stripes. What does this indicate about the spreading rates on either side of the ridge?

Figure 9–5 Magnetic profiles across the East Pacific Rise. See Map 9–3 for the locations of the four profile lines.

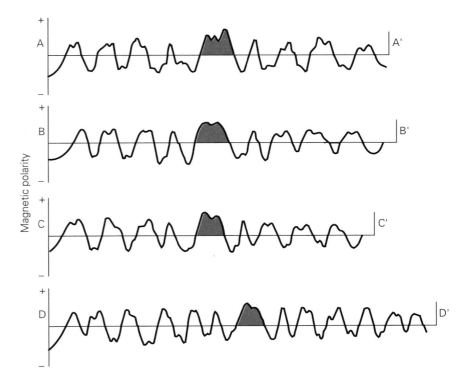

The continents of Africa, South America, Australia, India, and Antarctica were once part of Gondwanaland, the ancient supercontinent which formed sometime in the Proterozoic and rifted apart during the early Mesozoic. Africa and South America formed the western part of Gondwanaland until the middle Jurassic, when they rifted apart and opened the South Atlantic Ocean between them. The former contact between the two continents is still evident from the following:

- The shapes of their coasts and continental margins
- The late Carboniferous glacial till deposits in southern Africa (below 12°S), Uruguay, and southern Brazil
- The late Carboniferous glacial-marine deposits (mostly ice-rafted gravels) in central Africa between 0 and 12°S, Argentina, Paraguay, southern Bolivia, and the eastern coast of Brazil
- The late Permian fold belts in northern Morocco and Algeria (the Atlas Mountains) and northern Venezuela (the Cordillera de Merida)
- The mid-Jurassic evaporites along the eastern coast of Brazil between Rio de Janeiro and Recife, and the western coasts of Cameroon, Gabon, and Angola

1. Map 9–1 shows the outlines of the continents and continental margins of Africa and South America. Outline the outcrop belts of the Paleozoic and Mesozoic sedimentary deposits and the structural features listed above. (You may need to refer to a world atlas to locate them.) When you have finished, cut out the outlines and reunite the two continents as they were once joined in western Gondwanaland.

2. Examine the distribution of glacial-till deposits in southern Africa. What does this indicate about the location of western Gondwanaland relative to the South Pole during the Carboniferous?

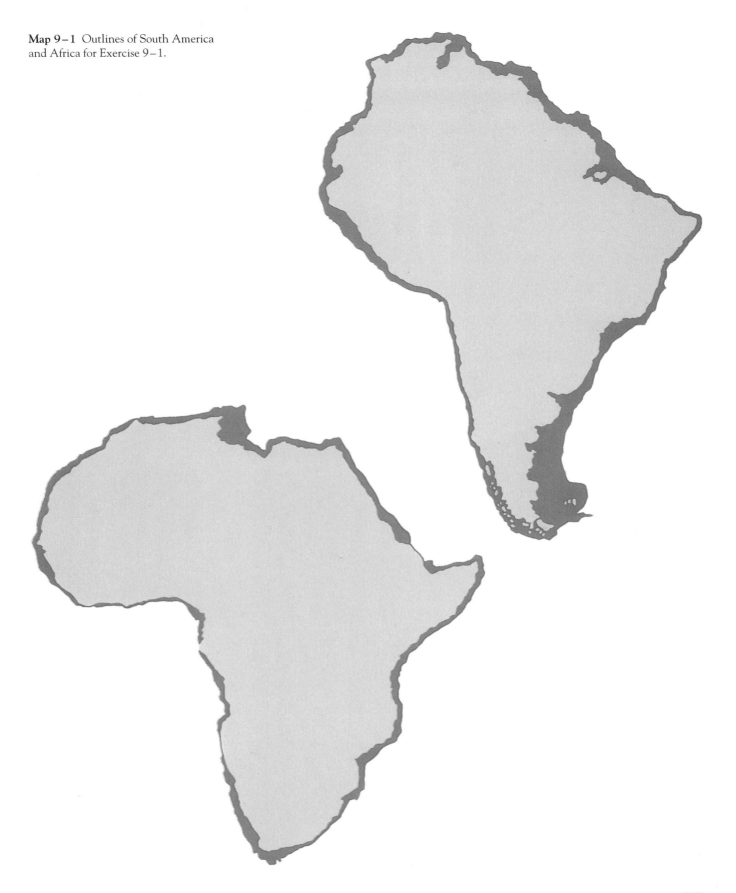

Map 9–1 Outlines of South America and Africa for Exercise 9–1.

MAP 9-2 Bathymetry of the northern Atlantic Ocean

Elevations and depths in meters

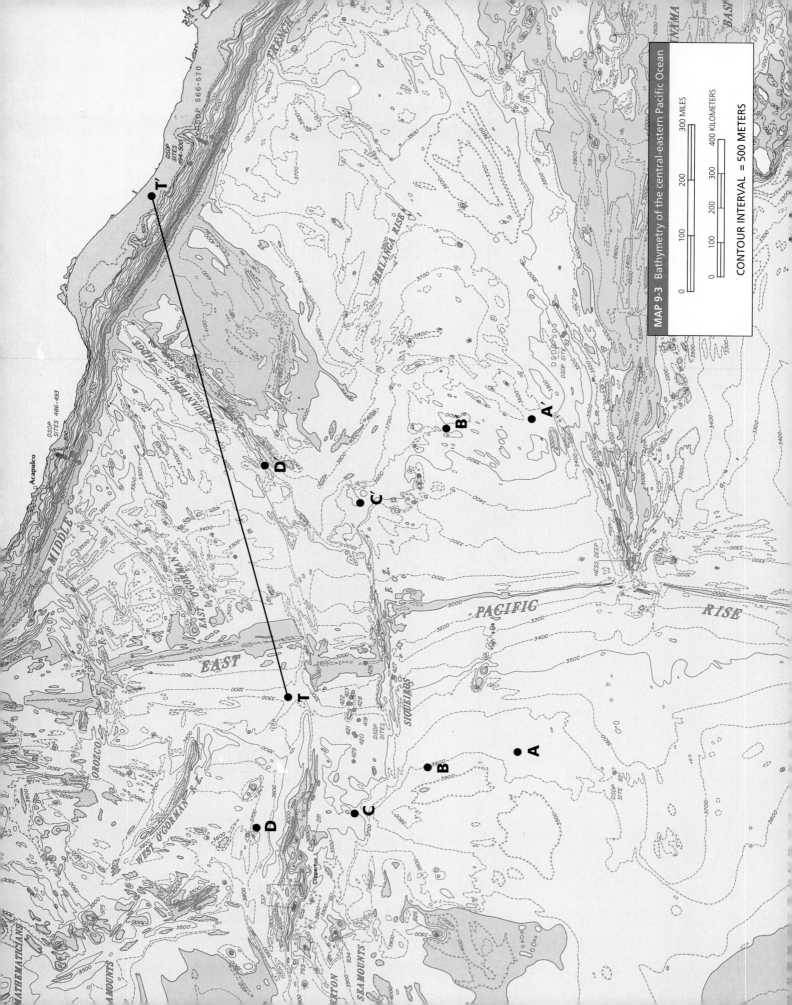

MAP 9-3 Bathymetry of the central-eastern Pacific Ocean

CONTOUR INTERVAL = 500 METERS

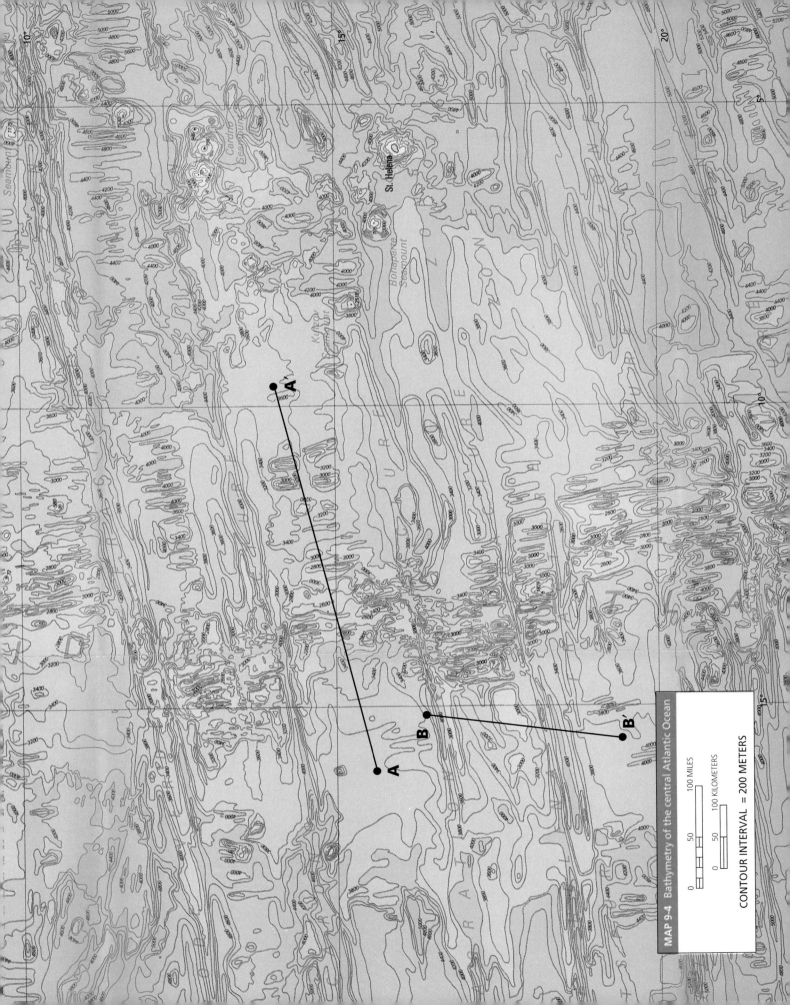

MAP 9-4 Bathymetry of the central Atlantic Ocean

CONTOUR INTERVAL = 200 METERS

0   50   100 MILES

0   50   100 KILOMETERS

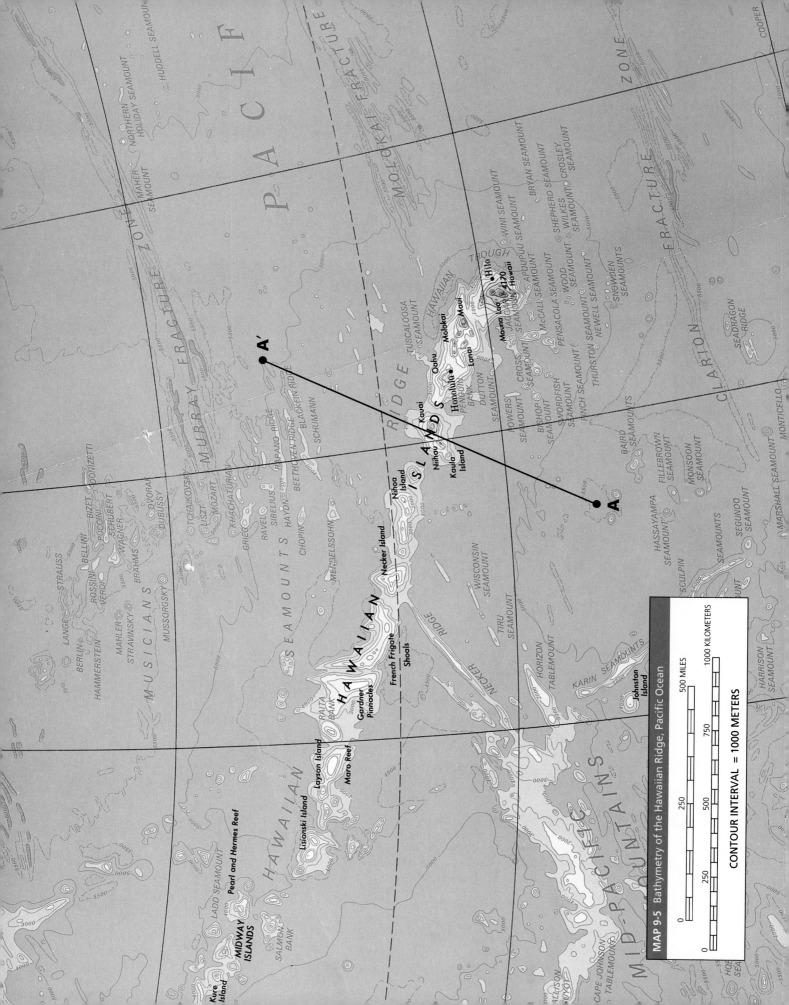

MAP 9-5 Bathymetry of the Hawaiian Ridge, Pacific Ocean

CONTOUR INTERVAL = 1000 METERS

T he present theory of plate tectonics holds that the crust of the Earth is fragmented into 14 major lithospheric plates and that these plates are moving in different directions and at different velocities across the Earth's surface. According to the theory, the formation and deformation of igneous, sedimentary, and metamorphic rocks, the origin and evolution of continents and oceans, and volcanic and seismic activity can all be largely explained by this motion and particularly by the interaction of lithospheric plates at **divergent, convergent,** and **transform plate boundaries** (Fig. 9–6).

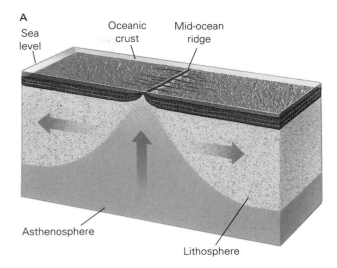

DIVERGENT BOUNDARIES

Divergent boundaries are those which form when two plates move or **rift** apart. The spreading centers between two diverging plates are marked physiographically by a mid-oceanic ridge, where volcanism plugs the ever-widening gap between the plates with new oceanic crust. This oceanic crust is composed of basaltic magmas, which are derived from the partial melting of the asthenosphere below the rift zone. When rifting begins in the interior of a continent, the spreading center begins as a deep **continental rift valley** and gradually widens into a narrow sea and eventually a large ocean with a fully developed mid-oceanic ridge.

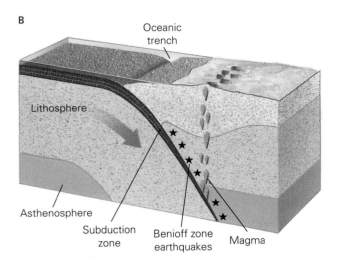

CONVERGENT BOUNDARIES

Convergent boundaries are those which form when two plates move toward each other. The boundary between two converging plates is marked by the **fold-and-thrust orogenic belts** of **collision zones** and the trenches and magmatic arcs of **subduction zones.**

Collision zones form between two converging plates which are both composed of continental crust. The great compressive force of collision folds, thrust-faults, and metamorphoses the rocks along the margins of the plates and creates high **mountain chains** with deep crustal roots.

Subduction zones form when one or both of the converging plates are composed of oceanic crust. When they converge, the dense oceanic crust is subducted beneath the second (overriding) plate and eventually consumed by melting in the Earth's hot interior. The path of the descending plate can be traced by the foci of earthquakes generated by friction in the Benioff zone between the two plates. The magma which is produced by the melting of the subducted plate rises upward into the overriding plate, where it is emplaced in crustal plutons and erupts on the surface at volcanic islands and mountains forming a **magmatic arc.**

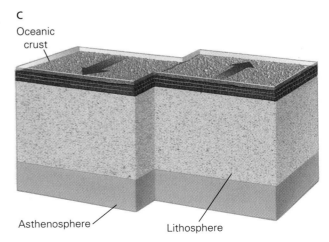

Figure 9–6 Plate interactions at divergent, convergent, and transform boundaries.

TRANSFORM BOUNDARIES

Transform boundaries are those which form when two plates move in parallel but opposite directions. They are marked by one or several strike-slip faults, which slice the two plate margins into many tiny microblocks and open deep pull-apart basins between them. The San Andreas fault of California is one of several such faults which mark the transform boundary between parts of the North American and Pacific plates.

INTRAPLATE SETTINGS

Tectonism and its associated rock-forming and rock-deforming processes are less important in the interiors of plates, but they are not entirely unknown there. The principal tectonic process in intraplate settings appears to be the product of the upwelling of magma in the mantle, which produces deep **hot spots** or **plumes** beneath the lithospheric plates.

When mantle plumes occur beneath continental interiors, they result in extensive crustal upwarping, the reactivation of old faults, and the formation of basins, domes, and basement uplifts. When they occur beneath oceanic crust (which is considerably thinner than continental crust), the upwelling magma breaks through to the surface to form volcanic mountains and seamounts. The movement of oceanic plates over such hot spots creates long chains of volcanic seamounts such as the Hawaii–Emperor Chain in the Pacific Ocean.

Digital satellite photograph of the southwest Pacific Ocean. The shallow parts of the ocean, including volcanic islands, seamounts, and continental margins, are colored light blue. Deeper parts, including the oceanic trenches, are colored dark blue. *(Photo courtesy of National Oceanic and Atmospheric Administration)*

EXERCISE 9-9   Spreading Rates in the Atlantic and Pacific Oceans

Map 9–6 shows the mid-oceanic ridge, the major transform faults, and selected magnetic anomalies in the eastern Atlantic. The age of each magnetic anomaly (in millions of years) is indicated in parentheses.

1.   Measure the distance between each pair of anomalies, and calculate the time interval between them. Record these data on the table below.

2.   Calculate the spreading rate during each time interval by dividing the distance by the time (see example).

3.   What is the age of the oldest oceanic crust in this part of the Atlantic Ocean? _____

	DISTANCE	AGE	TIME INTERVAL	SPREADING RATE
Ridge–5	150 km	0–10 my	10 my	15 km/my
5–6	225 km	10–20 my	10 my	22.5 km/my
6–13				
13–21				
21–25				
25–30				
30–32				
32–33				
33–34				
34–M0				
M0–M4				
M4–M10				
M10–M16				
M16–M21				
M21–M25				

4.   Construct a graph which shows the change in the spreading rate in the eastern Atlantic over geologic time. Label the horizontal (X) axis "Time" and the vertical (Y) axis "Spreading Rate." Plot the midpoint of the age vs. the spreading rate for each pair of anomalies in the table.

5. Calculate the average spreading rate for the eastern Atlantic by measuring the distance between the mid-ocean ridge and the oldest magnetic anomaly. Across your graph draw a horizontal line which intersects the Y axis at this value.

6. Has the spreading rate in the eastern Atlantic Ocean been constant over geologic time? _____

7. When did the slowest spreading occur? _____ the fastest? _____

8. Return to Figure 9–5, which shows the magnetic anomalies surrounding the East Pacific Rise.
   a. Refer to the magnetic time scale in Figure 9–3. What is the age of the oldest oceanic crust traversed by the magnetic profiles? _____
   b. What is its distance from the ridge crest? _____
   c. Given these two values, calculate the average spreading rate for this ridge.

   d. Compare and contrast the spreading rates of the Mid-Atlantic Ridge and the East Pacific Rise. Speculate on the origin of their differences.

EXERCISE 9-10   Global Tectonics

This exercise examines the tectonic structure of the world. You will need to refer to two resources which are provided by your instructor: the Heezen–Tharp World Ocean Floor map and a world atlas. Whenever possible, record your answers on Figure 9–7.

1. Locate and label the major plates on Figure 9–7. The major plates are:

Eurasian   Indian–Australian   Antarctic   Philippine   Pacific

Juan de Fuca   North American   Cocos   Nazca   Scotia

South American   Caribbean   African   Arabian

2. Use a red pencil to outline the mid-ocean ridge system on Figure 9–7. (Create a key in the lower left corner of the figure which shows the symbols for this and other features in the map area.) Locate and label the following segments of the ridge, and indicate their spreading directions with arrows:

East Pacific Rise   Mid-Atlantic Ridge   Juan de Fuca Ridge   Reykjanes Ridge

Carlsburg Ridge   Southwest Indian Ridge   Southeast Indian Ridge

3. Use a blue pencil to outline the oceanic trenches in the figure. Locate and label the trenches which are listed in the following table. Identify the principal islands, ridge, or mountain belt which forms the magmatic arc of each trench, and indicate their locations with this symbol: —V—. Indicate with arrows the relative directions of plate motion and subduction along the trenches.

TRENCH	MAGMATIC ARC	TRENCH	MAGMATIC ARC
Peru–Chile		Java	
Middle America		South Solomon	
Aleutian		New Hebrides	
Kuril–Kamchatka		Kermadec–Tonga	
Japan–Bonin		Hjort	
Mariana		Puerto Rico–Cayman	
Ryukyu		South Sandwich	
Philippine		South Shetland	

4. The Juan de Fuca plate is being subducted eastward beneath the North American plate. However, there is no discernible trench in this subduction zone. Examine the ocean floor map, and speculate on the reason for this. What is the name of the magmatic arc in this subduction zone?

5. Use a green pencil to outline the trends of the following mountain belts:

Rockies   Appalachians   Andes   Pyrenees   Alps   Carpathians   Caucasus   Ural   Himalaya
Hindu Kush   Tien Shen   Khingan   Cherskiy   Kolyma   Chukchi   Brooks

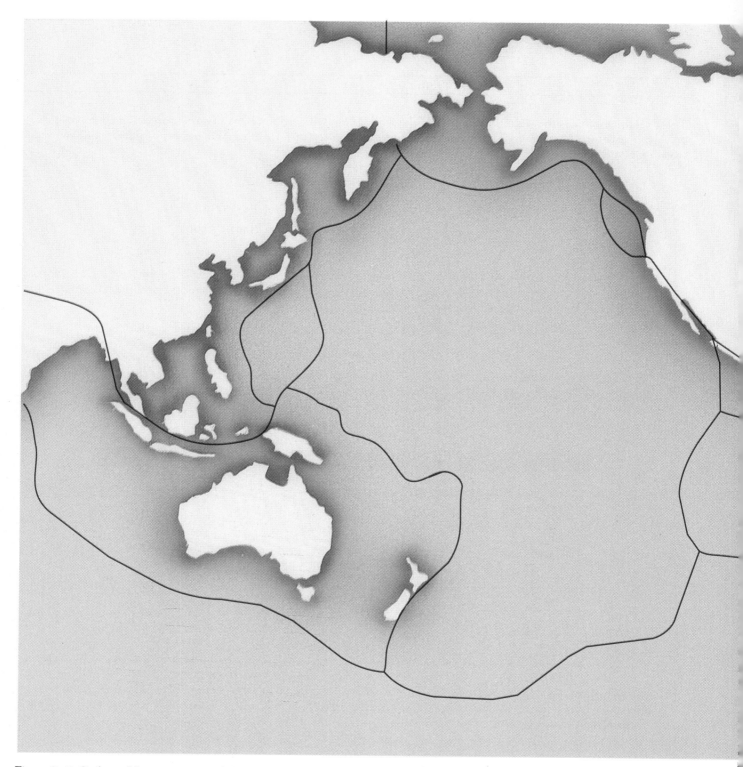

Figure 9–7 Outline of the continents and the major plate boundaries.

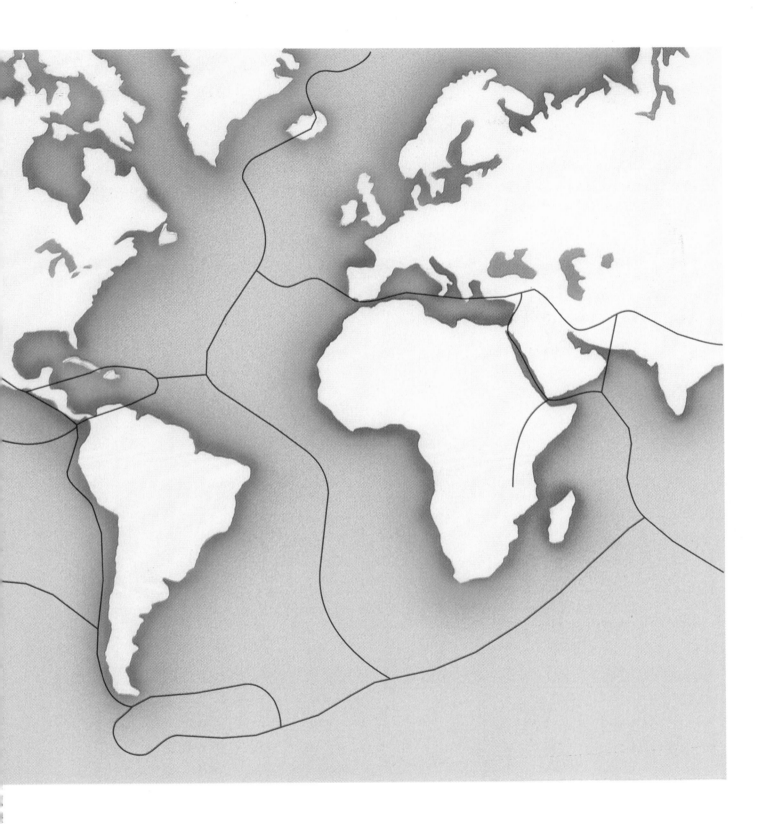

6. The Alpine–Himalaya mountain belt marks the collision zone between Africa, India, and Eurasia. Locate this collision zone on Figure 9–7. Indicate the relative directions of plate movement with arrows along its length from Gibraltar to Tibet. What are the names of the major mountain belts in this collision zone?

7. Identify the transform fault which forms the boundary between the North American and Pacific plates. _____ Indicate with arrows the relative directions of plate movement along this boundary.

8. Locate the following submarine fans, and mark their locations on Figure 9–7. Identify the river which supplies sediment to each fan and where this sediment comes from.

FAN	RIVER	SOURCE
Amazon		
Congo		
Orange		
Indus		
Ganges		
Nitanat		
Monterey		
Mississippi		

9. What kind of tectonic setting is most conducive to the formation of large submarine fans: divergent, subduction, collision, transform, or intraplate? Explain your answer.

10. The following table summarizes the compositions of several volcanic islands and mountains in and around the Pacific Ocean. Some of these volcanoes occur in the subduction zones along the margins of the Pacific plate, whereas others occur in intraplate settings. Locate them on the Heezen–Tharp map, and identify their tectonic settings.

What is the relationship between their tectonic settings and compositions, and what does this imply about the sources of magma in subduction zones and intraplate settings?

VOLCANO	COMPOSITION	SETTING	VOLCANO	COMPOSITION	SETTING
WESTERN PACIFIC			**EASTERN PACIFIC**		
Guam	Andesite		Mount St. Helens	Andesite	
Truk	Basalt		Parícutin	Andesite	
Marcus I.	Basalt		Galapagos I.	Basalt	
New Solomon I.	Andesite		Chimborazo (Ecu)	Andesite	
Fiji	Andesite		San Pedro (Chile)	Andesite	
Eniwetok	Basalt		Easter I.	Basalt	
Fujiyama	Andesite		Mauna Loa	Basalt	

11. The northeastern continental margin of North America and the northwestern continental margin of Africa are both examples of **passive continental margins**—that is, they occur on the trailing edges of two diverging continents. However, there is a great difference in the structures of these two margins: The **prerift** (pre-Jurassic) continental basement rocks of the North American margin are overlain by a thick wedge of Mesozoic and Cenozoic **postrift** sedimentary rocks, whereas the prerift continental basement rocks of the African margin are overlain by a thin, discontinuous veneer of postrift sedimentary rocks.

Examine the physiography of northwest Africa and its continental margin on the Heezen–Tharp map. Why are postrift sedimentary rocks relatively scarce there?

12. Describe the bathymetry of the northwestern continental margin of Africa. How is the relative paucity of postrift sedimentary rocks there reflected in the bathymetry?

13. Locate the spreading center between the African and Arabian plates. What physiographic feature occurs in this rift zone? _____

14. A second spreading center recently opened in the continental interior of eastern Africa. What physiographic features occur in this rift zone? _____ Mark the location of the spreading center with a red pencil, and indicate the spreading directions with arrows. What kind of rock-forming process occurs there? _____

15. Locate the Hawaiian and Emperor Islands on the Heezen–Tharp map. This chain of volcanic islands and seamounts was formed by the movement of the interior of the Pacific plate over a mantle plume. What is the youngest island or seamount in this chain? _____ the oldest? _____ What is your evidence?

16. Notice the change in the trend of this chain. What does this suggest about the direction of movement of the Pacific plate during its formation? When did this change occur (i.e., between the formation of which two volcanoes)?

17. What was the direction of movement of the Pacific plate before this change? _____ after this change? _____

18. The following table shows the ages of some of the volcanoes in the southern part of the Hawaii–Emperor Chain. Return to Map 9–5, measure the distances between these islands, and record this information on the table. Calculate the rates of plate movement (in km per thousand years) during the time intervals between the formations of these islands.

VOLCANO	AGE (in millions of years)	DISTANCE	RATE
Kilauea	<1.0		
Oahu	3.6		
Kauai	5.6		
Nihoa	6.9		
Necker	10.4		
Midway	16.2		

19. The Hawaii–Emperor Chain is one of the most prominent aseismic ridges in the Pacific Ocean. Examine the Pacific and Indian Oceans on the Heezen–Tharp map, and identify two other examples of aseismic ridges in each area.

EXERCISE 9-11 Tectonic Settings of Earthquakes

Figure 9–8 shows the foci of all earthquakes which occurred in the Pacific Ocean between 1961 and 1967. Refer to Figure 9–7 and the Heezen–Tharp map, and determine the tectonic settings of these earthquakes.

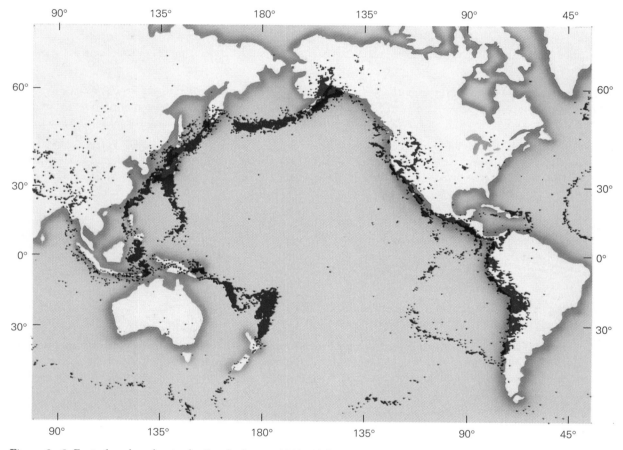

Figure 9–8 Foci of earthquakes in the Pacific Ocean, 1961–1967.

1. Use different-colored pencils to identify the seismic belts which occur in divergent, convergent, and strike-slip settings and in the interiors of plates.

2. What is the most common tectonic setting for earthquakes in this area? the least common?

Earthquakes are common in Benioff zones, where they are generated by the friction between the subducted and overriding plates. The foci of such earthquakes are shallow near the trench, but they increase in depth toward the volcanic arc. Consequently, it is possible to determine the geometry of a Benioff zone, and specifically the slope of descent of the subducted plate, by analyzing the depths of earthquake foci there. Map 9-7 shows part of the Tonga trench, where the Pacific plate is being subducted beneath the Indian-Australian plate, and part of the Peru-Chile trench, where the Nazca plate is being subducted beneath the South American plate. The following table summarizes the locations and depths of earthquake foci in these areas.

TONGA TRENCH			PERU TRENCH		
Latitude-Longitude		Depth (km)	Latitude-Longitude		Depth (km)
21.94S	174.92W	33	7.20S	76.07W	112
22.00S	175.03W	44	8.68S	79.88W	45
22.54S	174.70W	33	10.13S	78.62W	48
21.87S	175.06W	46	10.19S	78.27W	53
19.93S	175.88W	200	9.73S	74.63W	139
20.87S	176.30W	234	9.79S	74.70W	124
20.88S	177.93W	527	10.92S	78.02W	71
21.32S	173.35W	34	12.52S	75.11W	103
21.72S	173.90W	33	8.56S	79.81W	33
21.85S	175.19W	33	13.26S	76.15W	97
22.95S	175.65W	88	11.12S	79.01W	28
19.54S	175.67W	231	9.24S	76.74W	99
22.57S	174.60W	169	8.33S	74.62W	147
21.72S	174.03W	24			
21.25S	177.45W	331			

1. Calculate the slopes of the subducted plates in each area as follows:
    a. Locate the foci of the earthquakes on Map 9-7 and indicate their depths.
    b. Outline the edges of the Benioff zone; it runs parallel to the trench and encloses all the data points around it. Measure its width.
    c. Locate the two foci which define the two edges of the Benioff zone, and calculate the difference in their depths. The slope is equal to this difference divided by the width of the Benioff zone.

2. Examine the map of the age of the ocean (Fig. 9-2). What can you say about the relationship between the slope of a subducted plate and the age of the sea floor? What is the origin of this relationship?

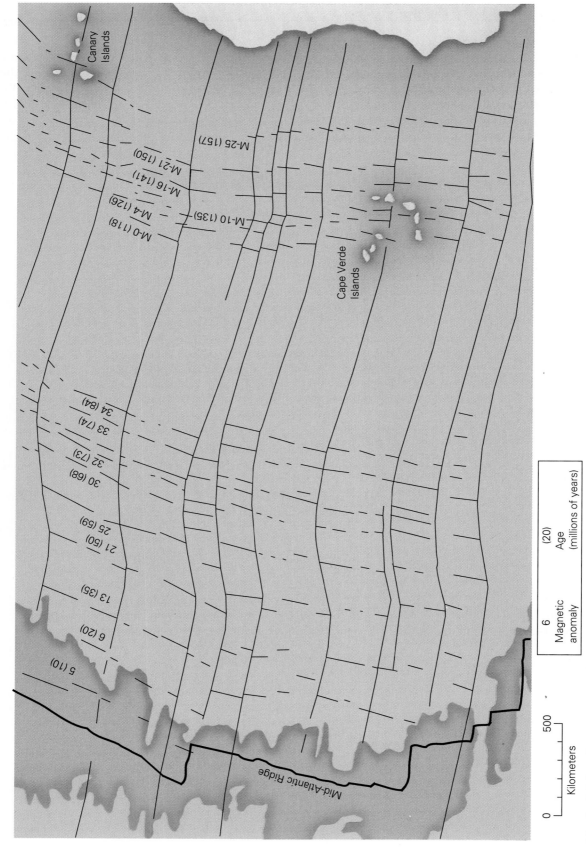

Map 9–6 Selected magnetic anomalies and their ages (in parentheses) in the eastern Atlantic Ocean.

Canary
Islands

M-25 (157)
M-21 (150)
M-16 (141)
M-10 (135)
M-4 (126)
M-0 (118)

Cape Verde
Islands

34 (84)
33 (74)
32 (73)
30 (68)
25 (59)
21 (50)
13 (35)
6 (20)
5 (10)

Mid-Atlantic Ridge

6   (20)
Magnetic   Age
anomaly   (millions of years)

Kilometers

0   500

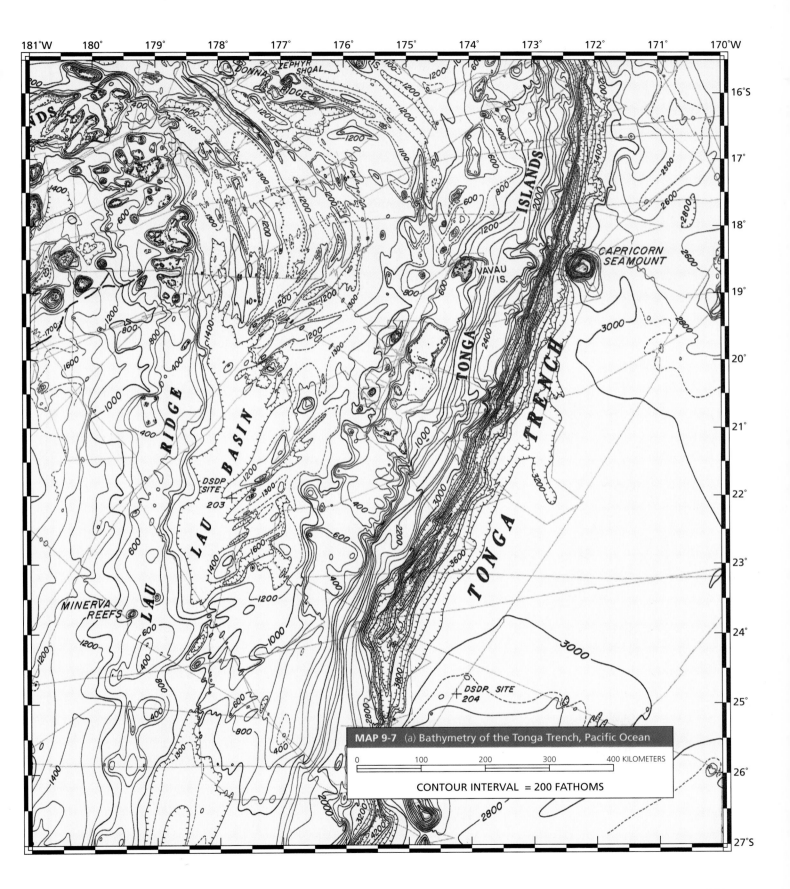

MAP 9-7 (a) Bathymetry of the Tonga Trench, Pacific Ocean

0   100   200   300   400 KILOMETERS

CONTOUR INTERVAL = 200 FATHOMS

231

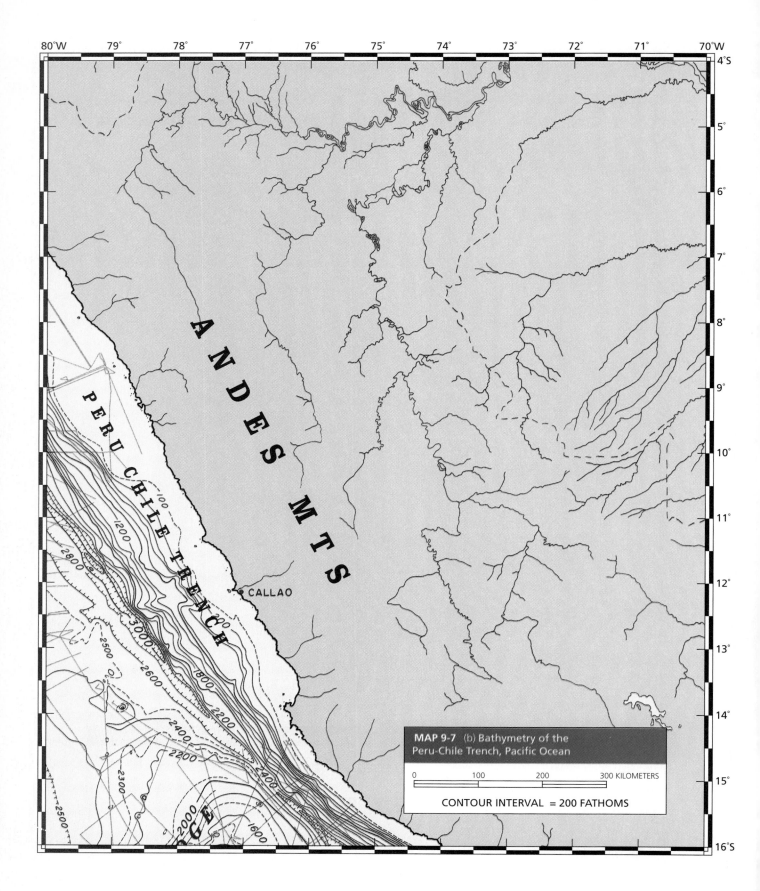

MAP 9-7 (b) Bathymetry of the
Peru-Chile Trench, Pacific Ocean

0   100   200   300 KILOMETERS

CONTOUR INTERVAL = 200 FATHOMS

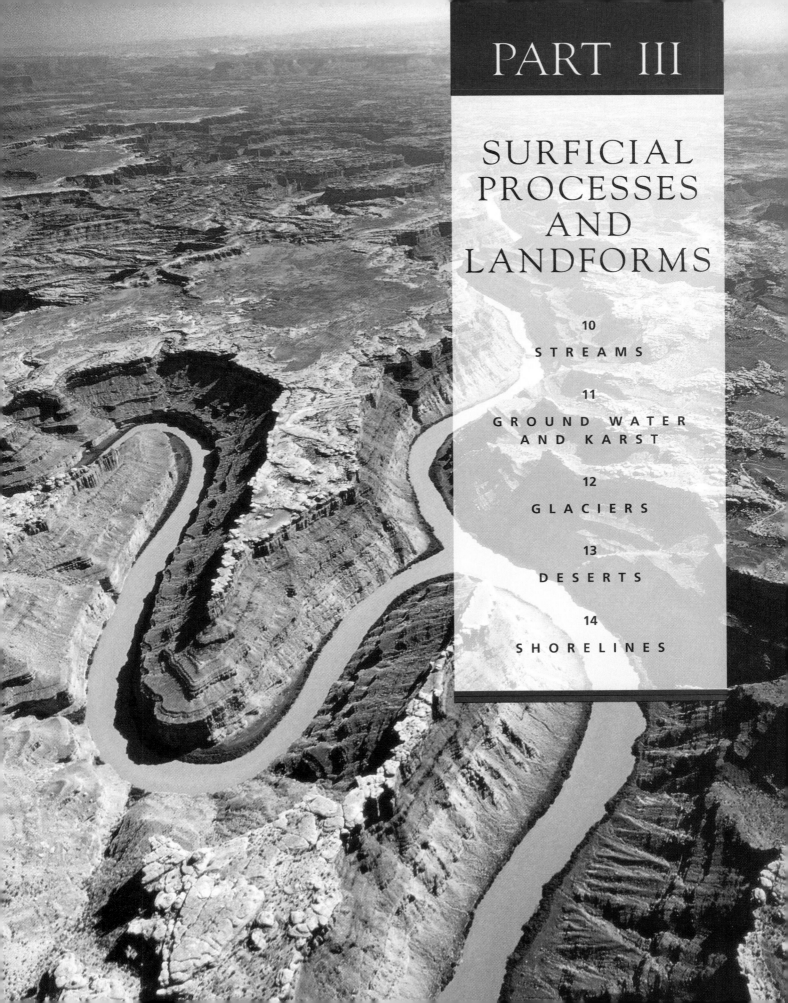

PART III

SURFICIAL PROCESSES AND LANDFORMS

10

STREAMS

**WATERFALL, McARTHUR-
BURNEY FALLS, CALIFORNIA**

© *Tony Stone Images, Renee Lynn)*

Materials Needed *Colored pens or pencils, lead pencils with erasers, ruler, calculator, stereoscope*

Streams receive a lot of attention from geologists, because they are by some estimations the most important agents of erosion and deposition on the continents. Streams erode and transport billions of tons of sediments each year, far more than glaciers and the wind, and they deposit this sediment in **alluvial fans, alluvial valleys,** and river-mouth **deltas.** Streams also carve deep, narrow **valleys** into the land and then slowly but inevitably widen and deepen them until the land is leveled.

This chapter examines streams and stream processes and the landforms which are created by them. It begins with an examination of stream drainage systems and the two basic components of all streams, **channels** and **flood plains.** The chapter continues with a consideration of stream erosion and the evolution of valleys from the steep canyons of youth to the flatlands of old age, and then concludes with an examination of the morphology of stream deposits.

ORIGIN AND COMPONENTS

The formation of streams and stream drainage systems requires excess surficial water—that is, there must be more precipitation of water on the land than there is loss of water to evaporation and infiltration into the ground. Where there is such excess **runoff water**, it flows over the ground as a shallow **sheetflood** and then collects into concentrated flows in the channels of **streams**. Individual streams then merge with one another to form a **drainage system**, which consists of a major **main** or **trunk stream** and its **tributary streams**. Every drainage system collects all the runoff from a specific area of land called a **drainage basin**. This basin, which is also called a **catchment area**, is separated from other basins by elevated **drainage divides** (Fig. 10–1).

The **headwaters** of drainage basins often lie in mountains, hills, and other topographic highs. The steep slopes of these **upper reaches** result in high stream-flow velocities and erosion, which tends to be concentrated along the fastest and least resistant paths to the sea (such as fault traces and outcrop belts of shales). Consequently, headwater streams assume areal arrangements and trends (**drainage patterns** and **densities**) which reflect the structure and lithology of the local bedrock (See Chapter 7).

The **lower reaches** of drainage basins flow across **coastal plains** and **interior basins** to the shores of ocean basins and lakes, where the main streams deposit their sediment loads at their **mouths**. The typically low slopes of coastal plains and interior basins also lead to low flow velocities and sediment deposition in **alluvial fans** and **alluvial valleys** there.

The two principal components of all streams are channels and **overbank** areas, or **flood plains**.

STREAM CHANNELS

Channels are the principal zones of flow in a stream. They are characterized by the highest flow velocities and the highest rates of erosion, and they transport and deposit the coarsest-grained sediments. Channels are generally described on the basis of their **gradient, cross-sectional area and shape**, and **sinuosity**.

Gradient

The gradient is the slope of a stream channel, or the vertical drop in the elevation of the channel bed over a given distance. It ranges from a high of 40 to 50 meters per kilometer, which is typical of mountain streams, to a low of a few centimeters per kilometer, which is typical of coastal-plain streams. The gradient of a stream channel can be represented graphically by a longitudinal profile, which is simply a topographic profile along the length of a channel (Fig. 10–2A).

Cross-Sectional Area and Shape

The size of a channel is measured by its cross-sectional area, the product of its average width and depth. This value is directly proportional to its discharge, which is the volume of water that passes through the channel in a given period of time: The higher the discharge, the greater the cross-sectional area.

The **cross-sectional shape** of a channel is determined by the texture of its banks. Wide and shallow channels typically form where the banks are composed of sand and gravel, which are not very cohesive and thus allow the channel to erode and expand laterally. Narrow and deep channels form where the banks are composed of mud, which is more cohesive and resistant to lateral erosion.

Sinuosity

Channels are not usually straight, but bend and curve to varying degrees along their lengths. This characteristic is called sinuosity, and it is measured by dividing the length of a channel by the horizontal distance between its two ends (Fig. 10–2B).

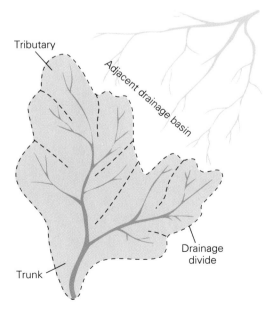

Tributary

Adjacent drainage basin

Drainage divide

Trunk

Figure 10–1 Components of a drainage basin.

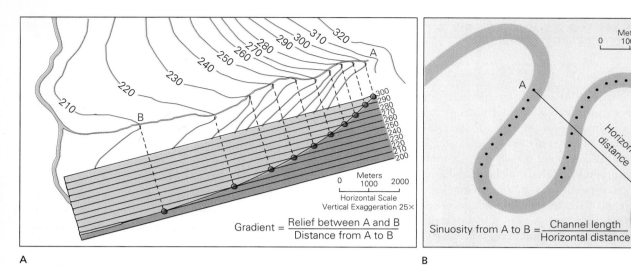

Figure 10–2 (A) The construction of a longitudinal profile of a stream. The gradient of the channel between A and B is equal to the vertical distance between these two points divided by the horizontal distance between them. (B) The sinuosity of the stream between A and B is equal to the channel length between these two points divided by the horizontal distance between them.

Two types of channels are defined on the basis of sinuosity: **straight** and **meandering**. Straight channels have sinuosity values between 1.0 and 1.5 and are most common to streams with high gradients. Meandering channels have sinuosity values greater than 1.5 and are most common to streams with low gradients.

FLOOD PLAINS

Channels are typically bordered by broad, low-relief areas called flood plains. Flood plains are inundated only when the discharge of a stream exceeds the capacity of its channels, causing them to overflow their banks. The resultant overbank flow deposits sand and silt in levees adjacent to the channel and a fertile blanket of clay and organics in a thin sheet across the rest of the flood plain.

Flood plains are tempting places for humans to settle, because they have an abundant water supply and fertile soils that are enriched periodically by floods. However, these same floods can also have catastrophic effects, destroying lives and property in a matter of days. The average flood can usually be contained by a system of dams and artificial levees, but major floods (called the **10-year** and **100-year floods** because of their statistical frequency) are often less containable and more destructive. Such great floods swept through the upper Mississippi Valley in the summer of 1993, following an extremely wet spring in the upper Midwest, and caused billions of dollars' worth of damage to cities, farms, livestock, and property.

EXERCISE 10 – 1   Drainage Basins of the United States

This exercise examines the morphology of the drainage systems in the United States. You may want to refer to the Heezen–Tharp world ocean map, a geologic map of the United States, geographic atlas of the United States, and other such resources to answer some of the following questions.

1. Locate and outline the drainage basins of each of the drainage systems on Map 10–1. Describe the relative area of each drainage basin as *small*, *moderate*, or *large*.

DRAINAGE SYSTEM	AREA	TECTONIC SETTING	DRAINAGE SYSTEM	AREA	TECTONIC SETTING
Columbia			San Joaquin–Sacramento		
Colorado			Rio Grande		
Humboldt			Mississippi		
Brazos			Mobile		
Appalachicola			Suwanee		
Savannah			Santee		
Peedee			Roanoke		
James			Potomac		
Susquehanna			Delaware		
Hudson			Connecticut		

2. Describe the tectonic setting of each drainage system with terms such as *stable interior*, *plateau*, and *active margin*.

3. What is the general relationship between the area of a drainage basin and its tectonic setting?

4. Locate and label these regions on the map. Trace the drainage pattern in each region, and describe it below.

	DRAINAGE PATTERN	STRUCTURE
Atlantic and Gulf coastal plains		
Valley and Ridge (Appalachians)		
Appalachian Plateau		
Upper Mississippi–Ohio Valley		
Great Plains		
Colorado Plateau		
Basin and Range		
Columbia–Snake Plateau		
Great Valley of California		

5. The surficial bedrock in each region consists largely of sedimentary rocks with the exception of the Columbia–Snake Plateau, which is composed of layered basalt. What does the drainage pattern of each region indicate about the structure of its bedrock?

6. Why is there such poor drainage in the Basin and Range of northern Nevada?

EXERCISE 10 – 2   Ste. Genevieve, Missouri

The city of Ste. Genevieve is built on the western banks of the Mississippi River a few miles south of St. Louis. During the floods of 1993, the river rose to an elevation of 390 feet above sea level along this stretch of river.

1. Construct a northeast–southwest topographic profile through the city and across the Mississippi River (Map 10–2). Indicate the topographic elevation of the flood water at its height. How high did the river rise above its banks?

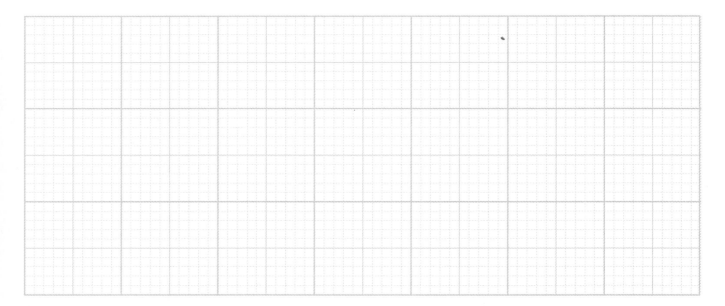

2. Examine the topographic map, and locate the contour line which marks this elevation. Shade the areas which were flooded by the overflow of the river with a colored pencil.

3. What important facilities were inundated by the flood? What facilities and buildings would have been flooded if the water had risen another 10 feet?

EXERCISE 10-3   Red River, North Dakota

Table 10–1 summarizes the maximum (peak) discharge of the Red River near Grand Forks, North Dakota, for the years 1882 to 1979.

TABLE 10–1 Peak Discharge* of the Red River, Grand Forks, ND					
YEAR	DISCHARGE	YEAR	DISCHARGE	YEAR	DISCHARGE
1882	25,000	1915	21,500	1948	34,200
1883	38,600	1916	29,000	1949	15,200
1884	20,600	1917	19,800	1950	54,000
1885	13,040	1918	4480	1951	23,600
1886	10,800	1919	13,600	1952	23,900
1887	7300	1920	30,300	1953	14,600
1888	19,000	1921	11,500	1954	9620
1889	3000	1922	19,000	1955	15,400
1890	3470	1923	16,200	1956	21,400
1891	6000	1924	2530	1957	14,700
1892	23,000	1925	9690	1958	7500
1893	53,300	1926	7720	1959	6300
1894	16,450	1927	10,600	1960	17,200
1895	2000	1928	12,200	1961	3400
1896	21,600	1929	17,100	1962	26,600
1897	85,000	1930	9610	1963	10,800
1898	4500	1931	1630	1964	13,200
1899	9000	1932	9400	1965	52,000
1900	4000	1933	4360	1966	55,000
1901	14,000	1934	3210	1967	28,200
1902	15,000	1935	2920	1968	9420
1903	18,800	1936	9500	1969	53,500
1904	33,000	1937	4180	1970	23,700
1905	16,800	1938	6600	1971	15,800
1906	27,800	1939	6720	1972	31,400
1907	30,400	1940	10,000	1973	11,300
1908	20,500	1941	13,400	1974	34,300
1909	9260	1942	11,000	1975	42,800
1910	18,500	1943	28,200	1976	23,600
1911	3520	1944	10,400	1977	2190
1912	4730	1945	21,300	1978	54,200
1913	17,200	1946	22,000	1979	82,000
1914	8240	1947	35,000		

* In cubic feet per second (cfs).   *Source:* U.S. Geological Survey.

1. Plot these data (with discharge on the vertical axis) on the accompanying graph paper on following page.

2. a. What decade witnessed the lowest discharges along this stretch of the river? _____
   b. What was the range of the discharges during this decade? _____
   c. What other events occurred in the United States during the same decade which might be related to these low discharges and might explain their cause?

3. a. Locate and label the two 100-year floods on your graph.
   b. What were the discharges of these two floods?

   c. What was the actual length of time which elapsed between them? _____

4. a. Locate and label the 10-year floods on your graph.
   b. What is the range of the discharges of these floods? _____
   c. What is the actual average frequency of these 10-year floods? (The average frequency equals the number of 10-year floods divided by the length of time over which they occurred.)

   d. Which decades witnessed the highest-discharge 10-year floods? _____
   e. Which decades witnessed the lowest-discharge 10-year floods? _____

5. Describe the overall cyclic nature of flood discharges from 1882 to 1979, and speculate on the cause of this cyclicity. Illustrate the approximate trend of this cyclicity on your graph.

6. Calculate the average frequency of floods with the following discharges.

Less than 15,000 cfs	
15,001–30,000 cfs	
30,001–45,000 cfs	
45,001–60,000 cfs	
60,001–75,000 cfs	
Greater than 75,001 cfs	

PACIFIC OCEAN

Principal Islands of
HAWAII
SCALE 1:7,500,000

MAP 10-1 Shaded relief map of the United States

ELEVATION TINTS

FEET		METERS
12,000		3,658
9,000		2,743
5,000		1,524
2,000		610
1,000		305
500		152
0		0 (sea level)

LAKE SUPERIOR

LAKE MICHIGAN

LAKE HURON

LAKE ERIE

LAKE ONTARIO

ATLANTIC OCEAN

0   100   200   300   400 MILE

0   100   200   300   400   500   600 KILOMETERS

MAP 10-2 Ste. Genevieve, Missouri

N

CONTOUR INTERVAL = 20 FEET

STREAM EROSION

Streams erode valleys into the land surface by the combined processes of **downcutting** into the valley floor and **lateral erosion** and widening of the valley walls. Stream erosion is most rapid and intensive in uplifted terranes, where the channel gradients are steep, the flow velocities are high, and the streams are far above base level.

The morphology of the landscape evolves through three successive stages during the course of stream erosion: **youth**, **maturity**, and **old age**. Each stage is distinguished by the sinuosity and gradient of the stream channels, their erosional style (i.e., whether they are downcutting into the valley floor or laterally eroding its walls), and the cross-sectional shape of the stream valley (Fig. 10–3).

The youth stage is characterized by streams with straight channels and steep gradients, rapid downcutting, and deep, **V-shaped valleys** with steep slopes and narrow flood plains. Downcutting by young streams continues until they begin to approach base level (usually sea level), at which point they evolve into mature streams with more sinuous channels, lower gradients, and valleys which have been widened considerably by lateral erosion. Eventually, this lateral erosion completely levels the land to form a flat, broad **peneplain**, which characterizes old streams.

However, this progression can be interrupted at any point by tectonic uplift, which increases the relief of a region, **rejuvenates** its streams, and reinitiates this sequence of stream erosion and valley evolution. Rejuvenation creates such features as **incised meanders**, which are formed when mature or old streams are uplifted and resume downcutting; and **terraces**, the uplifted and abandoned flood plains of streams.

Streams can also increase the areas of their valleys by a combination of **headward erosion** and **stream piracy**. Headward erosion describes the erosion and retreat of the drainage divides in the upper reaches of drainage systems, which expands the stream valleys in their sourceward direction. Stream piracy occurs when headward erosion and expansion of a stream progress to the point where the stream intersects a second stream, capturing its water flow.

STREAM DEPOSITION

The principal continental environments of stream deposition are **alluvial fans**, which form along the margins of inland basins, and **braided** and **meandering streams**, which typically fill the alluvial valleys of their lower reaches. Streams also deposit their sediment in deltas along the shores of oceans and lakes, as we will discuss in Chapter 14.

Alluvial Fans

When a stream flows out of a mountain range and into an inland basin, its gradient decreases sharply and its channel splits into progressively smaller channels. This causes a decrease in the flow velocity of the stream and the deposition of a cone-shaped **alluvial fan** at the base of the mountain range (Fig. 10–4).

The surface of an alluvial fan is characterized by a radiating drainage pattern of **distributary channels** separated by low-relief **interdistributary flats**. The distributary channels increase in number and decrease in flow velocity in the downstream direction, which results in a decrease in the texture of their deposits between the inner fan (where they are gravelly) and the outer fan (where they are sandy and muddy). The surface of the inner fan may also contain some tongue-shaped sheets of **debris flow deposits** that were derived from the adjacent mountain front.

Braided Streams

Braided streams contain a large number of channels which split and rejoin around **channel bars** to form a distinctive braidlike pattern (Fig. 10–5). They are commonly found in high-gradient mountain valleys, where the flow velocities are high and the sediment load contains abundant gravel and sand. The channels of braided streams are typically straight to slightly sinuous in plan view and wide and shallow in cross section, and they tend to migrate rapidly across the entire valley floor. The flood plains are usually very narrow and small in area, and their deposits are rarely preserved because they are usually eroded by the migrating channels.

Meandering Streams

A meandering stream consists of a single large channel which flows through a broad flood plain. Meandering streams are typically formed in low-gradient coastal-plain valleys, where the flow velocities are low and the sediment load is composed largely of sand and mud.

The channels of meandering streams are highly sinuous in plan view and narrow and deep in cross section (Fig. 10–5). The maximum depths and highest flow velocities in the channels occur along the outer banks of the meander bends, which are the erosive cutbanks. The shallowest depths and lowest flow velocities occur along the inner banks of the meander bends, which are the sites of deposition of sandy channel-margin **point bars**. Meandering channels migrate slowly in the lateral direction as the result of the simultaneous erosion and retreat of their cutbanks and the deposition and growth of their point bars. This migra-

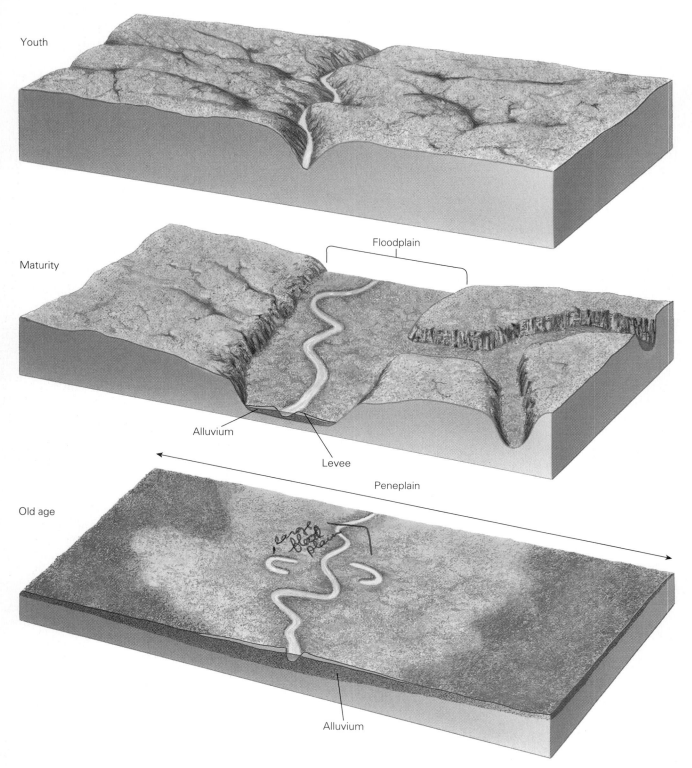

Youth

Maturity

Floodplain

Alluvium

Levee

Peneplain

Old age

Alluvium

Figure 10–3 The three stages in the evolution of a landscape by stream erosion. Youth is characterized by extensive downcutting and deep valleys with steeply sloping walls and narrow floors. Maturity is characterized by lateral erosion, which widens the valley floor to form a flood plain. This lateral erosion can eventually level the land to produce the broad peneplain which characterizes old age.

Figure 10–4 (at right) Cedar Creek alluvial fan, near Ennis, Montana. ▶

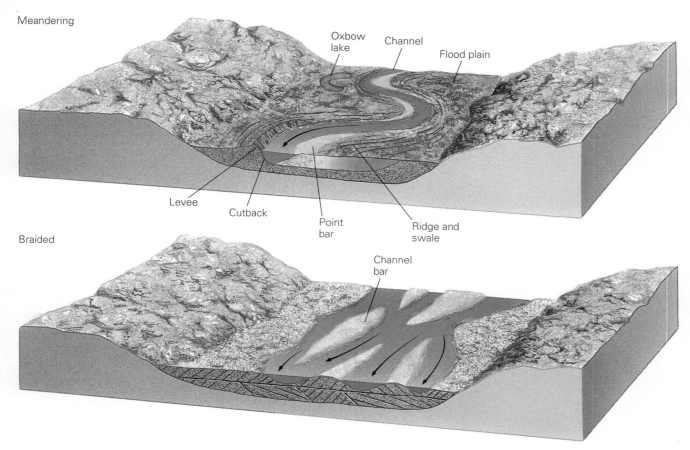

Meandering

Oxbow
lake

Channel

Flood plain

Levee

Cutback

Point
bar

Ridge and
swale

Braided

Channel
bar

Figure 10–5 The morphology of meandering and braided streams.

tion creates a distinctive morphology on the inner banks of meander bends which is called **ridge and swale topography**.

The erosion of a cutbank gradually increases the curvature of a meander bend. Eventually, the channel erodes through the narrow **meander neck** of such a bend as it attempts to find a more direct course toward its mouth. This meander cutoff isolates the former meander bend from the main flow of the channel, usually creating an **oxbow lake**.

The flood plains of meandering streams are flat, featureless plains interrupted only by **levees** and **swamps**. The levees are narrow, elongate ridges of sand and silt which are deposited next to the channel banks by overbank flow during floods; the swamps are vast, low-lying areas where stagnant water often accumulates. The swamps are supplied with mud by overbank flows during floods, but they also contain significant amounts of peat and other organic deposits which are preserved in their stagnant waters.

EXERCISE 10–4   Marias River, Montana

1. Examine the topographic map of the Marias River near Loma, Montana (Map 10–3).
   a. In the central part of the map, construct a topographic profile across the valley.

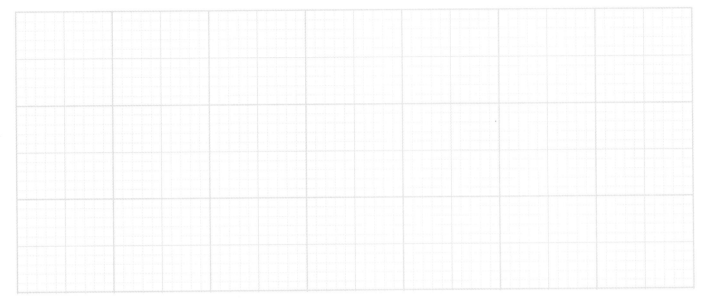

   b. Describe the shape of the cross-sectional profile of the valley. Measure the width of the flood plain and calculate the slopes of the valley walls.

   c. Calculate the sinuosity and gradient of the channel.

   d. Is this a young, mature, or old stream? Explain your answer.

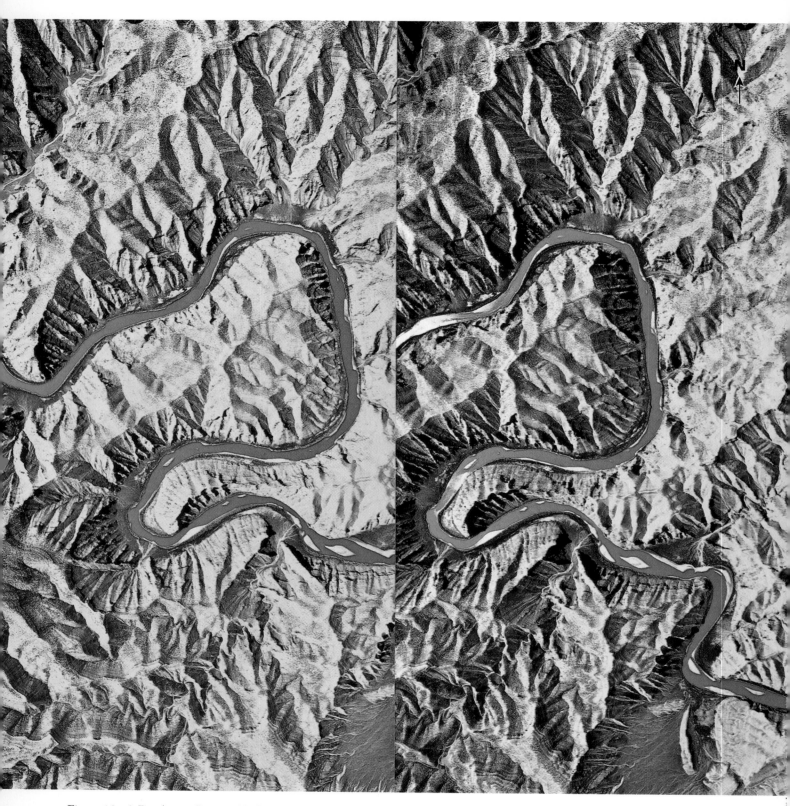

Figure 10–6 Desolation Canyon, Utah.

2. The Colorado Plateau is the only part of the western United States which was relatively unaffected by structural deformation and igneous activity during the Cordilleran orogeny. The plateau was, however, slightly uplifted in the late Tertiary, and this uplift began the downcutting by the Colorado River and its tributaries. This produced Grand Canyon, Glen Canyon, and many other scenic canyons in the plateau.

Examine the stereophotographs of Desolation Canyon (Fig. 10–6, at left) and the Rincon (Fig. 5–12) of the Colorado River.

a. Describe the shapes of the cross-sectional profiles of the two valleys, the slopes of their walls, and the widths of their flood plains.

b. What evidence do you see for uplift of the Colorado Plateau in these two sets of stereophotographs?

E X E R C I S E   1 0 – 5   Kaaterskill, New York

Map 10–4 is a topographic map of the Kaaterskill, New York area.

1. Examine the area near the town of Kaaterskill. Trace the drainage patterns of Gooseberry and Kaaterskill Creeks, and calculate their gradients.

2. North Lake, South Lake, and Spruce Creek were formerly the headwaters of Gooseberry Creek, but they are presently part of the Kaaterskill Creek drainage system. Trace the stream valley that once connected these two lakes and Spruce Creek to Gooseberry Creek. Where has Kaaterskill Creek captured their drainage? Why has this happened?

3. Trace the drainage patterns of Plattekill and Schoharie Creeks. Predict and explain any stream piracy that might occur between these two streams in the future.

E X E R C I S E   1 0 – 6   Ennis, Montana

Cedar Creek alluvial fan (Fig. 10–4 and Map 10–5) was deposited along the eastern side of the Madison River valley near Ennis, Montana, the town in the northwest corner of the map.

1. Construct topographic profiles across the alluvial fan in the strike and dip directions.

2. Describe the shapes of these profiles, and calculate the slope of the surface of the fan along the dip profile.

3. The alluvial fan is traversed by active and abandoned distributaries of Cedar Creek. Locate and label examples of each on the stereophotographs (Fig. 10–4). Describe and contrast their morphologies.

4. Construct a longitudinal profile of McDeed Creek, one of the active distributaries on the fan.
   a. How does the gradient of the creek change along the profile?

   b. What effects does this variation have on flow velocity and sedimentation on the fan?

5. The Madison River valley is presently being uplifted by tectonic activity.
   a. What evidence exists to support this statement?

   b. Approximately how far has the valley been uplifted? _____ How many separate episodes of uplift are apparent? _____ What is your evidence?

6. What kind of stream is the Madison? _____ Label examples of the depositional features which distinguish this kind of stream on the map. How wide is the Madison's flood plain at Ennis? _____ How far would the Madison River have to rise above its mapped level to flood the town? _____

```
EXERCISE  10 – 7     Mississippi River, Mississippi
```

The lower Mississippi River, which runs from Cairo, Illinois, to the Gulf of Mexico, is the classic example of a meandering stream. Map 10–6 shows a stretch of the river near Vicksburg, Mississippi.

1. Label examples of the following features on the map: meander bends, point bars, cutbanks, meander cutoffs, oxbow lakes, levees, and swamps.

2. Trace the present channel of the Mississippi River. Measure its length and calculate its sinuosity.

3. Lakes Palmyra and St. Joseph are remnants of the 1824 channel of the Mississippi River. Trace the path of the old channel of the river through these two lakes. Measure its length, and calculate its sinuosity.

4. What are the width and relief of the flood plain of the Mississippi in the map area? _____
How high above its elevation shown on the map would the river have to rise to inundate the flood plain? _____
_____

```
EXERCISE  10 – 8     Mississippi River
```

Figure 10–7 is a photomontage of the flood plain of the Mississippi River near Vicksburg, Mississippi.

1. Examine the photograph, and label the following features: abandoned channels (A), swamps (S), oxbow lakes (O), and ridge and swale topography (R).

2. Trace some of the older routes of the main channel of the Mississippi on the photo. How have the course and sinuosity of the river changed in recent times?

N

9-27-85   235-126   320912 NHA   9-27-85   235-33   320911 NHAP 2

9-27-85   235-124   320912 NH   9-27-85   235-35   320911 NHAP 2

Figure 10–7 Mississippi River near Vicksburg, Mississippi.

flood plain

MARIAS

RIVER

point bar

cutbank

gibson lake

Creek

MAP 10-3 Loma West, Montana

0   1/2   1 MILE

0   .5   1 KILOMETERS

CONTOUR INTERVAL = 20 FEET

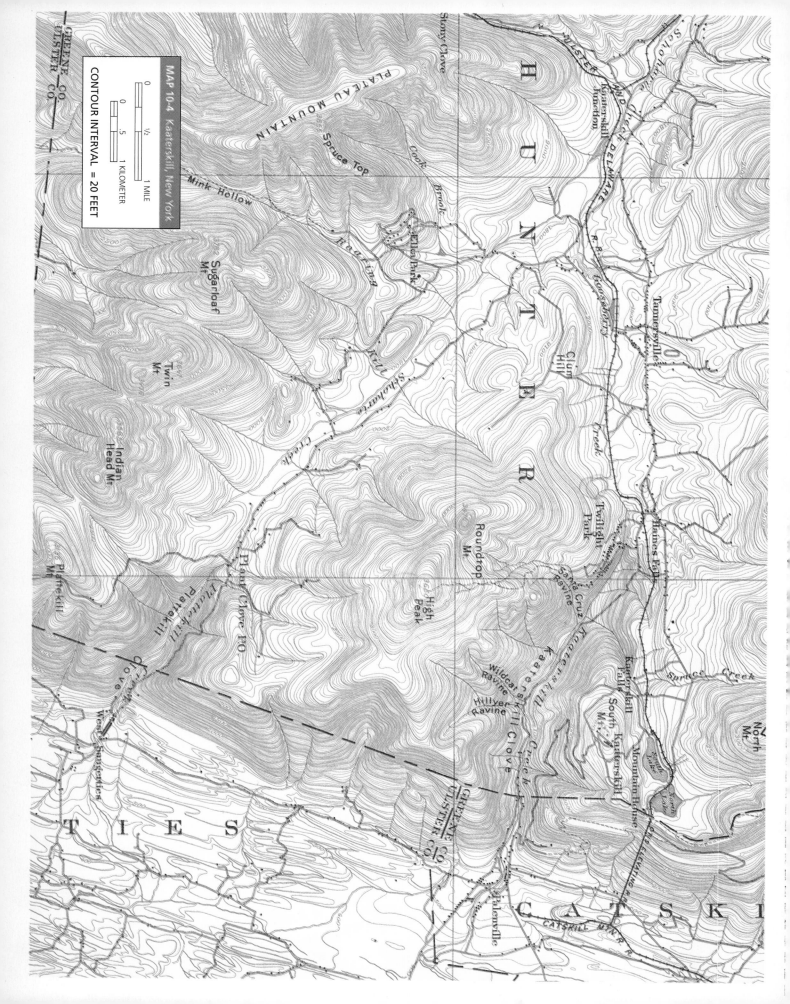

GREENE CO.
ULSTER CO.

H U N T E R

Stony Clove

PLATEAU MOUNTAIN

Spruce Top

Mink Hollow

Cook Brook

Sugarloaf Mt

Twin Mt

Indian Head Mt

Plattekill Mt

Hunting Kill

Elka Park

Schoharie Creek

Plattekill Clove

Plattekill

Plattekill PO

West Saugerties

Schoharie Junction
Kaaterskill Junction

Hoosahopa

Tannersville

Clum Hill

Creek

Twilight Park

Haines Falls

Roundtop Mt

High Peak

Santa Cruz Ravine

Wildcat Ravine

Hillyer Ravine

Kaaterskill Creek

Kaaterskill Clove

Kaaterskill Falls

Mountain House

South Kaaterskill Mt

Spruce Creek

North Mt

GREENE CO.
ULSTER CO.

T I E S

C A T S K I

Palenville

CATSKILL MTN. R.R.

OTIS ELEVATING R.R.

ULSTER AND DELAWARE R.R.

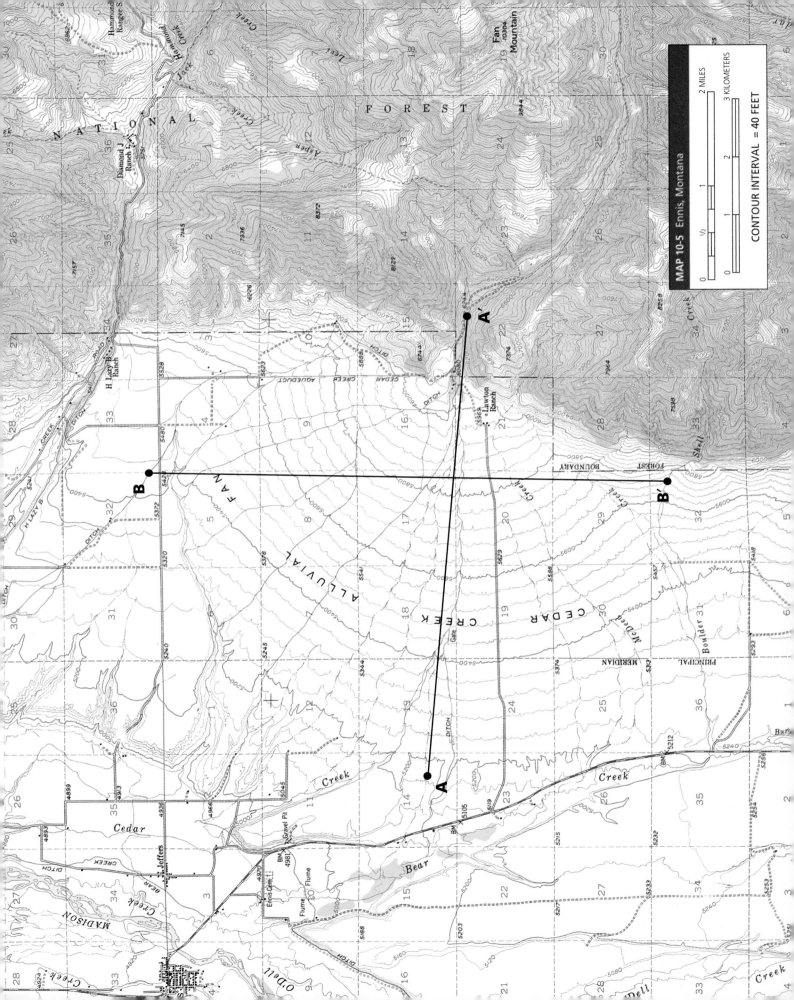

MAP 10-5 Ennis, Montana

CONTOUR INTERVAL = 40 FEET

MAP 10-6 Jackson, Mississippi

0   5 STATUTE MILES

0   5   10 KILOMETERS

CONTOUR INTERVAL = 50 FEET
WITH SUPPLEMENTARY CONTOURS AT 25 FOOT INTERVALS

11

GROUND WATER AND KARST

KEY WORDS

Ground water and hydrogeology . . . porosity, permeability, and hydraulic conductivity . . . aquifers and aquicludes . . . zones of saturation and aeration and the water table . . . recharge and discharge . . . hydraulic gradient, head, and Darcy's law . . . depletion and contamination . . . karst, caverns, and natural tunnels . . . sinking creeks, blind valleys, and haystack hills . . . sinks and rises, sinkhole ponds, and sinkhole plains

CARLSBAD CAVERNS, TWIN DOME AREA, NEW MEXICO (© *Tony Stone Worldwide, Ltd., Raymond G. Barnes*)

Materials Needed *Colored pens or pencils, lead pencils with erasers, ruler, calculator, stereoscope, protractor*

There is a vast and dynamic body of **ground water** beneath our feet. This water is derived from rain and melted snow which infiltrate the ground through fractures and pores in the soil and bedrock, and it flows underground through porous and permeable rock bodies called **aquifers**.

Ground water is a very important geologic agent, because it interacts with the aquifers through which it flows. In some cases, for instance, it is the source of the minerals which cement sedimentary rocks; in other cases, it is the fluid which dissolves pores and caverns in limestones. In addition, ground water is an important human resource. There is far more ground water stored in subsurface aquifers (approximately 4 million cubic km) than all the lakes, streams, and reservoirs in the world (160,000 cubic km), and it is the principal source of drinking and agricultural water in the United States (and elsewhere). However, it is also a finite and fragile resource, one that can be easily exhausted or contaminated if not properly protected and managed.

This chapter deals with the subject of **hydrogeology**, the scientific study of the occurrence, movement, and quality of ground water. It begins with a summary of the principles of this science, including a discussion of the hydraulic properties of aquifers and the movement of water through them, and then examines the effects of water withdrawal and pollution by humans on water quality and content. The chapter concludes with a study of **karst** and the unique set of landforms produced by the interaction of surface and subsurface waters with soluble bedrock.

HYDRAULIC PROPERTIES OF ROCKS

Aquifers are rock bodies which can act as conduits for the movement of ground water through the subsurface. They are also capable of storing significant quantities of water and yielding this water readily to pumping wells. Aquifers can function in this capacity because they are **porous** and **permeable**.

Porosity is a measure of the total volume of void space in a soil or rock. There are three major generic classes of pores (Fig. 11–1):

- **Intergranular pores**: The open spaces between grains in a sediment or sedimentary rock. Intergranular porosity is typically very high in unconsolidated surface sediments, but it is significantly reduced when they are buried, **compacted**, and lithified by natural **cements**.

- **Secondary pores**: Open voids created by the dissolution of soluble grains and cements in a rock. They are common to rocks which contain soluble minerals such as calcite, feldspars, and anhydrite in the form of grains or cements.

- **Fracture pores**: Narrow and elongate voids created by the breakage of brittle rock. Fractures are the principal pore types in igneous and metamorphic rocks, but they are also common to massive and brittle sedimentary rocks.

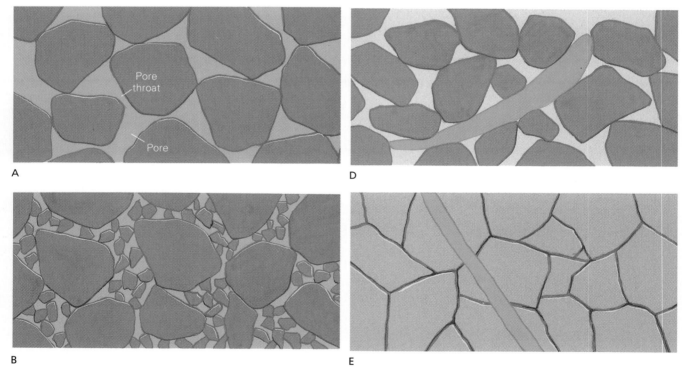

A

B

C

D

E

Figure 11–1 Common pore types. The pores are colored blue in all these figures. (A) Intergranular pores in a well-sorted sand. The pores are connected by unobstructed pore throats through which fluids can flow easily. (B) Intergranular pores in a poorly sorted silty sand. The pore throats are clogged by silt grains, which impede the flow of fluids from pore to pore and thus reduce the permeability of this sediment. (C) Reduced intergranular pores in a sandstone, formed by the partial filling of its intergranular pores by calcite cement. (D) A secondary moldic pore in a sandstone, formed by the selective dissolution of a calcareous shell fragment. (E) A fracture pore in a granite. Such pores are typically straight and cut indiscriminately across grains.

Permeability is a measure of the velocity at which any fluid (water, oil, gas, and so on) flows from pore to pore through a soil or rock. It reflects the degree of **connectivity** of pores, that is, whether they are linked by a network of passageways (**pore throats**) and to what degree the pores and throats are obstructed by silt and clay, cements, or anything else which might impede the flow of water through them (Fig. 11–1).

The term **hydraulic conductivity** is used by geologists specifically to describe the permeability of soil or rock to water. It is generally measured in units of meters or centimeters per second (m/sec or cm/sec), and it ranges from a high of approximately 1 to 3 cm/sec for gravels and porous limestones to a low of 1×10^{-7} cm/sec or less for shales and massive granites. The hydraulic conductivities of common Earth materials are known from laboratory analyses and field tests of rock and soil samples (Table 11–1).

Common Aquifers

The highest-quality aquifers are generally composed of unconsolidated sands and gravels; sandstones, conglomerates, and limestones; and basalts (Table 11–1).

Unconsolidated sand and gravel aquifers have intergranular porosities which range from 24% to 53% and maximum hydraulic conductivities on the order of several centimeters per second. The hydraulic conductivity is highest for well-sorted sands and gravels, which contain few silt and clay grains to obstruct their pore throats, and it decreases with decreased sorting (Figs. 11–1A and 11–1B).

TABLE 11–1	Hydraulic Conductivities of Common Earth Materials
MATERIAL	**HYDRAULIC CONDUCTIVITY (m/sec)***
Gravel	3×10^{-4} to 3×10^{-2}
Coarse sand	9×10^{-7} to 6×10^{-3}
Fine sand	2×10^{-7} to 2×10^{-4}
Silt	1×10^{-9} to 2×10^{-5}
Clay	1×10^{-11} to 5×10^{-9}
Limestone	1×10^{-9} to 2×10^{-2}
Sandstone	3×10^{-10} to 6×10^{-6}
Siltstone	1×10^{-11} to 1×10^{-8}
Shale	1×10^{-13} to 2×10^{-9}
Halite	1×10^{-12} to 1×10^{-10}
Fractured crystalline rocks	8×10^{-9} to 2×10^{-2}
Massive crystalline rocks	3×10^{-14} to 2×10^{-10}

* To convert from m/sec to cm/sec, multiply this value by 10^2.

Sandstone and conglomerate aquifers have porosities of 5% to 30% and maximum hydraulic conductivities on the order of 10^{-6} m/sec (0.0001 cm/sec). They may contain several different types of pores, including **reduced intergranular** pores, which are only partially compacted and cemented; secondary pores created by the dissolution of soluble grains and cements; and fracture pores, which are common to quartz-cemented sandstones (Figs. 11–1C through 11–1E).

Limestone aquifers have porosities of 5% to 50% and maximum hydraulic conductivities of several centimeters per second and more. Their pores are typically secondary in origin, ranging in size from the small **moldic pores** formed by the selective dissolution of fossil fragments to the vast **caverns** of karst terrains (see the section titled "Karst" later in this chapter). Chalks, micrites, and crystalline limestones, which are typically dense and brittle, may also contain fracture pores (Figs. 11–1D and 11–1E).

Basalt aquifers have porosities of 3% to 35% and maximum hydraulic conductivities of several centimeters per second. Their porosity is usually in the form of **shrinkage fractures** which formed when the rocks cooled and crystallized from molten lava.

Common Aquicludes

Aquicludes are soils and rocks which are not especially permeable to water. They are most commonly composed of clay, shale, and massive (nonfractured) igneous and metamorphic rocks, which have hydraulic conductivities on the order of 10^{-9} m/sec and less (Table 11–1).

Whereas aquifers are the conduits for the movement of water through the subsurface, aquicludes are hydraulic barriers which slow or impede the water's movement and thus cause it to pool and accumulate in aquifers. **Unconfined** and **perched aquifers** are formed wherever the downward percolation of surface water through a permeable rock is blocked by an underlying aquiclude. **Confined aquifers** are formed wherever water infiltrates a permeable rock which is bordered above and below by aquicludes (Fig. 11–2).

THE WATER TABLE

An aquifer can be divided into two zones: a lower **zone of saturation**, in which all pores are completely filled with water; and an upper **zone of aeration**, in which they contain both water and air. The boundary between these zones is called the **water table**. The depth of the water table is evident from wells, which often fill to the level of the water table, and from the elevations of springs, streams, and lakes (Fig. 11–2A).

It has long been understood that the elevation of a water table—that is, the volume of water in an aquifer—is not always constant but instead rises and falls over time. This fluctuation is due to changes in the balance between the rate of **recharge** and the rate of **discharge** of the aquifer.

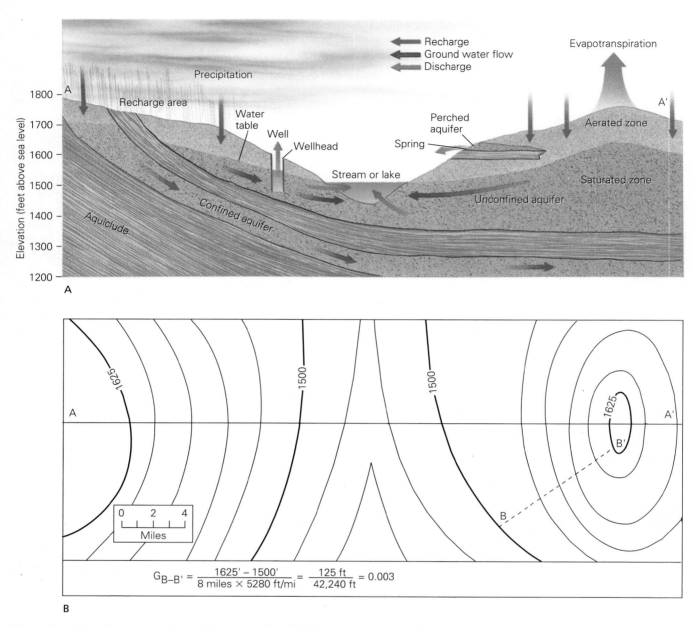

Figure 11–2 Aquifers, aquicludes, and the water table. (A) The structure of unconfined, confined, and perched aquifers and the occurrence and movement of ground water through them. (B) Contour map of the water table in A. The hydraulic gradient, G, between points B and B′ has been calculated.

Recharge and Discharge

Recharge describes the input or supply of water to an aquifer. An aquifer is typically recharged at its outcrops (or **recharge area**), where it is exposed to rain and snow. (Coastal aquifers are also recharged with seawater, which intrudes landward into them in **salt wedges.**) This recharge percolates downward through the aerated zone until it reaches the water table and saturated zone (Fig. 11–2A). The rate of recharge is thus dependent on the climate of the recharge area, which controls precipitation levels there, and the hydraulic conductivity of the aquifer, which controls the rate at which water flows from its recharge area to its water table.

Discharge is the withdrawal of water from an aquifer. A large volume of ground water is discharged naturally by **transpiration**, the process by which plants absorb water through their roots and transport it to the surfaces of their leaves, where it is evaporated into the atmosphere. When the water table intersects the surface, ground water is also naturally discharged from **springs** into lakes and streams, which rise to the level of the water table (Fig. 11–2A). Finally, ground water is discharged through water wells which penetrate the saturated zones of aquifers.

The elevation of the water table of an aquifer remains in a **steady-state condition** over a given interval of time if the rate of water discharged during that time equals the rate of recharge. When the balance shifts in favor of recharge—for example, when precipitation in the recharge area increases but there is no change in the rate of discharge—the water table rises. On the other hand, when the balance shifts in favor of discharge—for example, when the precipitation level in the recharge area is constant but there is increased pumping of water from wells—the water table falls.

MOVEMENT OF GROUND WATER

A water table is rarely a level planar surface (despite its name). Rather, it usually has a three-dimensional configuration that mimics the shape of the land above: It is elevated beneath topographic highs and depressed beneath topographic lows, and it slopes in the same general direction as the surface. The shape or subsurface topography of the water table can be represented graphically by a contour map which shows the elevation of its surface (measured in water wells and lakes), its relief (the vertical distances between points on the water table), and its slope or **hydraulic gradient** (Fig. 11–2B).

The hydraulic gradient between two points on a water table can be calculated with the general equation for slopes:

Hydraulic gradient
$$= \frac{\text{Vertical distance between the two points}}{\text{Horizontal distance between the two points}}$$

Generally, the vertical and horizontal distances between two such points are measured in the same units of measure (feet or meters), which cancel out each other in this equation. Thus, the hydraulic gradient is expressed as a **dimensionless number**.

Ground water exerts a certain amount of downward pressure on the saturated zone of an aquifer due to its mass and the pull of gravity. This pressure is directly proportional to the height of the water column: It is high where the water table is elevated and low where it is depressed. Thus, the relief of a water table creates differences in pressure or **head** throughout the saturated zone, and it is this head which drives the flow of ground water through the subsurface. This flow occurs down the hydraulic gradient of the water table—that is, from high points of the water table to low points—perpendicular to the contour lines of equal water elevation.

The velocity of ground water flow (V) is determined largely by the hydraulic gradient of the water table (G) and hydraulic conductivity of the aquifer (K). This relationship is described by **Darcy's law**,

$$V = KG,$$

named after the engineer Henry Darcy, who discovered it during his study of the ground water supply beneath Dijon, France.

USE AND ABUSE OF AQUIFERS

Aquifers are obviously valuable resources, for they are abundant sources of fresh drinking and irrigation water. However, overpumping of water from aquifers and careless management of their recharge areas can cause their **depletion** and **contamination**.

Depletion

Depletion of aquifers occurs when the rate of discharge exceeds the rate of recharge, so that there is a net loss of water from the aquifer over time. Depletion occurs whenever water wells are drilled into an aquifer, upsetting the natural balance between recharge and discharge. The rate of recharge (which is generally very slow) cannot keep pace with the rate of discharge (which is generally much faster, thanks to pumps), and the volume of water in the aquifer decreases. This causes the water table to drop and water wells to dry up, and it requires drilling deeper and deeper wells to reach the saturated zone (Fig. 11–3).

Depletion has its most severe effects on aquifers which are poorly recharged. For example, the principal aquifer in the western plains of the United States is the Ogallala formation, a thick accumulation of tertiary sands and gravels which blankets the foothills of the Rocky Mountains (Fig. 11–4). Water is presently being withdrawn from this aquifer to supply irrigation water to farms from South Dakota to

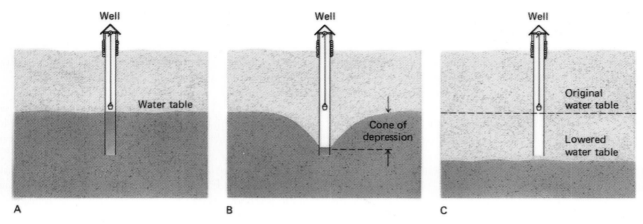

Figure 11–3 (A) When a well is drilled into the saturated zone of an aquifer, the water level rises to the water table. (B) If water is withdrawn from the well more rapidly than it can be recharged, a cone of depression forms around the well. (C) If the high rate of withdrawal continues, the water table falls below the bottom of the well, and the well runs dry.

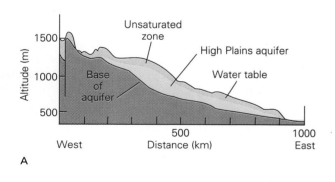

Figure 11–4 (A) Geologic cross section through the Ogallala aquifer. (B) The aquifer is the major source of agricultural water for the western plains of the United States.

Texas. However, this water was derived from the melting of mountain glaciers in the Central and Southern Rockies during the late Pleistocene and Holocene. The loss of this "fossil water" is not presently being balanced by any significant recharge, because the recharge area of the Ogallala lies in the arid rainshadow of the Rockies. In other words, this aquifer is a **nonrenewable resource** which will eventually be exhausted, with disastrous effects on agriculture in the western plains.

The rapid withdrawal of ground water from an aquifer can also cause the **subsidence** of the land surface above it. The pore water in a sandy or gravelly aquifer exerts **pore pressure**, which hinders compaction of the aquifer by the weight of overlying sediments (Fig. 11–5). The rapid removal of ground water reduces pore pressure, allowing compaction to reduce pore space and the total volume of the aquifer. The resultant subsidence of the land above the aquifer can occur at rates of a few centimeters or inches per year, depending on the rate of water withdrawal.

Contamination

The rapid withdrawal of ground water from a coastal aquifer can provide an opening through which seawater can penetrate that aquifer and contaminate it with salt (Fig. 11–6). This phenomenon, called **salt-water intrusion**, is a problem in the coastal plains of the eastern and southern United States, where the pumping of fresh water from coastal aquifers has caused the **fresh water–salt water interface** to migrate several miles to tens of kilometers landward.

The greater risk of ground water contamination, however, comes from pollution in the recharge areas of aquifers. There are many substances which can pollute the soil and bedrock in recharge areas, including agricultural pesticides; livestock waste; fuel and oil from leaky tanks and engines; garbage and toxic-waste dumps; mining wastes, which often contain arsenic, lead, and other toxic metals; car and truck exhaust, which is deposited on roads and then washed into the ground by rain; and sewer and septic systems. Recharge waters can dissolve many of these substances and transport them downward into aquifers, where they contaminate the water supply. The rate and direction of this downward flow or **advection** of contaminants are the same as the rate and direction of ground water flow.

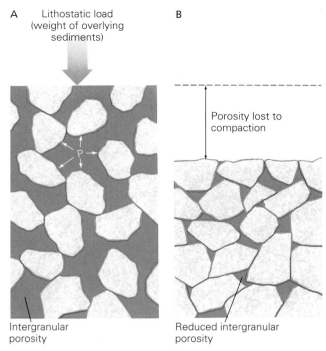

Figure 11–5 The effects of ground water withdrawal from a sandy aquifer. (A) The pore pressure, P, of water supports the fabric of the sediment and prevents its compaction. (B) When this water is removed, the weight of the overlying sediments compacts the grains and reduces the porosity of the aquifer.

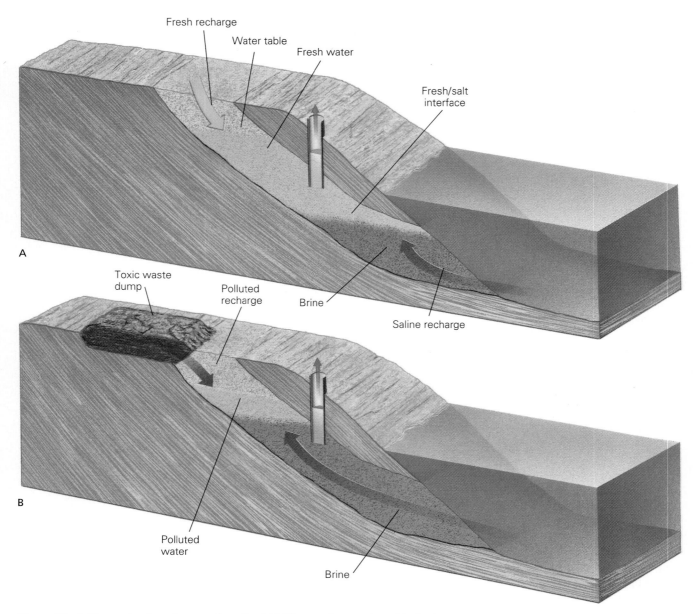

Fresh recharge

Water table

Fresh water

Fresh/salt interface

A

Toxic waste dump

Polluted recharge

Brine

Saline recharge

B

Polluted water

Brine

Figure 11–6 Contamination of a coastal aquifer. (A) The aquifer is recharged by both precipitation and seawater. The well is drilled above the fresh water–salt water interface and discharges fresh water. (B) The rapid withdrawal of ground water from the aquifer (with no change in the fresh water recharge) allows more salt water to enter it. This raises the fresh water–salt water interface and contaminates the well water. An equally serious threat to the quality of the ground water in the aquifer comes from pollution in the recharge area.

EXERCISE 11–1   Contouring a Water Table

Figure 11–7 is a topographic map of an imaginary land surface which is incised by a stream and underlain by a shallow, unconfined aquifer.

1. Construct a topographic profile of the surface from the northwest corner to the southeast corner of the map.

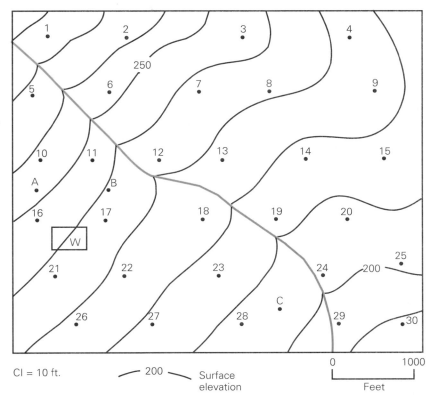

Figure 11–7 Topographic map for Exercise 1.

CI = 10 ft.   —— 200 ——   Surface elevation

0   1000

Feet

2. The locations of several water wells are indicated on the topographic map. Table 11–2 gives the depth of the water level below the wellhead (i.e., below the surface) for each well. Determine the elevation (in feet above sea level) of each wellhead from the topographic map, and calculate the elevation (in feet above sea level) of its water level. Plot these data on the map, and contour them at a 10-foot interval. Use a red or green pencil for these contour lines. See well number 1 for an example.

TABLE 11–2 Well Data (in feet)

WELL NUMBER	WELLHEAD ELEVATION*	WATER DEPTH	WATER LEVEL ELEVATION†	WELL NUMBER	WELLHEAD ELEVATION*	WATER DEPTH	WATER LEVEL ELEVATION†
1	275'	48	227'	16		39	
2		42		17		36	
3		38		18		31	
4		32		19		28	
5		46		20		26	
6		41		21		38	
7		36		22		34	
8		33		23		29	
9		30		24		23	
10		44		25		25	
11		39		26		35	
12		35		27		30	
13		34		28		35	
14		28		29		43	
15		30		30		49	

* From topographic map.
† Wellhead elevation minus water depth.

3. On the same graph paper, and with the same scales as the surface profile, construct a topographic profile of the water table from the northwest corner to the southeast corner of the map.

4. Describe the relationship between the topography of the land surface and the topography of the water table.

5. Ground water flow occurs down the hydraulic gradient of the water table — that is, from high points of the water table to low points — perpendicular to the contour lines of equal water elevation. What are the directions of ground water flow in this aquifer? Indicate your answer with arrows.

6. Calculate the average hydraulic gradient between the northwest and southeast corners of the map area. Measure the vertical and horizontal distances between the two corners in feet, so that the hydraulic gradient will be a dimensionless number.

7. Use this average hydraulic gradient to calculate the flow velocity of ground water through rocks with the following hydraulic conductivities. Summarize your data in the following table.

HYDRAULIC GRADIENT	×	HYDRAULIC CONDUCTIVITY	=	FLOW VELOCITY
	×	1×10^{-1} m/sec	=	
	×	1×10^{-2} m/sec	=	
	×	1×10^{-3} m/sec	=	
	×	1×10^{-4} m/sec	=	
	×	1×10^{-5} m/sec	=	
	×	1×10^{-6} m/sec	=	
	×	1×10^{-7} m/sec	=	

8. Construct a graph that shows the change in flow velocity with variations in hydraulic conductivity. Use a horizontal scale of hydraulic conductivity which begins on the left with 1×10^{-6} m/sec and increases to the right. Use a vertical scale of flow velocity which increases upward.

9. Describe the effect of variations in the hydraulic conductivity on flow velocity.

10. The shaded area marked "W" is a sludge pond of a coal mine which was abandoned in the early 1980s. The EPA wants to drain the sulfur-rich waters of this pond because it suspects that the pond sprang a leak in 1990. There are three municipal wells (marked A, B, and C) in the immediate area of the sludge pond. Which well is most threatened by the leaky pond? Explain your answer.

11. What is the horizontal distance between the sludge pond and the threatened well? _____ Given the range of flow velocities calculated for question 7, estimate the minimum and maximum amounts of time required for the contaminated ground water to reach this well.

E X E R C I S E   1 1 – 2   Ground Water Flow, San Fernando Basin

---

Figure 11–8 is a contour map of the water table of a shallow, unconfined aquifer in the San Fernando basin of southern California. The aquifer consists of poorly consolidated arkosic sands with hydraulic conductivities on the order of 1×10^{-5} m/sec. Surface runoff and stream drainage from the surrounding mountains serve as the principal natural sources of water for the aquifer in the basin. In addition, the aquifer is recharged with injection wells which recycle water back into the subsurface.

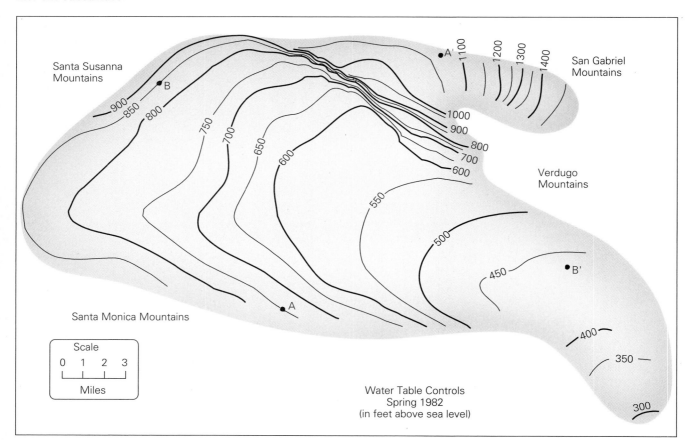

Figure 11–8 Contour map of a shallow unconfined aquifer in the San Fernando basin, California.

1. What is the maximum elevation of the water table in the aquifer? _____ the minimum? _____ its total relief? _____

2. What is the relationship between the height of the water table and its proximity to the recharge area?

3. Construct a topographic profile from point A to point A′ and a second profile from B to B′.

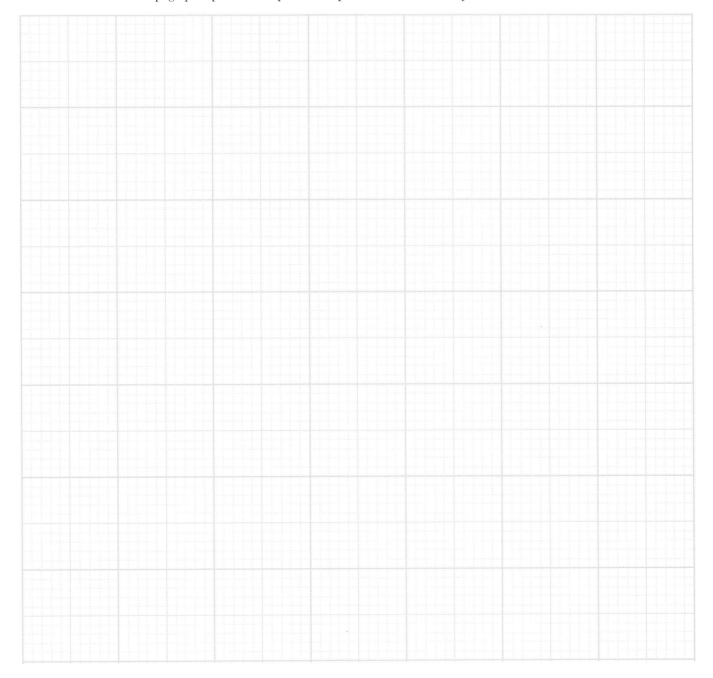

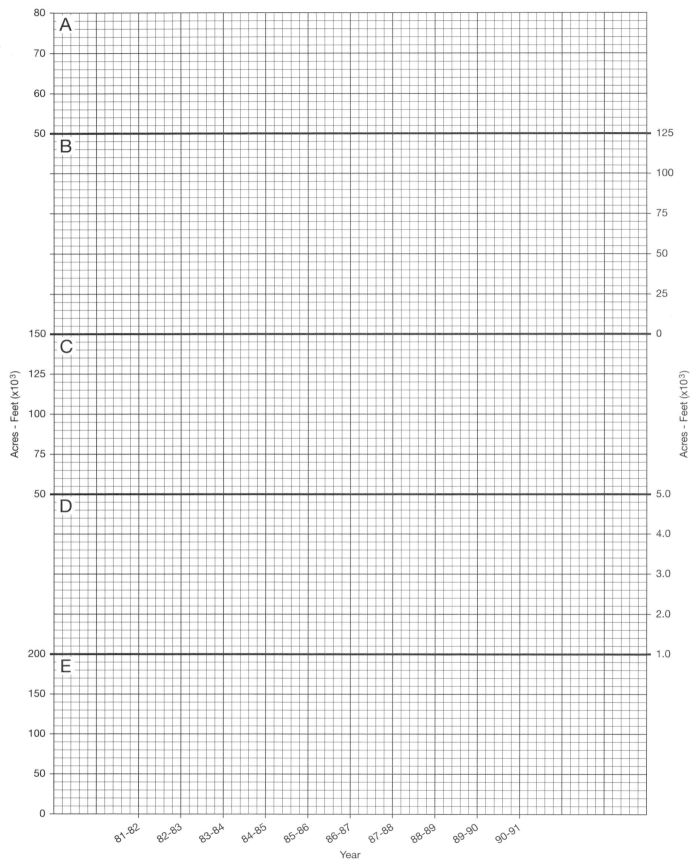

4. Describe the shape and attitude of the water table.

5. Use arrows to indicate the direction of ground water flow in the San Fernando basin. What is the net direction of ground water flow? _____ Where in the subsurface does ground water exit the basin? _____

6. Where along the two profiles does the water table have its steepest hydraulic gradient? Calculate this value and the flow velocity of ground water there.

7. Where along line B-B' does the water table have its gentlest hydraulic gradient? Calculate this value and the flow velocity of ground water there.

8. Table 11–3 is a ground water inventory summary of the aquifer for the period 1981 to 1990. The data summarize its natural recharge from streams and surface runoff and its natural discharge from surface flow (into streams) and subsurface flow (see question 5). Table 11–3 also summarizes the aquifer's recharge in injection wells (which return waste water to the subsurface) and its discharge from water wells. The data are given in units of **acre-feet**, which are a standard means of measuring large volumes of water. One acre-foot is the volume of water which would flood one acre of land with one foot of water; it is equal to 1234 cubic meters.

Construct six graphs of the following ground water data on the accompanying sheet of graph paper.

Graph A: natural recharge

Graph B: injection-well recharge

Graph C: well discharge

Graph D: ground and subsurface discharge

Graph E: First complete Table 11–3 by calculating (a) the total recharge for each year (streams and runoff + injection wells) and (b) the total discharge for each year (wells + surface and subsurface flow). Plot the total annual recharge (a) on the graph paper and connect the points with a red line. Plot the total annual discharge (b) on the same graph and connect the points with a blue line.

TABLE 11–3 Ground Water Inventory for the San Fernando Basin

					TOTAL IN ACRE-FEET					
	1981–82	1982–83	1983–84	1984–85	1985–86	1986–87	1987–88	1988–89	1989–90	1990–91
RECHARGE From streams and runoff	60,000	74,000	65,000	62,000	70,000	62,000	69,000	61,000	62,000	57,000
From injection wells	29,000	113,000	41,000	25,000	33,000	10,000	28,000	8,000	6,000	23,000
Total recharge	_____	_____	_____	_____	_____	_____	_____	_____	_____	_____
DISCHARGE From wells	88,000	71,000	120,000	106,000	91,000	97,000	109,000	133,000	87,000	76,000
From surface– subsurface flow	1,700	3,900	3,400	3,700	4,300	3,400	3,400	3,400	3,900	3,200
Total discharge	_____	_____	_____	_____	_____	_____	_____	_____	_____	_____

9. Examine graphs A through E and answer the following questions:
   a. Describe the changes in the total annual recharge for the last 10 years.

   b. When was the annual recharge highest? _____ Examine Graphs A and B. What was the cause of this peak?

   c. When was the annual recharge lowest? _____ Speculate on the cause of this low.

   d. Describe the changes in the total annual discharge for the last 10 years.

   e. When was the annual discharge highest? _____ Examine Graphs C and D. What was the cause of this peak?

   f. When was the annual discharge lowest? _____ Speculate on the cause of this low.

   g. A **water deficit** in a ground water inventory occurs when the annual discharge from an aquifer exceeds its annual recharge. What was the year of the greatest water deficit for the San Fernando aquifer? _____ What was the volume (in acre-feet) of this deficit? _____

   h. A **water surplus** occurs when the annual recharge from an aquifer exceeds its annual discharge. What was the year of the greatest water surplus for the San Fernando aquifer? _____ What was the volume of this surplus? _____

   i. Was there a significant surplus or deficit in the last year of the inventory? _____

   j. Discuss with your classmates the total effect of human activity on this aquifer. Do you think that the people responsible for this resource manage it well? What recommendations would you make regarding its future management?

EXERCISE 11 – 3   Subsidence Around Galveston Bay

The southeastern coast of Texas has been slowly subsiding due to the rapid withdrawal of large volumes of ground water from shallow aquifers and oil from deeper reservoirs. This subsidence has caused billions of dollars' worth of damage to buildings and roads, as well as the loss of coastal land to flooding by the waters of Galveston Bay. Figure 11–9 shows the subsidence which occurred in this region between 1950 and 1970.

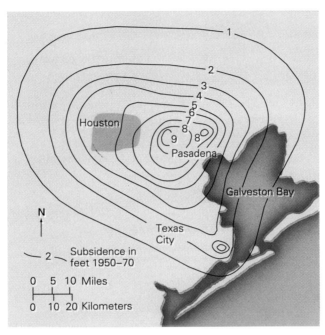

Figure 11–9 Subsidence along the southeast Texas coast.

1. Where are the areas of maximum subsidence?

2. What was the average rate of subsidence per year for each of these areas?

3. According to the 1970 topographic map, Texas City was approximately 10 feet above sea level and Pasadena was approximately 40 feet above sea level. Assuming no changes in the rates of water and oil withdrawal and subsidence after 1970, determine the decade in which each town would be flooded by Galveston Bay.

4. The city of Houston (population in excess of 3 million) was also approximately 40 feet above sea level in 1970. Given the present rate of subsidence there, when would this major urban area have to be abandoned to the waters of Galveston Bay?

This is a hands-on experiment on the effects of grain size and sorting on permeability. The basic apparatus for the experiment (which will be supplied by your instructor) consists of a funnel (3 to 6 inch diameter), a paper coffee filter, and a graduated cylinder. Place the filter paper inside the funnel, and place the funnel at the top of the cylinder.

Your instructor will provide several 50 to 60 gram (approximately 2 ounces) samples of different types of dry unconsolidated sediment, including fine gravel, well sorted sand, poorly sorted sand, silt, and clay. Follow this procedure for each sample:

1. Examine each sample, and describe its grain size and sorting.

2. Pour each sample in the paper-lined funnel, and smooth its top.

3. Gently pour 50 ml of water into the funnel.

4. Measure the amount of time required for this water to flow into the graduated cylinder.

5. Record your data in tabular form in the space provided below.

You can run this experiment separately for each sediment sample on the same apparatus, but be sure to empty and dry the cylinder and replace the filter paper after each run. If the supplies are available, you can also set up several apparatuses and run the experiment on all the samples at the same time.

When you are finished, discuss the effects of grain size and sorting of the flow rate with your classmates.

enerally, the movement of ground water through aquifers is very slow. The major exception to this generalization exists in **karst** terrains, where the large-scale dissolution of bedrock by ground water creates fast-moving streams which flow through subterranean **caverns** and **natural tunnels** (Fig. 11–10).

Karst is most common in limestone and dolomite formations, which are composed of soluble calcareous minerals (calcite and dolomite). The **karstification** of such formations requires the flow of large volumes of ground water through the bedrock, because the solubilities of these minerals are very low. Consequently, karstification usually occurs in humid and tropical climates, where there are high levels of precipitation and ground-water recharge. However, the solubilities of calcite and dolomite can be increased by the introduction of carbon dioxide (CO_2) into the ground water, which makes it more acidic according to this reaction:

$$CO_2 + H_2O \longrightarrow H_2CO_3 \longrightarrow H^+ + HCO_3^-.$$

| Carbon dioxide | Water | Carbonic acid | Hydrogen ion (acid) | Bicarbonate ion |

The carbon dioxide in ground water can come from atmospheric sources, particularly volcanoes and the combustion of fossil fuels. It can also be derived from sources within the crust itself, such as plutons and hydrocarbon

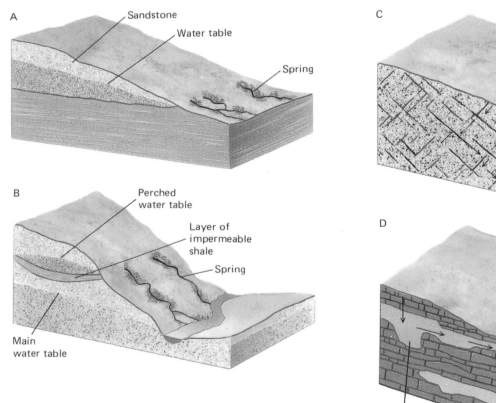
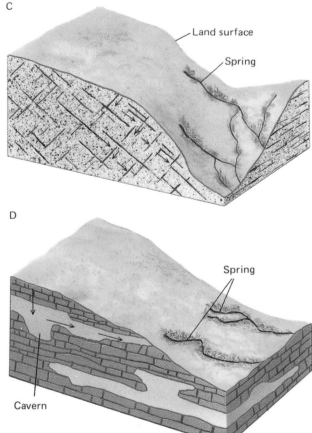

Figure 11–10 The occurrence and movement of ground water. (A) Ground water in a sandstone aquifer underlain by a shale aquitard. The ground water is discharged from springs wherever the saturated aquifer intersects the land surface. (B) Ground water above a perched aquitard is discharged from springs; the main ground water body discharges directly into a stream. (C) Ground water in fractured bedrock. The ground water is discharged from springs wherever the fractures intersect the land surface. Thus, the surface distribution of springs is a good indicator of structural trends in fractured bedrock. (D) Ground water in soluble limestone bedrock. The ground water flows in subterranean streams through caverns and tunnels. The caverns and tunnels are typically formed along the most permeable horizons in the bedrock such as a fracture plane or a bed of poorly cemented skeletal limestone. The ground water is discharged from springs and rises wherever such permeable horizons intersect the land surface.

reservoirs, which release CO_2 as a byproduct of oil and gas maturation.

Karstification is promoted by the presence of fractures, joints, and bedding planes in the bedrock, which provide permeable pathways through which ground water can circulate and dissolve a network of caverns and tunnels (Fig. 11–10). It is also favored by the entrenchment of the main stream of the drainage system, because this increases the gradients of its tributaries and promotes their downcutting into the bedrock.

KARST LANDFORMS

Karst terrains are drained by streams which flow for varying distances both across and below the surface. They are distinguished on topographic maps by low density, poorly integrated surface drainage patterns, and the presence of **sinking creeks** and **blind valleys**, **rises** and **sinks**, and **haystack hills** (Fig. 11–11).

The surface drainage pattern of a typical karst terrain consists of several isolated sinking creeks which emerge from rises and run for short stretches through blind valleys before vanishing abruptly into sinks. The streams continue to flow in the subsurface through natural tunnels and caverns until they reemerge at rises or reach the main stream. Together, sinking creeks and subterranean streams form fully connected or integrated drainage systems with drainage patterns which reflect the local bedrock structure. However, this is not usually apparent from the surface drainage patterns of karst terrains, because their sinking creeks and blind valleys often have no obvious connections to each other or to the main stream and its valley.

Sinks (or **sinkholes**) are circular and elliptical depressions in a karst surface. They can be formed by two means: the collapse of cavern roofs (**collapse sinkholes**) and the dissolution of fractures and bedding planes by downward-percolating ground water (**dolines**). Wherever the water table is high, they fill to form **sinkhole ponds**. Haystack hills (also called **pepinos**) are mound-shaped erosional remnants on the karst surface. They are often composed of rocks which are less soluble than the surrounding bedrock and thus dissolve more slowly.

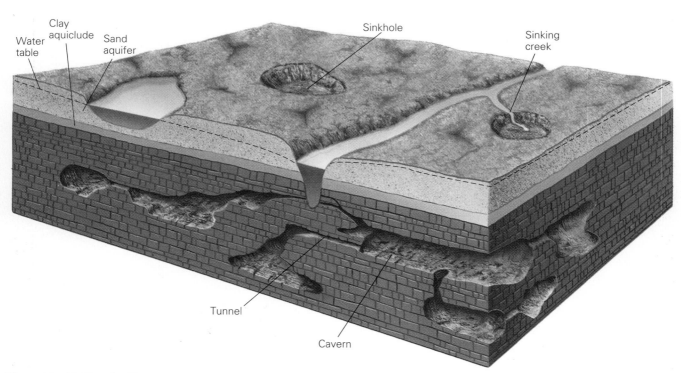

Figure 11–11 Karst landforms.

EXERCISE 11–5   Interlachen, Florida

Map 11–1 is a topographic map of the area around Interlachen, Florida. The bedrock consists of a thick section of porous limestone which is overlain by interbedded sandstone and shale. The land surface contains many sinkhole ponds which were created by the dissolution of the limestone bedrock. The surfaces of these ponds are exposures of the water table.

1.   Describe the shapes and dimensions (width and depth) of some of the larger sinkholes in the map area.

2.   Would you buy real estate in the towns of Interlachen or Mannville? Explain your answer.

3.   Use the spot elevations of the sinkhole ponds (where they are recorded) and the surface contour lines (where they are not) to determine the approximate elevations of the water table in the larger sinkhole ponds. Record these data directly on the topographic map.
     a. What is the maximum relief between the top of a sinkhole and the water table? _____
     b. What is the minimum relief between the top of a sinkhole and the water table? _____

4.   Use a colored pencil to contour the water table at a contour interval of 5 feet.

5.   What is the:
     a. maximum elevation (above sea level) of the water table in the map area? _____
     b. minimum elevation of the water table? _____
     c. total relief of the water table in the map area? _____
     d. direction of slope of the water table? _____
     e. hydraulic gradient of the water table? _____
     f. direction of ground water flow in the aquifer (indicate with arrows)? _____

6.   Given a hydraulic conductivity of 10 cm per sec, what is the velocity of ground water flow through the aquifer?

7.   Draw topographic profiles of the land surface and water table along line A-A'. Indicate the water levels in the lakes which lie on the line. Use colored pencils to shade the aerated and saturated zones of the aquifer. Indicate the direction of water flow in the saturated zone with an arrow.

8.   Locate point B on the profile line. If you wanted to drill a water well at this spot, and if the cost of such drilling were $3 per foot, what would be the minimum cost of the well?

9.   Locate point C. Suppose there is a gasoline station at this point, and one of its storage tanks has ruptured. Which might be polluted by the spill: Trotting Pond or Grassy Lake? Why?

| EXERCISE 11-6 | Mammoth Cave, Kentucky |

Some of the best examples of karst landforms can be seen around Mammoth Caves National Park in central Kentucky. The local bedrock there consists of two sedimentary rock formations of Carboniferous age: the Mauch Chunk, which is a resistant sandstone unit, and the underlying St. Louis, which is a karstified limestone. The sandstone caps a gently inclined plateau which extends northward from Dripping Springs escarpment toward the Green River (Map 11–2). This plateau is eroded by the tributaries of the river, some of which are sinking creeks. The limestone is exposed throughout a sinkhole plain south of the escarpment as well as in the deeper valleys in the plateau.

1. Examine the sinkhole plain on the topographic map. Describe its general morphology.

2. Use the topographic map to determine
    a. the maximum elevation of the sinkhole plain _____
    b. its minimum elevation _____
    c. its total relief _____

3. Measure the widths and depths of some of the largest sinkholes in the sinkhole plain. Compare their dimensions to the sinkholes in Interlachen, Florida (Fig. 11–12). Speculate on the origin of any differences between the two groups.

4. Locate Dripping Spring. What is the origin of this feature? _____ What is its elevation? _____

5. Locate the sinking creek south of Rocky Hill. Trace its path on the topographic map, and indicate its flow direction with an arrow.

    a. What is the flow direction of Sinking Creek? _____
    b. Where does it terminate? _____
    c. Trace the other sinking creeks in the sinkhole plain. Describe the overall drainage pattern and density of the sinkhole plain.

6. Construct a topographic profile along line A-A'. Label the following features: Dripping Springs, Dripping Springs escarpment, Brownsville Pike, Double Sink, and the Green River. Record the elevations of Dripping Springs and the surface of the river. Examine the slope of the escarpment, and mark the approximate location of the contact between the St. Louis and the Mauch Chunk.

7. Examine and trace (with a red pencil) the drainage pattern on the plateau between Brownsville Pike and the escarpment.
   a. Describe the drainage pattern. _____
   b. Do you see much evidence for karst formation in this area? _____
   c. What do the drainage pattern and topography indicate about the lithology of the bedrock there? Explain your answer.

8. Double Sink is an example of a blind valley of a sinking creek. Trace the drainage pattern and tributaries of this sinking creek with a blue pencil.
   a. Where does the sinking creek terminate? _____
   b. Which formation must be exposed at the bottom of Double Sink? _____
   c. Which formation is exposed in the high ground around Double Sink? _____

9. What is the approximate elevation of the contact between the St. Louis limestone and the Mauch Chunk sandstone in Double Sink? _____ Indicate this contact on your topographic profile.

10. Construct a geologic cross section on your profile by connecting the contacts between the two formations at Double Sink and Dripping Springs escarpment. Label the two formations on your profile.

11. Locate other blind valleys in the plateau, and trace the drainage patterns of their creeks with your blue pencil. Compare these drainage patterns with the pattern of the plateau between Brownsville Pike and the escarpment. How are they similar?

12. What does this similarity in drainage pattern indicate about
   a. the erosional history of the blind valleys?

   b. the future evolution of the topography between Brownsville Pike and the escarpment?

13. Mammoth Caves is the third largest cave system in the world, with greater than 250 km of interconnected subterranean caverns and natural tunnels. These caverns and tunnels have developed along two intersecting sets of joints in the St. Louis limestone. Examine the topography around the entrance to the caves (in the northeast corner of the map), and locate and trace surface features which reflect this jointing pattern. What are the approximate strike directions of the joints? _____

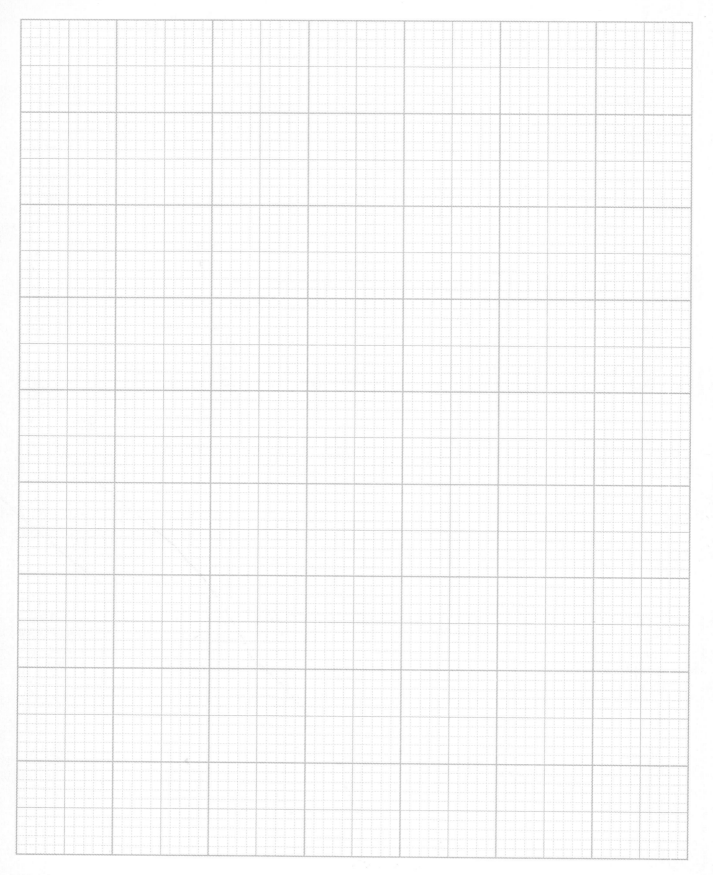

MAP 11-1 Interlachen, Florida

MAP 11-2 Mammoth Cave, Kentucky

CONTOUR INTERVAL = 20 FEET

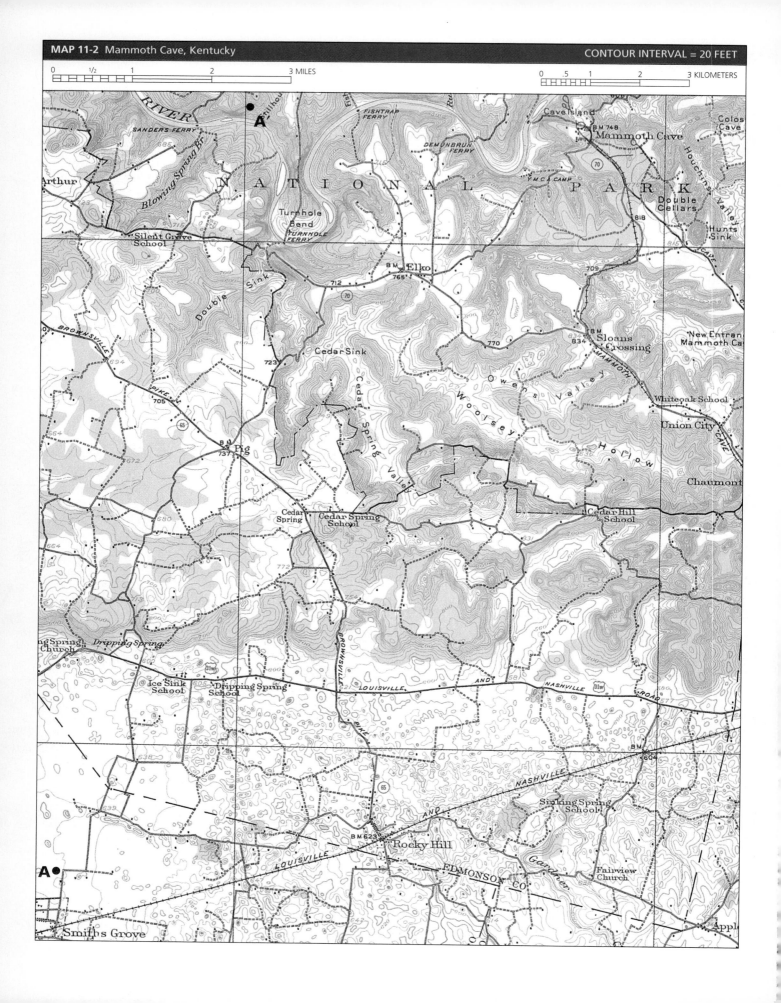

GLACIERS

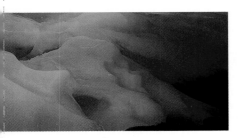

GLACIAL ICE FORMATIONS IN AN ICE CAVE. *(© Tony Stone Images/Stuart Westmorland)*

Materials Needed *Colored pens or pencils, lead pencils with erasers, ruler, calculator, stereoscope*

Glaciers are permanent masses of flowing snow and ice. They form wherever a combination of short, cool summers, low temperatures, and high levels of precipitation allows for the net accumulation of winter snow over the years. They range in size from the thick **continental ice sheets** which blanket Antarctica and Greenland to the small **alpine** or **mountain glaciers** which flow through the valleys of high mountain ranges.

Glaciers are dynamic geologic agents which advance and retreat across mountains and continents in response to changes in the Earth's climate. This movement results in intensive erosion of the landscape, deposition of enormous volumes of sediment, and the creation of some of the most spectacular and distinctive landforms in the world. In this chapter we first consider the conditions that lead to the formation and growth of glaciers and then examine some of their erosive and depositional landforms on topographic maps and aerial photographs.

MOUNTAIN GLACIERS

Mountain glaciers form in the summits of high mountains, where cold temperatures, high rates of precipitation, and short, cool summers lead to the formation of a permanent ground cover of ice. The largest mountain glaciers are several hundred meters thick and cover several hundred square kilometers of land. Most, however, are considerably thinner and less extensive in area.

Mountain glaciers typically begin to form in spoon-shaped depressions called **cirques** in the **zone of accumulation**, which lies above the snowline. Eventually, however, the ice overflows these cirques and forms **ice streams** which flow down the steep mountain slopes through preexisting valleys (usually fluvial in origin). The individual ice streams often merge to form a **glacial drainage system** consisting of a **main** or **trunk glacier** and its smaller **tributary glaciers**. Glacial flow ends at the **glacial front** in the zone of wastage (or **ablation**) below the snowline, where warm temperatures and less precipitation lead to a net loss of glacial ice every year from **melting, sublimation,** and **calving**. The glacial fronts are often situated on land, where they are drained by fast-flowing **outwash** or **meltwater streams**. The glacial fronts of **tidewater glaciers**, which advance from mountains into the sea, constantly calve to form loose blocks of drifting ice called **icebergs**.

The elevation and areal extent of a mountain glacier are limited by the topographic elevation of the snowline, which is itself fixed by the balance between precipitation (the source of glacial ice) and temperature (which controls its wastage). Glaciers are confined to the peaks and upper slopes of mountains in the warmer lower latitudes, where the snowline lies at elevations as high as 7000 meters. They can flow downslope to lower elevations and spread over larger areas in the colder higher latitudes, where the snowline is at or near sea level.

However, the topographic elevation of the snowline also fluctuates in response to global and local changes in climate. This climatic fluctuation produces corresponding fluctuations in the extents of mountain glaciers. Glaciers expand and advance when the snowline is lowered by an increase in precipitation and/or a decrease in temperature, and they contract and retreat when the snowline is raised by a decrease in precipitation and/or an increase in temperature.

CONTINENTAL ICE SHEETS

Continental ice sheets form in high latitudes, where low temperatures create a permanent ground cover of ice over most of the landscape. They also form in middle latitudes by the expansion and coalescence of mountain glaciers during glacial periods. Ice sheets are vast in area and many hundreds to thousands of meters thick, and often cover all but the highest mountain peaks on a continent. There are presently two ice sheets on the surface of the Earth: the Antarctic ice sheet, which covers about 13 million sq km of Antarctica (including some of its continental shelves) and the Greenland ice cap, which covers about 2 million sq km of land.

Continental ice sheets expand and contract in response to global changes in the climate of the Earth. For example, they expanded over much of North America during four separate **stages** in the Pleistocene Epoch (from oldest to youngest): the Nebraskan, Kansan, Illinoian, and Wisconsinian. During each stage, the Laurentide ice sheet formed around Hudson Bay and then expanded over eastern and central Canada and the northern United States, whereas the mountain glaciers in the western part of the continent expanded and merged into the Cordilleran ice sheet. We are now living in an **interglacial** stage between the Wisconsinian and the next glacial stage (and it is likely that there will be a next one), during which most mountain glaciers are confined to the heights of mountains and ice sheets to the high latitudes.

GLACIAL EROSION

Glaciers are powerful agents of erosion. They are often heavily laden with sediment, particularly along their margins and bases, where they come into contact with bedrock. Such ice acts like sandpaper to **abrade** and erode the bedrock and grind it into fine **rock flour**, and it leaves behind a polished surface with elongate, flow-parallel **striations** carved into it. This erosion also undercuts and oversteepens the slopes on both sides of a mountain glacier, which leads to their erosion by slope failures and debris flows. The exposed slopes are also susceptible to mechanical breakup by **plucking**, which occurs when water seeps into cracks in bedrock and refreezes there.

Erosion by Mountain Glaciers

The two basic erosive landforms of mountain glaciers are **cirques** and **U-shaped valleys** (Fig. 12–1). Cirques are formed by the plucking and mass wastage of the bedrock surface beneath the glacial ice that accumulates at the heads of the glaciers. They are typically spoon-shaped, steep-walled depressions on the high slopes of mountain ranges, and they cluster together to form sharp, jagged ridges called **arêtes** and isolated peaks called **horns** (the most famous example of which is the Matterhorn of Switzerland).

U-shaped valleys are evidence of the tremendous erosive capacity of glaciers. Mountain glaciers often flow through older stream valleys, which are typically V-shaped in profile. However, this profile is greatly altered by intensive lateral erosion along the margins of the glaciers and basal erosion beneath them. This widens the valleys and flattens their floors to form the distinctive U-shaped profile. The main ice stream of a glacier drainage system usually carves the deepest and widest valley, and its tributaries erode smaller **hanging valleys** which are perched high above its floor. **Fjords** are glacial valleys which have been flooded by the sea.

Glacial erosion also creates many depressions in the bedrock which become lakes when the glaciers retreat. The most common types are **tarns**, which are small subcircular lakes in abandoned cirques, and **paternoster lakes**, which form beadlike strings on the floors of glacial valleys.

Erosion by Continental Ice Sheets

Continental ice sheets are considerably thicker than mountain glaciers, and their great weight multiplies their erosive capacity many times over. Continental glaciers level mountains and hills as they grow and advance, leaving behind a vast plain which is striated and polished, dotted by lakes, and covered by a thin and patchy veneer of stony glacial debris. The low relief of this plain is only occasionally interrupted by elongate, streamlined hills of resistant bedrock called **roche moutonnées**, which have been smoothed and rounded by glacial abrasion. The advancing lobes of continental glaciers also widen and deepen older stream valleys to form huge elongate lake basins such as the Great Lakes of the central United States and the Finger Lakes of New York.

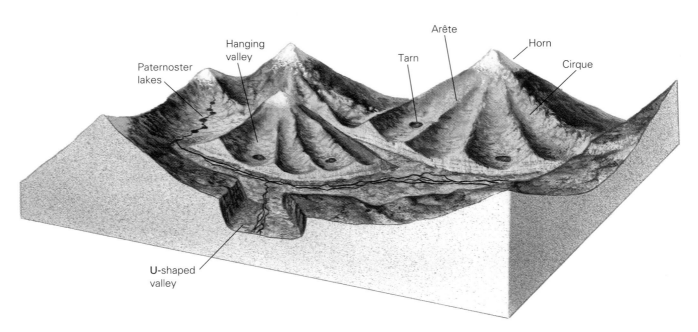

Figure 12–1 Erosive landforms of mountain glaciers.

Name _____ Section _____ Date _____

EXERCISE 12–1   Kenai Peninsula, Alaska

Examine the topographic map of the Kenai Peninsula near Seward, Alaska (Map 12–1).

   1. What is the elevation of the snowline in this region? _____

   2. There are many separate glacial drainage systems in this area. Outline the drainage pattern of each system, and indicate its flow direction. Describe the drainage pattern. _____

   3. Locate and label examples of the following features on the map: (a) ice-filled cirques, (b) abandoned cirques, (c) tidewater glaciers, (d) tarns, (e) arêtes, (f) horns, (g) U-shaped valleys, (h) fjords, and (i) paternoster lakes.

EXERCISE 12–2   Holy Cross, Colorado

The central Rocky Mountains near Holy Cross, Colorado, contain excellent examples of the erosive landforms of mountain glaciers, including cirques, arêtes, horns, tarns, U-shaped and hanging valleys, and paternoster lakes.

   1. Locate an example of each of these landforms on the stereophotographs (Fig. 12–2) and on the topographic map (Map 12–2) of this region.

   2. Construct topographic profiles across an arête, a horn, and a U-shaped valley.

   3. Many of the cirques in this area are on the northern and eastern slopes of the mountains. What might have caused this uneven ice accumulation?

Hemlock and Canadice Lakes (Map 12–3) are two small members of the Finger Lakes of New York State. The lakes were carved by lobes of the Laurentide ice sheet during the Pleistocene glaciation of North America.

    1.   Construct a topographic profile across the two lakes.

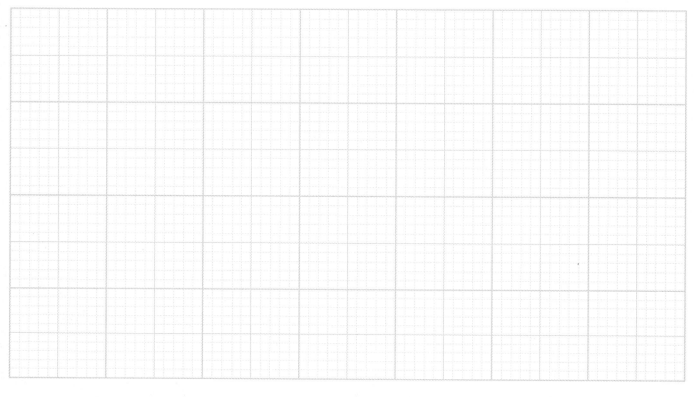

    2.   Calculate the slope and relief of the valley wall of one of the two lakes.

    3.   What evidence do you see in the geometry of the lakes to indicate a glacial (rather than fluvial) origin?

    4.   Describe the shape and morphology of Bald Hill. What kind of landform is it?

    5.   Locate and label the numerous **fan deltas** which are slowly filling the two lakes. Why are they called fan deltas?

Figure 12-2 Holy Cross, Colorado.

MAP 12-1 Seward, Alaska

10 STATUTE MILES

15 KILOMETERS

CONTOUR INTERVAL = 200 FEET

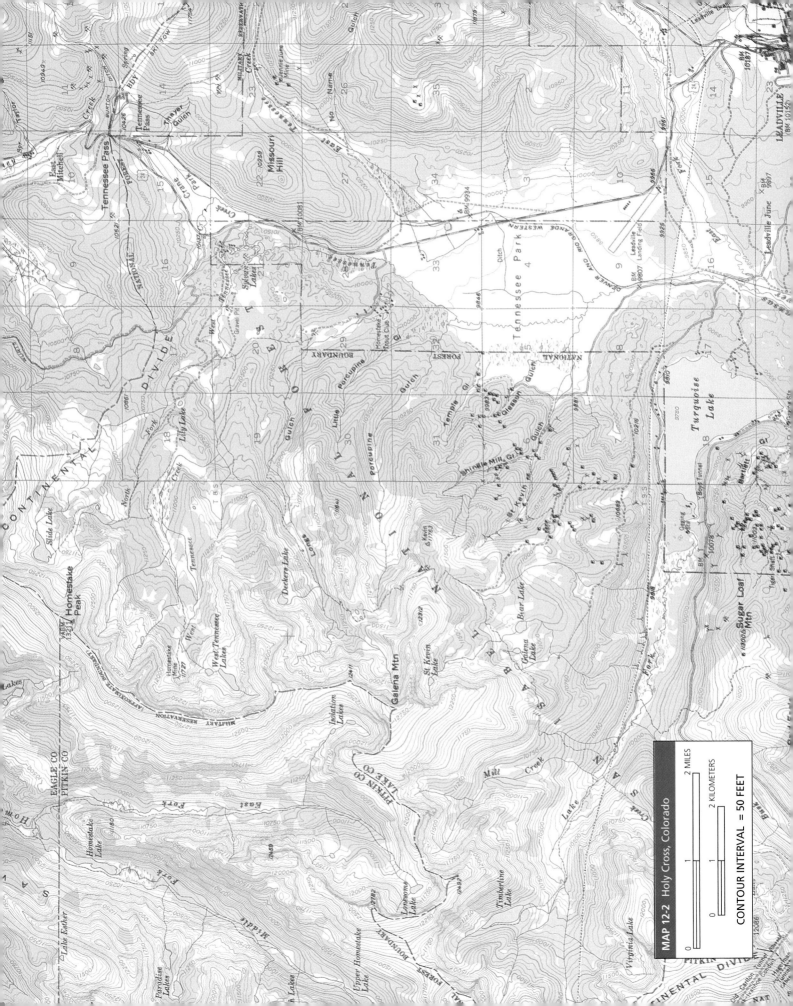

MAP 12-2 Holy Cross, Colorado

CONTOUR INTERVAL = 50 FEET

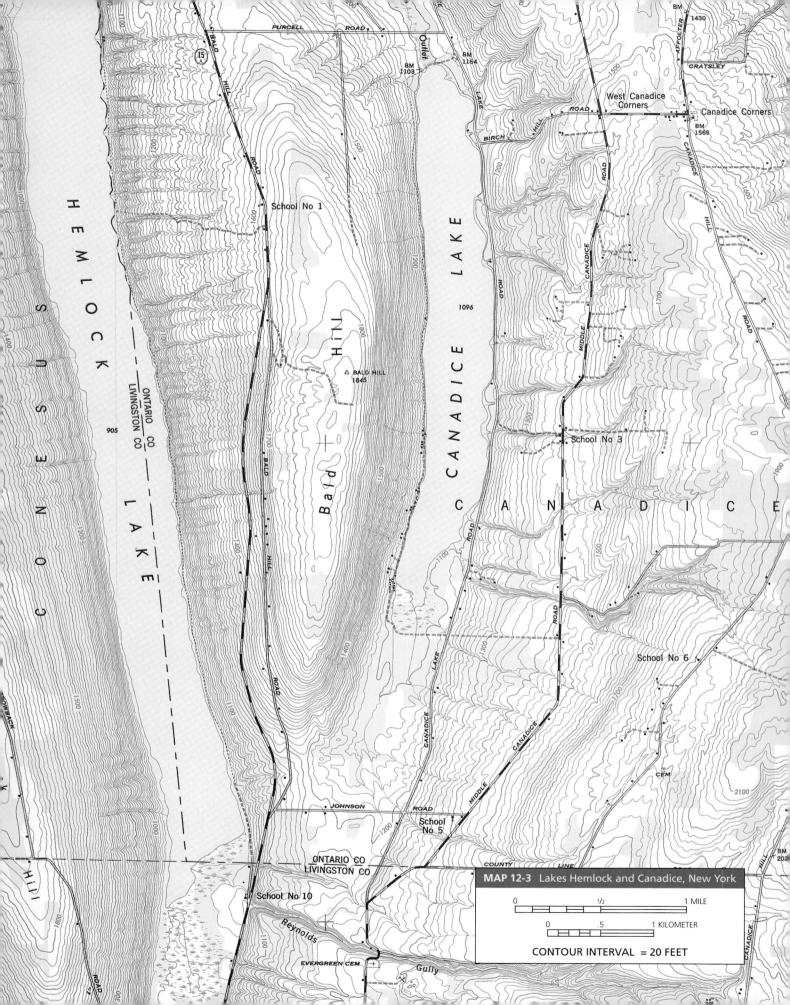

MAP 12-3 Lakes Hemlock and Canadice, New York

0   ½   1 MILE

0   .5   1 KILOMETER

CONTOUR INTERVAL = 20 FEET

TILLS AND MORAINES

Glaciers transport large volumes of sediment. The bulk of this sediment is concentrated beneath the glaciers and along their margins, and it flows downslope with the ice toward the glacial front. When glaciers retreat, this sediment (**till**) is deposited by the melting ice to form **moraines**, which are the principal depositional landforms of continental and mountain glaciers. There are several types of moraines (Fig. 12–3), including

- **Lateral moraines**: Ridges of till that accumulate between the margins of mountain glaciers and their valley walls.
- **Medial moraines**: Ridges of till that accumulate between two mountain glaciers as their lateral moraines merge.

- **Ground moraines**: Thin blankets of till that are dispersed over once-glaciated surfaces. Ground moraines of continental ice sheets are particularly widespread. They are also noted for their poor water drainage and abundance of small lakes and swamps.
- **End moraines**: Ridges of till that accumulate at glacial fronts. End moraines of mountain glaciers are relatively small and limited in areal extent, whereas those of continental ice sheets are thicker and more widespread. **Terminal moraines** are end moraines which form at the outermost limit of glacial advance; and **recessional moraines** are end moraines which form during **stillstands**, when retreating glaciers stabilize in one position for a period of time.

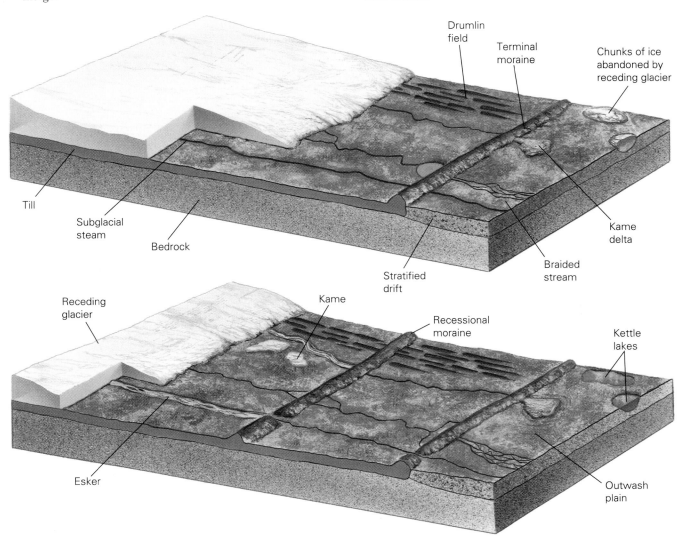

Figure 12–3 Till, moraine, and stratified drift deposits of a retreating glacier.

Glacial deposits are often eroded and reworked into other types of depositional landforms by the same glaciers which initially deposited them. The most common landforms of this type are **drumlins**, which are formed when glaciers readvance over their deposits and mold them into elongate, streamlined hills.

STRATIFIED DRIFT

Glacial deposits can also be reworked and redeposited by glacial meltwaters that flow through crevasses and beneath glaciers and drain the glacial front. Glacial-meltwater deposits, which are called **stratified drift** or **outwash**, form a variety of depositional landforms (Fig. 12–3), including

- **Kames**: Mounds of stratified drift deposited by streams that flow through and beneath glaciers.
- **Eskers**: Long, often sinuous ridges of stratified drift that are deposited by streams flowing beneath glacial ice. They are most common to continental ice sheets.
- **Valley trains**: Long, narrow bodies of outwash stream deposits which extend downdip from the terminal moraines of mountain glaciers.
- **Outwash plains**: Broad, flat sheets of outwash stream deposits which extend beyond the terminal moraines of continental ice sheets.
- **Kettles** and **kettle lakes**: Shallow subcircular depressions on the surfaces of till and stratified drift that are formed by the melting of blocks of buried glacial ice.

An end moraine of a mountain glacier.

EXERCISE 12-4 Mountain Glaciers, Alaska and Colorado

1. Return to the topographic map of the Kenai Peninsula of southern Alaska (Map 12–1), and locate the valley trains of the mountain glaciers there.
    a. What kind of streams are found there: braided or meandering? _____
    b. What does this indicate about the texture of the sediment in the valley trains? Explain your answer.

2. Return to the aerial photographs and map of the Holy Cross, Colorado, area (Fig. 12–2 and Map 12–2). Notice that Lake Fork Creek has been dammed to form Turquoise Lake.
    a. Draw a topographic profile from the gaging station on the west bank of the lake to the Leadville Landing Field, and a second profile from the center of the lake into Tennessee Park.
    b. What natural landforms created this lake? _____
       Label them on your profiles.

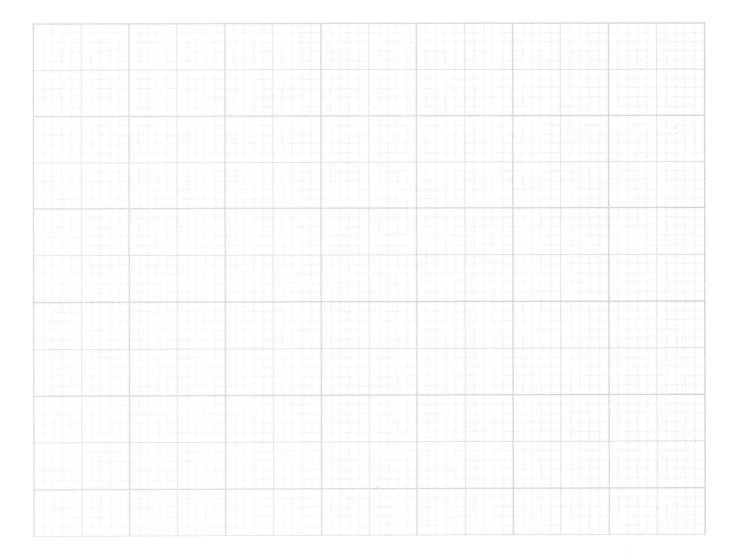

Map 12–4 is a topographic map of Kingston, Rhode Island. The bedrock here is buried beneath ground and terminal moraine deposits in the northern part of the map, and by outwash plain and beach deposits in the south.

1.   Locate and label the ground moraine, terminal moraine, and outwash plain (use G for the ground moraine, T for the terminal moraine, and O for the outwash plain). Trace the northern and southern edges of the terminal moraine. What is its maximum width in the map area? _____

2.   Construct a topographic profile along line A-A′. Describe and compare the reliefs of the terminal moraine and outwash plain.

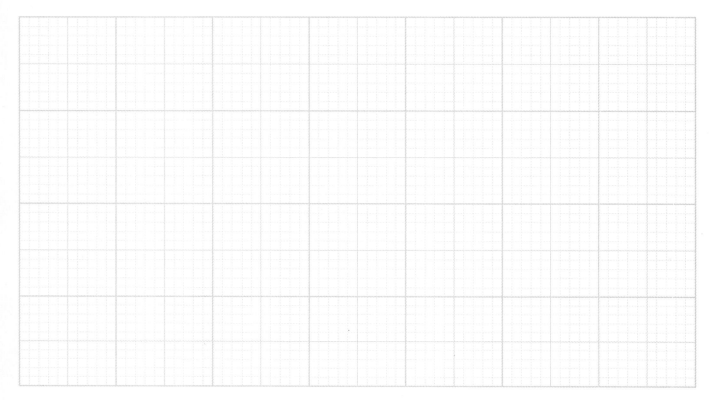

3.   Compare the areal extent and relief of this terminal moraine to the terminal moraine in the Holy Cross region (Map 12–2). What does the difference indicate about the relative sediment transport capabilities of mountain and continental glaciers?

4.   Trace and describe the drainage patterns of the ground moraine and outwash plains. How and why are they different?

5.   Locate examples of kettles, and label them with the letter K.

| **E X E R C I S E   1 2 – 6**   Williamson, New York |

The landscape in the central-western part of New York State contains an estimated 10,000 drumlins, some of which are nearly 30 m high and 2 km long. Figure 12–4 and Map 12–5 show a swarm of these drumlins near Palmyra, New York.

1. Examine the stereophotographs, and describe the geometry of the drumlins.

2. Construct a topographic profile through the length of a drumlin.

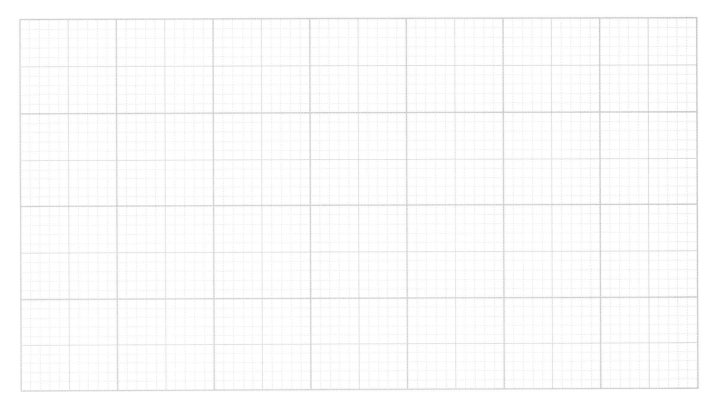

3. What was the flow direction of the glaciers that molded these drumlins? _____ What evidence do you have to support this conclusion?

4. How have the drumlins affected the placement of homes, roads, and railroad tracks in the map area?

Map 12–6 is a topographic map of the Whitewater, Wisconsin, area, which contains many of the common depositional landforms of continental ice sheets.

1. Construct a topographic profile along line A-A'. Label the end moraine on the profile.

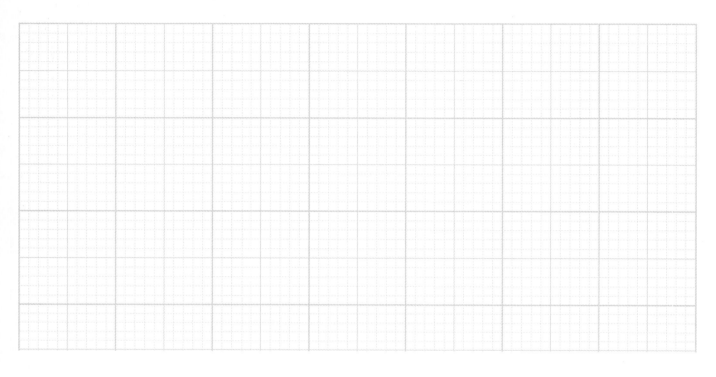

2. What is the probable origin of the hills in the northwestern corner of the map? _____ What do they indicate about possible directions of glacial flow in the area? _____

3. What is the origin of the low mound of gravel which appears to the north of the end moraine on your topographic profile? _____ Label this feature on your profile. Locate and label other similar features in the map area.

4. Compare and contrast the drainage patterns of the areas to the northwest and southeast of the end moraine.

5. The end moraine is bordered to one side by ground moraine, and to the other side by a pitted outwash plain. Identify and label these two glacial deposits on the topographic map and the profile. What evidence did you use to distinguish them?

6. What is the origin of the pits on the outwash plain? _____

EXERCISE 12-8 Pleistocene Glaciation, Northern United States

Map 12–7 shows the distribution of the Pleistocene deposits of the Laurentide ice sheet in the eastern United States. The geologic map shows the ages of the deposits, the major glacial depositional landforms, and the orientations of bedrock striations and streamlined landforms (such as drumlins) which indicate the flow directions of the ice sheet's lobes.

1. Examine the New York–New England segment of the map. During the Wisconsinian glaciation, this region was covered by two large lobes of the Laurentide ice sheet: the Lake Champlain–Hudson River lobe of New York, and the New England lobe. Draw a line on the map which shows the maximum southern extent of these two glacial lobes.

2. Label the terminal moraine (TM) and recessional moraine (RM) of these two lobes.

3. What kind of glacial deposits form:
   a. the northern and southern forks of Long Island? _____
   b. the southern part of Long Island? _____
   c. Cape Cod? _____
   d. Martha's Vineyard and Nantucket Island? _____

4. It is fairly common for large glacial lobes to diverge into several smaller flows, or **sublobes**, each flowing in a slightly different direction.
   a. Examine the orientations of the glacial striations and landforms throughout New England and New York State. Do you see any evidence for such divergence? Indicate the approximate flow directions and extents of any sublobes you see.

   b. Now examine the map area west of the Ohio River. There were three major glacial lobes in this area during the Wisconsinian stage: the Green Bay (GB) lobe of Wisconsin, the Lake Michigan (LM) lobe of Illinois, and the Lake Erie (LE) lobe. The Lake Erie lobe has four distinct sublobes (from east to west): the Grand River, Scloto, Miami, and White River. Locate and label these lobes and sublobes, and identify their respective terminal moraines.
   c. How many recessional moraines of the Scloto sublobe can you identify within the state of Ohio? _____
   d. Radiocarbon dates indicate that the terminal moraine of the Scloto was deposited approximately 21,000 years ago, and that its youngest recessional moraine in Ohio was deposited about 13,000 years ago. What was the rate of retreat of this glacial sublobe (in km per century)?

5. What are the origins of Lakes Pontiac, Wauponsee, and Watseka in northern Illinois? These lakes were drained after a catastrophic break in their natural dams. Locate and label this breach (as B).

6. Identify and label the outwash streams of the Scloto (So), Miami (Mo), and White River (WRo) sublobes of the Erie lobe.

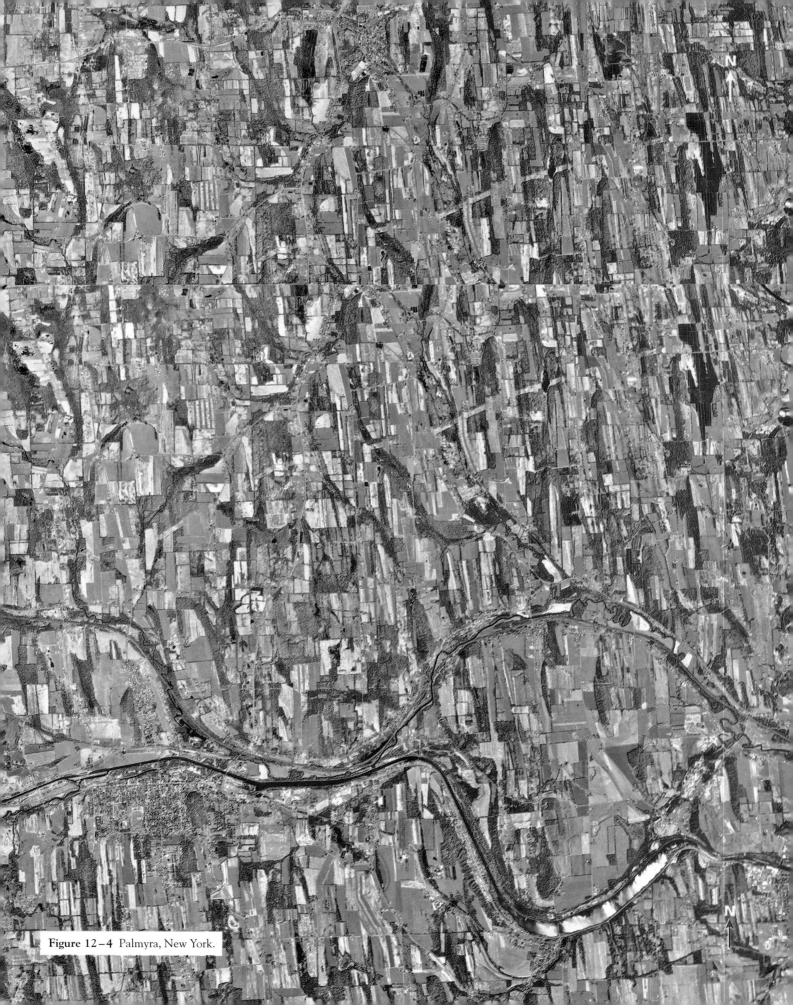

Figure 12−4 Palmyra, New York.

MAP 12-4 Kingston, Rhode Island

CONTOUR INTERVAL = 10 FEET

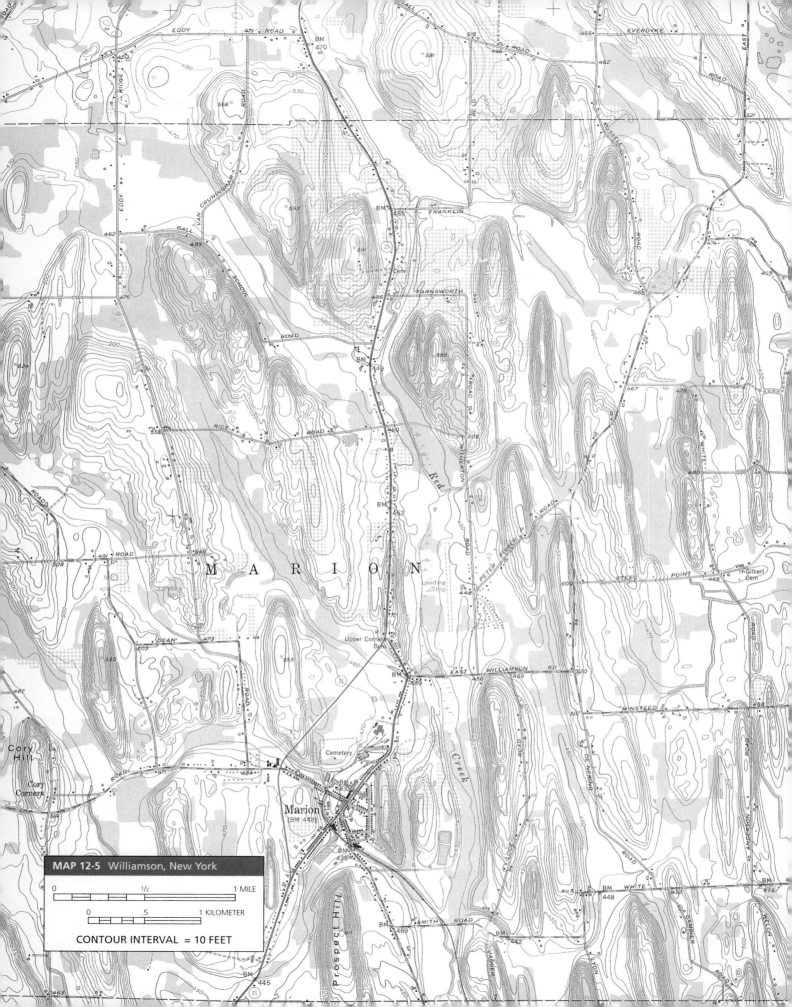

MAP 12-5 Williamson, New York

0 ⸻⸻⸻ 1/2 ⸻⸻⸻ 1 MILE

0 ⸻⸻⸻ .5 ⸻⸻⸻ 1 KILOMETER

CONTOUR INTERVAL = 10 FEET

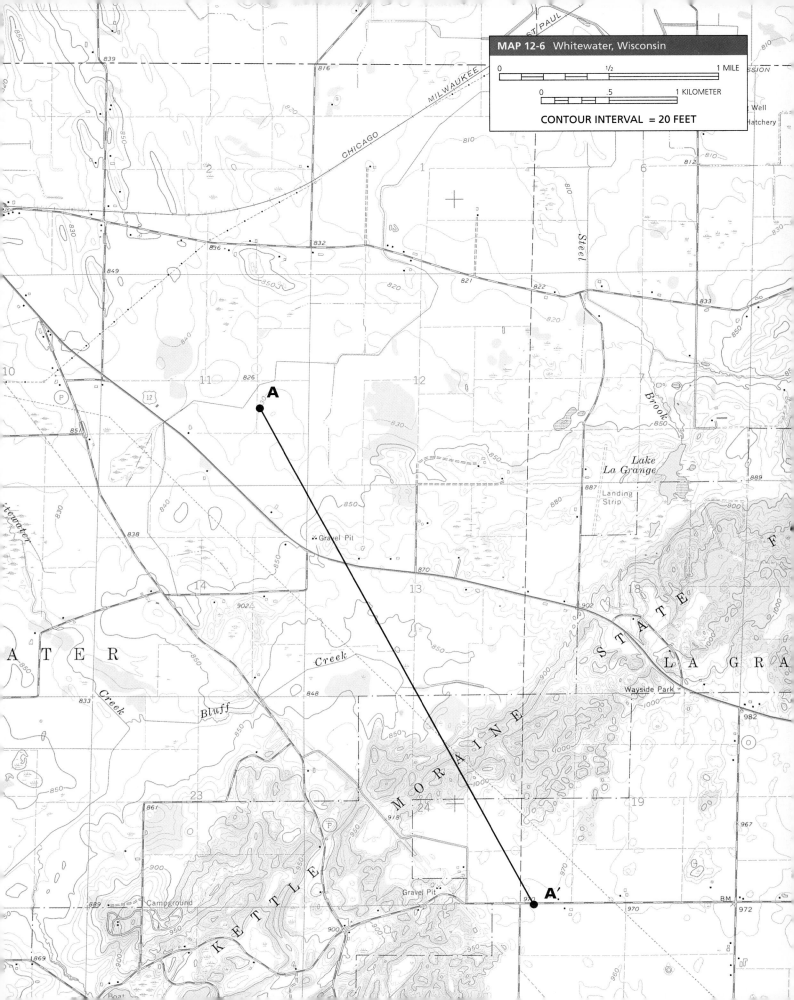

MAP 12-6 Whitewater, Wisconsin

CONTOUR INTERVAL = 20 FEET

A

A'

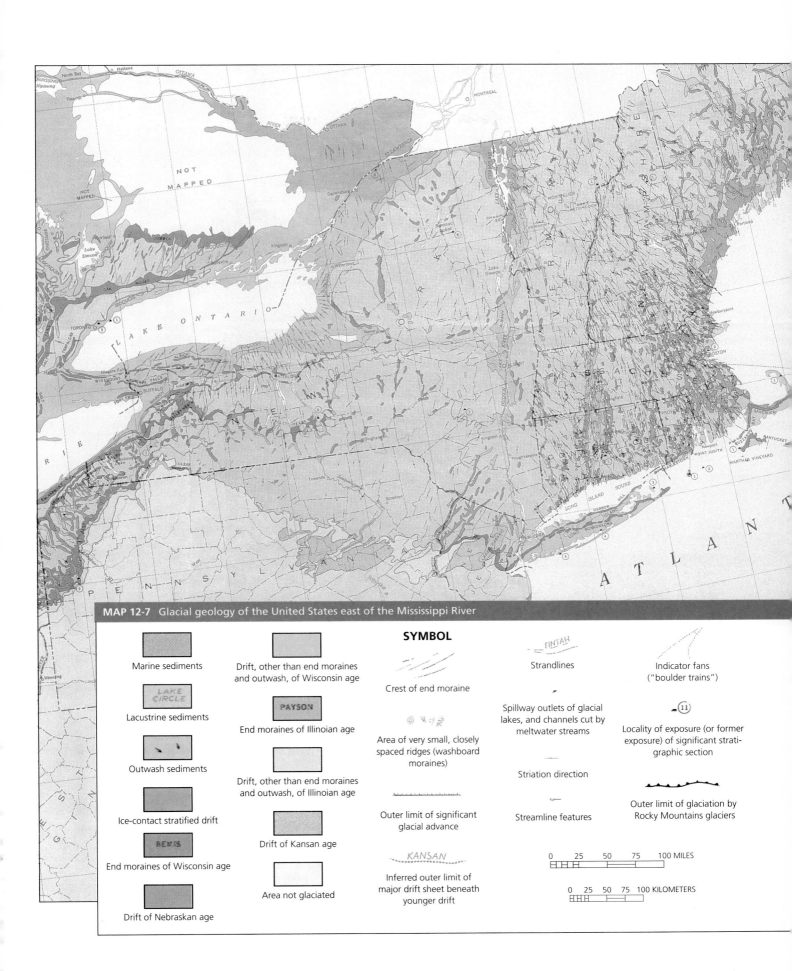

MAP 12-7 Glacial geology of the United States east of the Mississippi River

SYMBOL

Marine sediments

Lacustrine sediments

Outwash sediments

Ice-contact stratified drift

End moraines of Wisconsin age

Drift of Nebraskan age

Drift, other than end moraines and outwash, of Wisconsin age

End moraines of Illinoian age

Drift, other than end moraines and outwash, of Illinoian age

Drift of Kansan age

Area not glaciated

Crest of end moraine

Area of very small, closely spaced ridges (washboard moraines)

Outer limit of significant glacial advance

Inferred outer limit of major drift sheet beneath younger drift

Strandlines

Spillway outlets of glacial lakes, and channels cut by meltwater streams

Striation direction

Streamline features

Indicator fans ("boulder trains")

Locality of exposure (or former exposure) of significant stratigraphic section

Outer limit of glaciation by Rocky Mountains glaciers

0   25   50   75   100 MILES

0   25   50   75   100 KILOMETERS

13

DESERTS

A RIPPLE PATTERN ON SAND DUNE IN DESERT, NAMIBIA.

(©Tony Stone Worldwide/Willi Dolder)

Materials Needed *Colored pens or pencils, lead pencils with erasers, ruler, calculator, stereoscope*

Deserts are defined as land surfaces with sparse plant cover. It may seem strange to define deserts on the basis of a biological variable, but geologists recognize that plants have a significant influence on erosion, deposition, and the morphology of land surfaces. Plants increase the resistance of sediment to wind and water erosion by adding large amounts of cohesive organic matter to it and binding it with their roots. Deserts lack a dense soil-binding plant cover, and so their sediment is more susceptible to erosion by wind and water.

The major agent of erosion in many deserts is the wind, which winnows desert sediment and molds it into distinctive depositional landforms such as dunes. However, deserts are also prone to occasional rainfalls and flash floods, which leave their own distinctive marks on their landscapes. Consequently, the surfaces of deserts are complex and constantly changing mosaics of both eolian and fluvial landforms.

This chapter examines the many different landforms of deserts and the processes which create them. It begins with a consideration of the origins of deserts, the effects of wind erosion, and the deposition of windblown sediment. The chapter continues with a discussion of desert streams, and concludes with a study of desert fluvial landforms and depositional features in the Mojave Desert of California.

ORIGINS OF DESERTS

The principal cause of the lack of plant cover in deserts is aridity. Arid climates and large **low-latitude deserts** are most common in the high-pressure belts 30° north and south of the equator, where precipitation levels are extremely low (less than 25 cm/yr) and the potential for evaporation is considerably higher. Arid climates and small **periglacial** ("around glaciers") **deserts** are also common in the polar regions, where the air is cold and dry and most available moisture is locked up in glaciers or permafrost. Finally, arid climates and deserts are also found on the downwind sides of high mountain ranges (**rainshadow deserts**) and in the interiors of large continents, two areas which are often deprived of atmospheric moisture by physiographic barriers.

However, aridity is not the sole factor that contributes to the formation of deserts. Some deserts are products of the high salinities of the local stream or ground water. Small coastal deserts are particularly common, for example, because there are few plants which can tolerate the salty and brackish waters of beaches and other coastal environments. Similar **saline deserts** can also be found in inland settings where the local stream waters are rich in dissolved salt.

Finally, some deserts are the unfortunate result of human activity, which often destroys the native plant cover and leads to **desertification**. The careless and excessive cultivation of prairie grasslands and soils by farmers, the felling of large forests and the failure to plant wind-breaks, and overgrazing of plants by sheep, goats, and cattle, for instance, contributed to the collapse of many ancient civilizations in the Old and New Worlds as well as the more recent expansion of the Sahara Desert, and the American Dust Bowl of the 1930s, and the desertification of Bangladesh and the Amazon Basin.

EROSION AND DEPOSITION BY THE WIND

Deserts are distinguished from vegetated land surfaces by their susceptibility to erosion by the wind as well as by water. However, wind erosion is more size-selective than fluvial erosion; that is, streams are generally capable of eroding nearly all grain sizes (except perhaps the largest boulders), but the wind can erode only sand, silt, and clay. This process of selective wind erosion, or **deflation**, leaves thin mantles of gravel called **deflation lags** or **desert pavements** over topographic highs, alluvial fans, and other sources of windblown desert sediment. Wind deflation can also create shallow de-

pressions in the desert surface called **deflation basins** or **blow-outs**. These basins often fill with water to become deflation **ponds** when the local water table rises or the amount of rainfall increases.

Once the wind has eroded sand, silt, and clay grains from a sediment source, it quickly separates or sorts them on the basis of their sizes. The silt and clay fractions are lifted high into the atmosphere and transported in suspension far from their source, whereas the sand bounces, or **saltates,** more slowly across the desert floor. The sand eventually spreads over the desert to form large **sand seas** or **ergs**, where it is molded into the most distinctive eolian landform of deserts, **dunes.**

Dunes

Dunes are migrating ridges of windblown sand. They are asymmetrical in profile with a gently sloping upwind **stoss side** and a steeply sloping downwind **slip face** that meet at **crest** (Fig. 13–1). Dunes migrate across desert surfaces as a result of the deposition of bed after bed of sand on the slip face, which thereby advances in the downwind direction.

Dunes can actively migrate if there is little or no plant cover to bind the sand to the desert surface. They are only

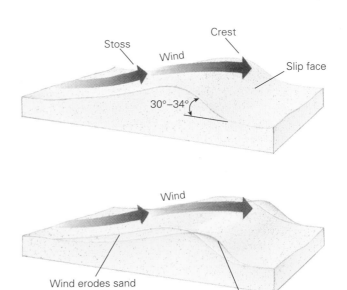

Figure 13–1 Morphology and migration of a sand dune.

hindered from migrating in any particular direction by topographic highs and vegetation. Dunes become **stabilized** when they migrate from poorly vegetated into more vegetated areas or whenever the climate changes sufficiently to support a plant cover.

Sand dunes vary in geometry and orientation due to variations in wind velocity, the supply of sand, the prevailing wind direction, and the nature of the desert surface. The most common types of sand dunes are (Fig. 13–2):

- **Barchan** dunes: Crescent-shaped dunes with horns pointed downwind. They form on rocky desert surfaces with low sand supplies.

- **Parabolic** dunes: Crescent-shaped dunes with horns pointed upwind. They are most common in coastal settings, where the slight plant cover anchors the horns of the dunes and causes them to move more slowly than the central part of the dune. The areas between the horns are erosive **blow-outs**, which function as the immediate source of sand for the central dune ridge and its horns.

- **Transverse** dunes: Straight-crested or linear dunes with crests oriented normal to the wind. Transverse dunes are the result of high sediment supply and typically cover large areas to form sand seas.

- **Longitudinal** or **seif** dunes: Straight linear dunes with crests oriented parallel to the wind. They are most common in warm deserts where the prevailing winds blow strongly from one direction.

- **Star** dunes: Complexly shaped dunes with multiple crests and slip faces. They form wherever the wind is constantly shifting in direction.

Loess

Loess is an accumulation of well-sorted windblown silt. This silt is derived from the deflation of desert and glacial deposits by the wind and transported downwind, until it is eventually deposited in thick, massive sheets which blanket large areas of land. Generally, the coarsest silt grains are deposited fairly close to their source, whereas the finest grains are deposited farther away, sometimes far into the oceans. Loess deposits are known for the richness and fertility of their soils, and they often serve as agricultural breadbaskets.

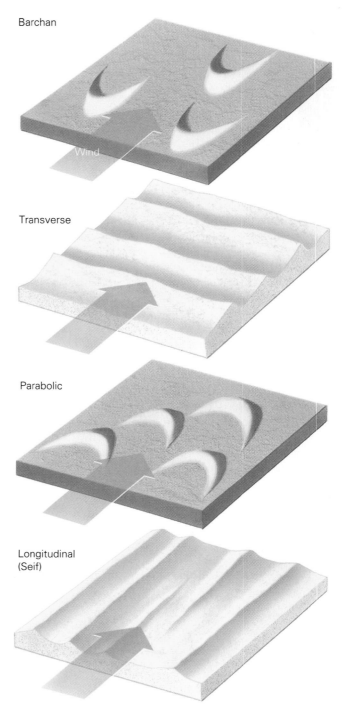

Figure 13–2 Common types of dunes.

EXERCISE 13–1   Deserts of the World

This exercise examines the distribution of deserts around the world. You will need to refer to a World Atlas and a Heezen-Tharp World Ocean Map to answer some of the following questions.

1.   Locate and label the following deserts on Map 13–1.

DESERT	TYPE
Sahara	
Namib	
Kalahari	
Rubᶜ al-Khali (Arabian)	
Turkestan	
Sonoran	
Mojave	
Atacama	
Patagonian	
Simpson	
Gobi	
Takla Makan	

2.   Locate and trace the 30°N and 30°S lines of latitude on the map. Which deserts are intersected by these lines? Indicate your answer with the letter "L" (for low latitude) on the table.

3.   The Gobi and Takla Makan deserts are dominated by northerly winds from central Asia, but they are also influenced by southerly winds from the Indian Ocean during the summer months. Warm ocean winds are typically moisture rich, but these southerly winds are dry by the time that they reach the Gobi and Takla Makan.
    a.   Locate the Gobi (G) and Takan Makan (T) on the Heezen-Tharp map. What prevents the moisture of the southerly winds from reaching the Gobi and Takan Makan? _____
    b.   What kinds of deserts are the Gobi and Takan Makan? (Indicate your answer on the table above.)

4.   The Patagonia and Mojave are two large rainshadow deserts in the Western Hemisphere. Locate them on the Heezen-Tharp map. What geologic features form the precipitation barriers for them? _____

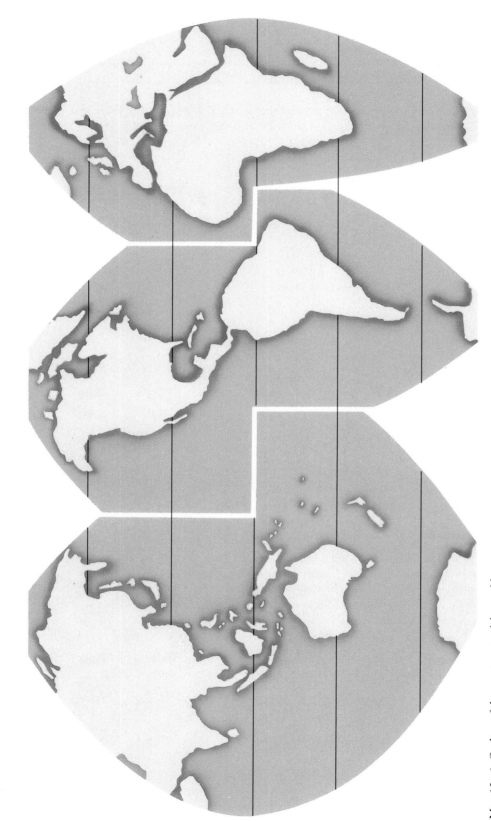

Map 13-1 Outline of the continents of the world.

EXERCISE 13-2   Desert Dunes

Figure 13–3 consists of high-altitude photographs of the deserts of Africa and Australia which were taken by the space shuttle astronauts. The prevailing wind direction in each desert is indicated with an arrow.

   1. Examine the photographs, and identify the type or types of dunes which are present in each desert. Illustrate each dune type.

   2. What are the gray-colored patches of land in the Kalahari and Simpson desert photographs?

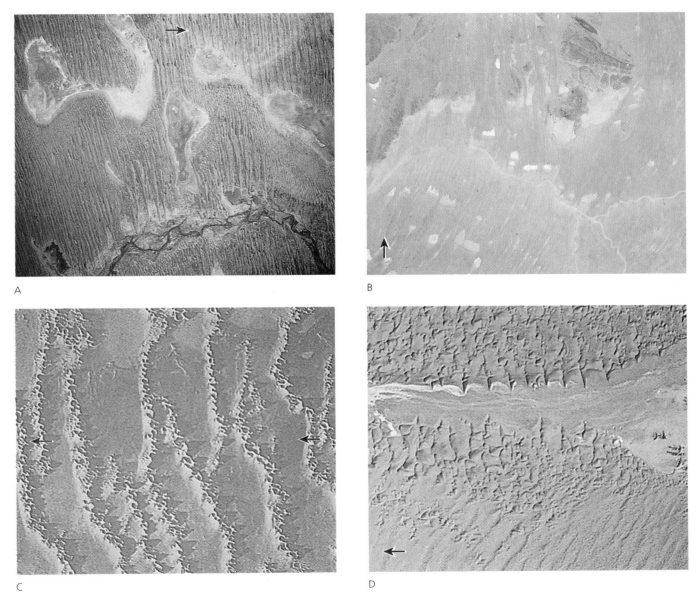

Figure 13–3 Satellite photographs of sand dunes. (A) Simpson Desert, Australia. (B) Kalahari Desert, Africa. (C) Sahara Desert, Africa. (D) Namib Desert, Africa.

One of the largest and most beautiful sets of coastal dunes is found near Coos Bay in southern Oregon, in the Oregon Dunes National Recreation Area. This area contains active migratory dunes as well as older dunes which have been stabilized by vegetation.

1. Examine the recent stereophotographs and the 1985 topographic map for the coastal area between the northern edge of the map and the Tenmile Creek (Fig. 13–4 and Map 13–2).
    a. What kind of dunes are found there? _____
    b. Construct a topographic profile from A to A′, and label the stoss side, crest, and slip face of each dune.
    c. What is the maximum relief of the dunes? _____ What is the direction of their migration? _____

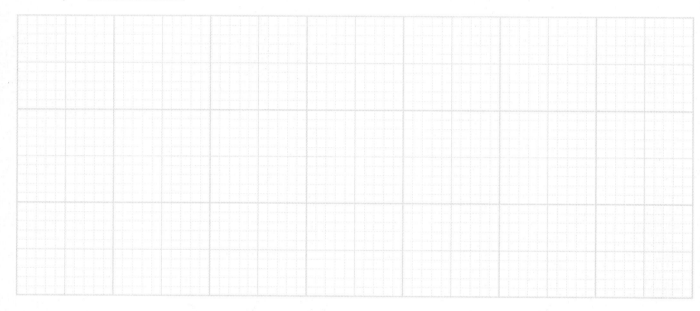

2. Construct a topographic profile from B to B′.
    a. What is the origin of the low divide and ponds between the two parts of the dune field?
    b. What does this tell you about the history of the dunes in this area?

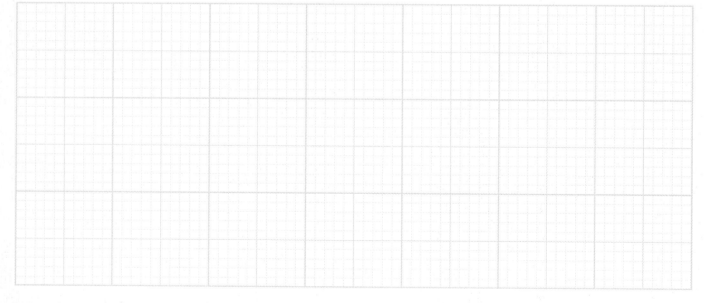

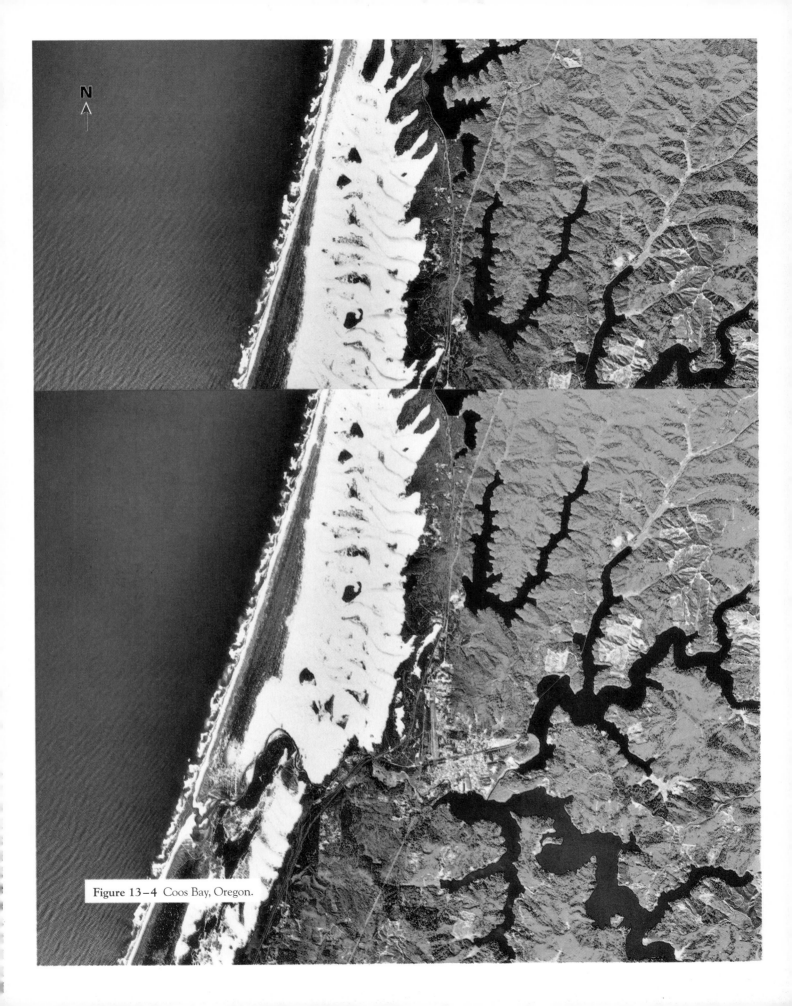

Figure 13-4 Coos Bay, Oregon.

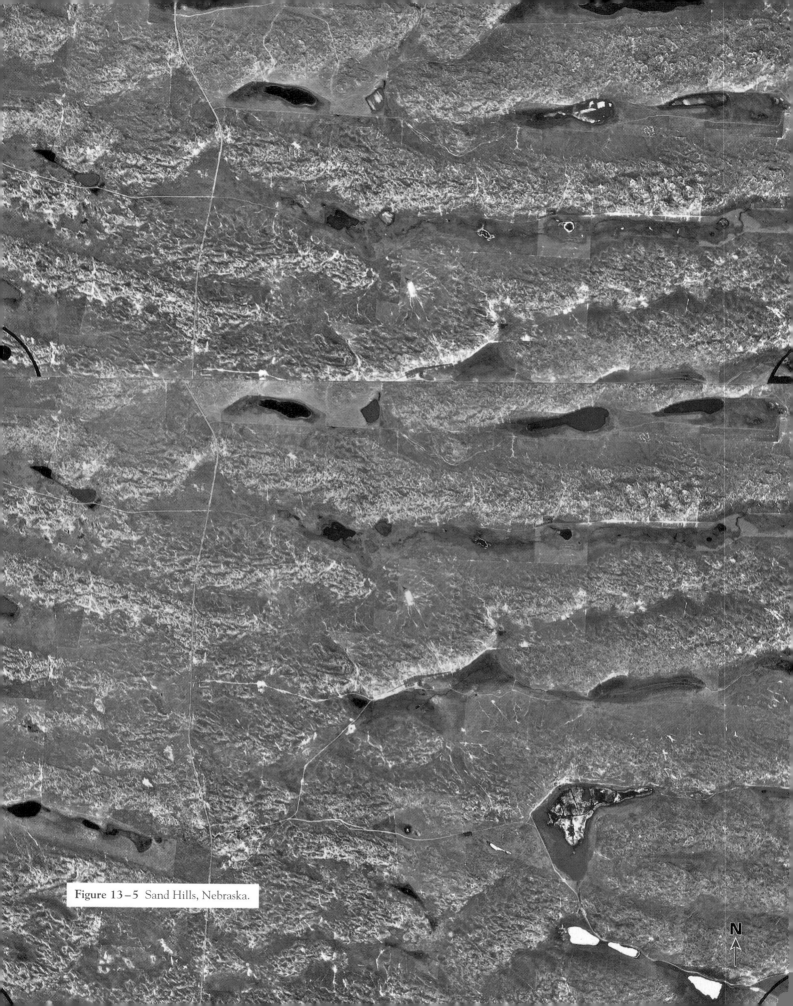

Figure 13–5 Sand Hills, Nebraska.

N

EXERCISE 13 – 4   Sand Hills, Nebraska

The Sand Hills make up one of a number of large dune fields in the Mid-Continent of the United States and Canada. These dune fields were active throughout the late Pleistocene and early Holocene, when the climate of this area was relatively arid, but they are presently stabilized by prairie grasses which flourish in the modern semiarid climate.

1. Examine the topographic maps and stereophotographs of the Sand Hills (Map 13–3 and Fig. 13–5). What kind of dunes are present in this area? _____ What is the origin of the lakes between these dunes? _____

2. Draw a topographic profile along line A-A′, and label the stoss side, crest, and slip face of the dunes.

3. What was the prevailing wind direction in western Nebraska during the late Pleistocene and early Holocene? Explain your answer.

The largest accumulation of loess in the world is found in the Loess Plateau of northern China, a 300-meter-thick deposit which covers over 800,000 square kilometers of land. Figure 13–6 shows the mean grain sizes of samples which were collected from the central part of the plateau. (The data are presented in units of microns, with 1000 microns equal to 1 millimeter.)

1. Contour these data in 10-micron intervals.

2. Indicate with an arrow the prevailing wind direction for the Loess Plateau.

3. Examine Map 13–1. Locate the Loess Plateau. What is the most probable source of the loess? Explain your answer.

Figure 13–6 Grain size variation in the Loess Plateau of northern China.

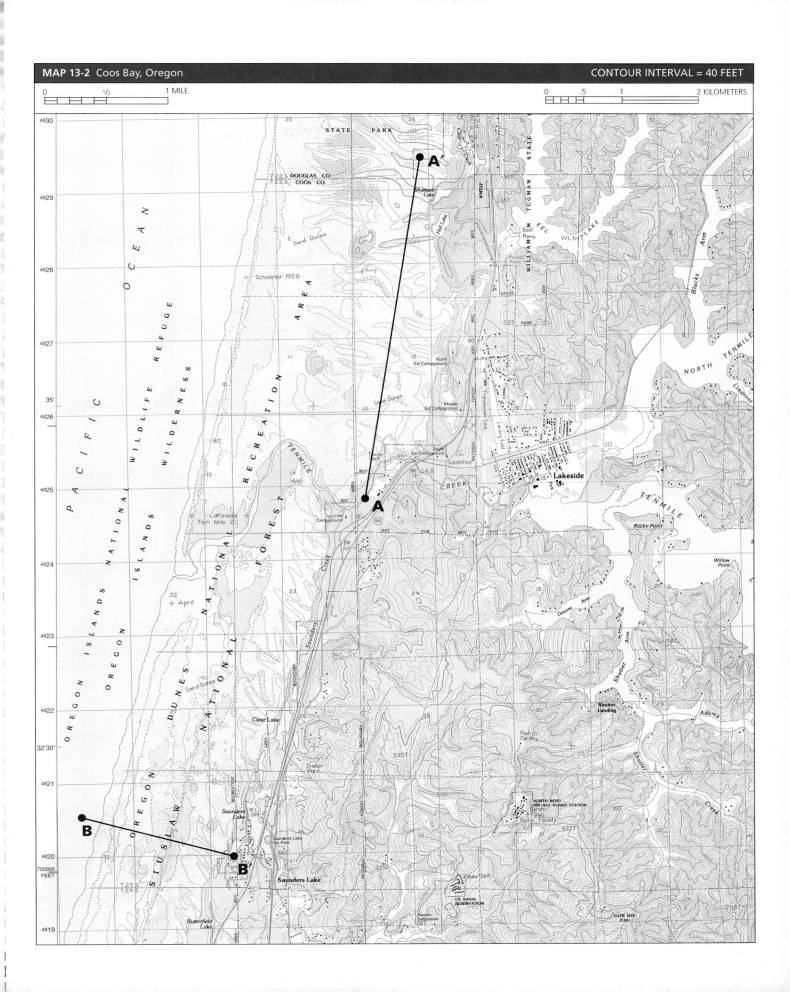

0   1/2   1 MILE

0   .5   1   2 KILOMETERS

MAP 13-3 Steverson Lake, Nebraska

0   1/2   1 MILE

0   .5   1   2 KILOMETERS

CONTOUR INTERVAL = 20 FEET

Deserts are generally perceived as places where the wind plays the major role in erosion and deposition. There are some deserts where this is true, but there are others in which streams play the principal role in erosion, deposition, and the shaping of the landscape. Two types of streams are important in this regard: **perennial** and **ephemeral** streams.

PERENNIAL STREAMS

Perennial streams flow and erode sediment throughout the year, barring any interference by humans. Perennial flow of a stream through a desert, where water is lost to evaporation, requires that the stream have either a large drainage basin or headwaters in more humid regions outside the desert.

The Colorado River is an example of such a perennial stream. The Colorado is presently eroding the many spectacular canyons and gorges which dominate the desert landscapes of the Colorado Plateau. Its perennial flow is due to the fact that it drains a very large part of the southwestern United States, including the semiarid western slopes of the Rocky Mountains (see Map 10–1).

EPHEMERAL STREAMS

Ephemeral, or intermittent, streams flow only during and immediately after desert rains. Rainfalls are rare in low-latitude and continental-interior deserts but more common in coastal and rainshadow deserts. When they do occur, however, they are usually torrential, falling within a few minutes or hours and causing sudden, catastrophic **flash floods** and intense erosion.

Ephemeral desert streams consist of channels called **wadis**, **dry washes**, or **arroyos**, and broad, low-relief flood plains called **sand flats** and **mud flats** (Fig. 13–7). The channels often flow from steep-walled mountain valleys onto the desert floor, where they form braided and parallel drainage patterns across the sand flats and mud flats. Ephemeral streams are sometimes the tributaries of larger perennial streams, but they can also flow into ephemeral lakes, or **playas,** which form in grabens, deflation basins, and other topographic lows on a desert floor.

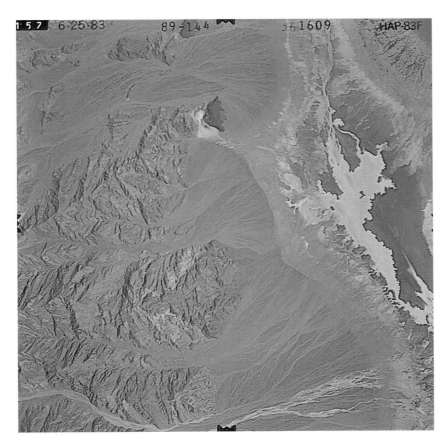

Tucki Mountain and central Death Valley, California.

The erosion of mountains and other topographic highs by ephemeral streams is very intense and rapid. Mountains are gradually reduced and dissected into small, isolated erosional remnants called **inselbergs**, and mountain fronts are eroded into gently sloping surfaces called **pediments** (Fig. 13–7). The pediments are then buried by the coarse clastic sediments of **alluvial fans**, which expand over time and coalesce into broad mantles or aprons of sediment called **bajadas**. Finer clastic sediments are usually transported past the alluvial fans to the desert floor, where they are deposited in channels, sand flats, mud flats, and playa-margin deltas.

Evaporites are often found with desert stream deposits. For example, the evaporation of playas results in the deposition of thin, widespread layers of halite, gypsum, and other salts on their bottoms. Such playa evaporites underlie the Bonneville salt flats, a testing ground for high-performance cars, and Edwards Air Force Base, the landing field for the NASA space shuttles. Evaporite nodules are also common within the clastic deposits of sand flats and mud flats, which are typically saturated with saline water.

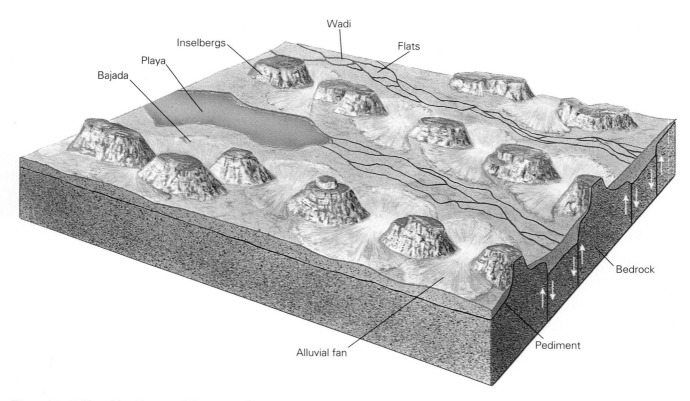

Figure 13–7 Fluvial landforms and deposits in deserts.

E X E R C I S E   1 3 – 6   Death Valley National Park, California

---

This exercise examines the topography of Death Valley National Park. Death Valley lies in the southern part of the Basin and Range province in the Mojave Desert of California. The valley is an elongate pull-apart basin which was formed by movement along several strike-slip faults. It receives runoff and sediment from two inclined fault blocks: the Panamint Range to the west and the Black Mountains to the east. Death Valley is filled largely with late Tertiary and Quaternary fluvial and lake sediments, and the topography of its floor, which is the lowest point in North America, is dominated by landforms of fluvial and lake origins.

1. Examine the 1:250,000 topographic map of Death Valley (Map 13–4).
   a. Locate and label examples of inselbergs and playas.
   b. The eastern margin of the Panamint Range is lined by alluvial fans which have coalesced into a broad bajada. Use a colored pencil to shade this bajada and others in the map area.

2. The principal drainage system in the map area consists of the Amargosa River and its tributaries in the Panamint Range and Black Mountains. The Amargosa River enters through the southern end of Death Valley and follows the trace of the South Death Valley fault zone until it reaches Mormon Point, where it splits into several small channels that flow into Badwater Basin.
   a. Locate and trace the channels of the Amargosa River and its mountain tributaries.
   b. Describe the regional drainage pattern of the Amargosa and its tributaries. _____
   c. Describe the channel pattern of the Amargosa River on the valley floor. _____
   d. Locate and label examples of the flats between the channels of the Amargosa.
   e. What kind of stream is the Amargosa in the southern part of the map: strike or dip? _____ perennial or ephemeral? _____

3. Construct a topographic profile along a line which runs from Wildrose Peak in the Panamint Range across the Badwater Basin to Coffin Peak.
   a. Label the mountain slopes, bajadas, and valley floor on the cross section.
   b. Calculate the slope and maximum relief of (1) the eastern front of the Panamint Range, (2) the bajada of the eastern front of the range, and (3) the valley floor.

   c. What is the approximate elevation of Badwater Basin? _____
   d. What kind of sediments might you expect to find in the center of the basin? _____

4. Construct a second topographic profile along the crest of the Greenwater Range from Ryan Mines to Browns Peak (in the southwest corner of the map).
   a. Locate the two passes through the Greenwater Range. What are the width and maximum relief of each pass?

   b. Explain the origins of the two passes.

   c. Examine the topographic profile of the Greenwater Range between Deadman Pass and Ryan's Mine. Identify locations where new passes are currently being eroded.

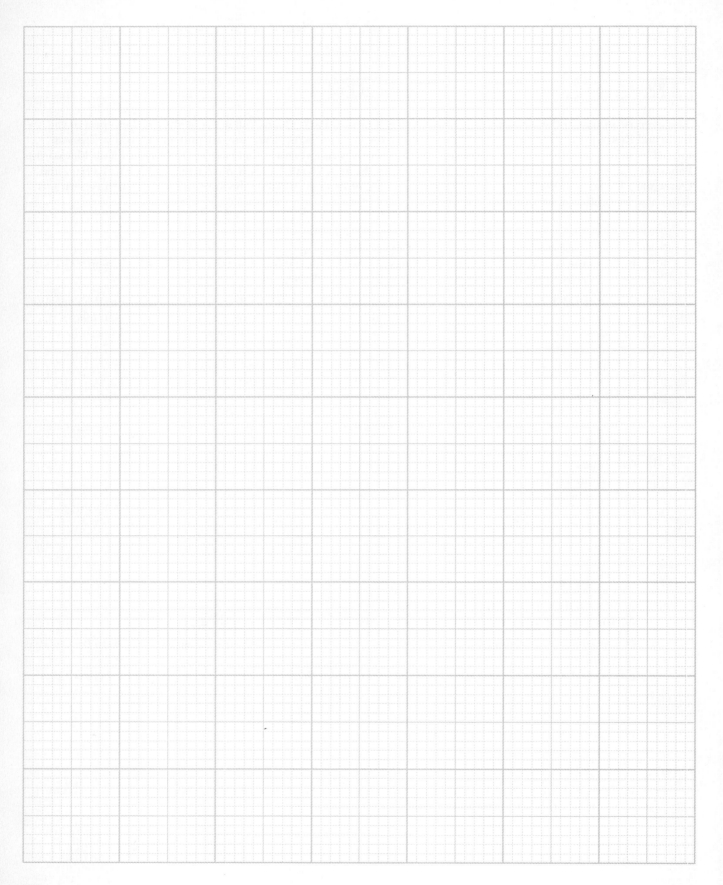

5. Examine the stereophotographs of Death Valley (Fig. 13–8).
   a. The photographs show the distal edges of some of the fans which form the western bajada of Death Valley, as well as the ephemeral streams of the valley floor. Locate and trace the boundary between them.
   b. Examine the distributary channels on the alluvial fans. Describe the change in their depths between the valley margin and the valley floor.

   c. Compare the Death Valley fans to the Cedar Creek alluvial fan of western Montana (Fig. 10–4). Death Valley has a hyperarid climate, whereas the Cedar Creek fan is in a semihumid to semiarid setting. Explain the effect of this difference in climate on the vegetation covers of the fans in the two regions.

   d. How has the difference in vegetation influenced fluvial erosion and the formation of distributary channels on the surfaces of the Death Valley and Cedar Creek fans?

   e. Examine the streams of the valley floor. Describe the channel pattern. _____ Describe the relief of the valley floor, particularly the depth of the channels, and speculate on the effects of this relief on flooding in Death Valley.

   f. Locate and label some examples of interchannel flats. Describe and illustrate their relief and shape.x

   g. Locate and label landforms of eolian origin in the photographs. What is the significance of your observations?

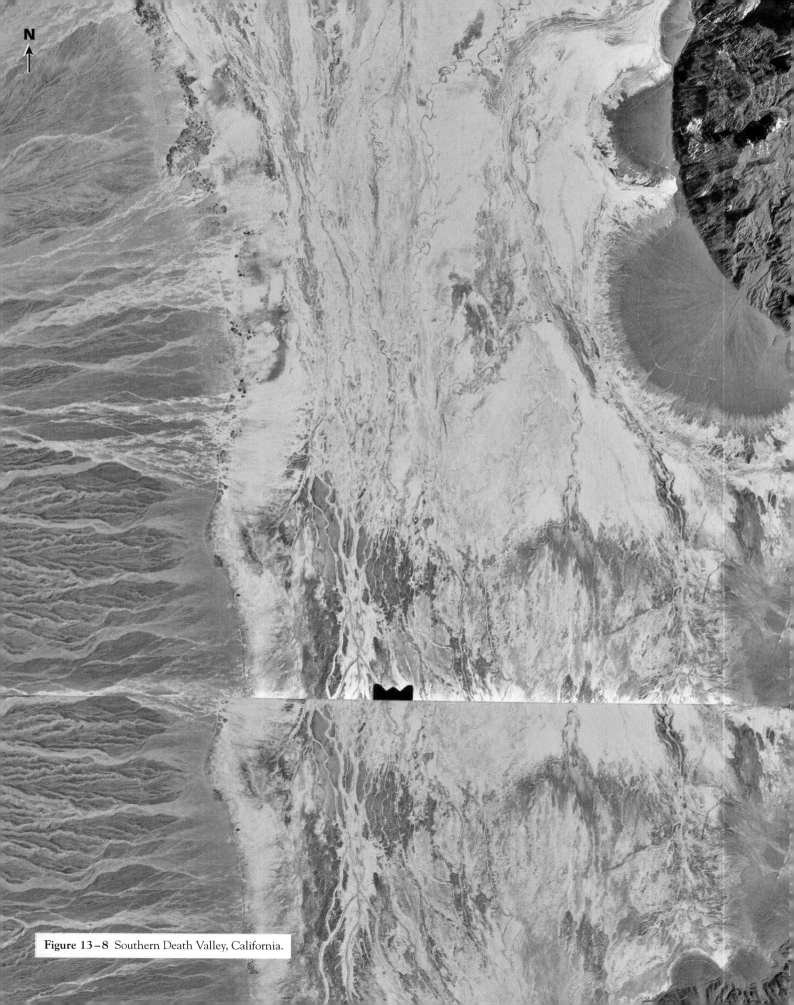

Figure 13-8 Southern Death Valley, California.

MAP 13-4 Death Valley National Park, California

| 0 | | | 5 | | 10 STATUTE MILES |

| 0 | | | 5 | | 10 KILOMETERS |

CONTOUR INTERVAL = 200 FEET

SHORELINES

THE SOUTH SHORE OF LONG ISLAND IS A BARRIER ISLAND WHERE SAND IS CONTINUOUSLY DEPOSITED BY LONGSHORE CURRENTS THAT FLOW PARALLEL TO THE COAST.

(Thompson/Turk)

Materials Needed *Colored pens or pencils, lead pencils with erasers, ruler, calculator, stereoscope*

The shoreline is the place where the land meets the sea. It is a dynamic and constantly changing environment where sediments eroded from the continents are deposited by streams, glaciers, and the wind and then reworked and dispersed to varying degrees by waves and tides.

The shapes of shorelines and the types of landforms which are present along them reflect the delicate balance between two competing sets of factors: the **rate of sediment supply** to the coast, and the **rate of reworking and redispersal** of coastal sediment by waves, tides, and sea-level fluctuations. Two general types of shorelines are defined by this balance: **constructive shorelines**, which are characterized by the net accumulation of sediment in barrier islands, deltas, and other depositional landforms; and **destructive shorelines**, which are characterized by the complete reworking of coastal sediment and the formation of rocky cliffs and other erosive landforms.

This chapter examines shoreline processes and the types of landforms which they produce. It begins with a discussion of the sources of coastal sediment and the roles of waves, tides, and sea-level variations in shaping the coast and its sediment, and continues with an examination of constructive and destructive coastal landforms. The chapter includes several exercises which address coastal landforms of the United States and the processes which led to their formation.

SEDIMENT SUPPLY

Rivers are perhaps the major source of shoreline sediments, depositing billions of tons of sediment at their mouths every year. The rate of sediment supply by a river to the shoreline is determined by three factors: its **discharge** and the **area** and **relief** of its drainage basin.

Discharge is largely a function of the climate of the drainage basin, particularly the amounts of precipitation and evaporation. The highest discharges are common to drainage basins with humid or tropical climates, the lowest to drainage basins with desert or polar climates.

The area and relief of a drainage basin are largely functions of its tectonic setting. Active tectonic settings have considerable relief, which contributes to high rates of sediment erosion. However, they also have many drainage divides and are often drained by many small rivers with modest discharges. Passive tectonic settings have lower relief and rates of erosion, but they are typically drained by a few rivers with large drainage basins and high discharges.

There are many other sources of shoreline sediments besides rivers. For example, glacial moraines and outwash deposits are the major source of sediment for the shorelines of New England, and desert dunes are the major source of sediment for the shorelines of northwest Africa and the Persian Gulf. In addition, shoreline sediments can derive from the erosion of rocky cliffs, deltas, beaches, carbonate reefs, and other local coastal sources by waves, tides, and gravity.

WAVES AND WAVE ACTION

Waves are oscillatory motions of water that are generated when the wind blows across the surface of an ocean or lake. They consist of high **crests** separated by low **troughs**, and they are described on the basis of **wavelength** (the horizontal distance between two successive crests), **wave height** (the vertical distance between the crest and trough), **wave base** (the lower limit of water oscillation), and **wave velocity** (Fig. 14–1).

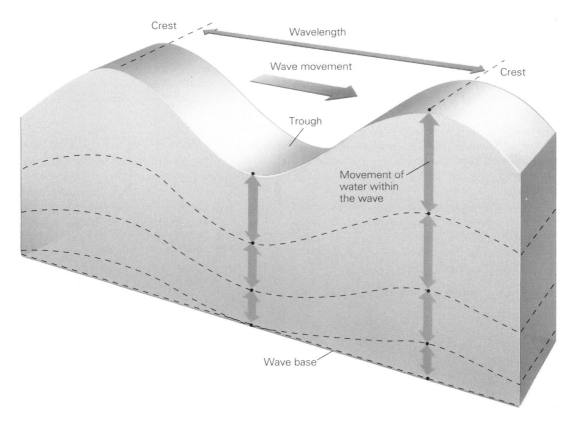

Figure 14–1 The basic components and descriptive terms for waves.

Waves are the major agent of sediment reworking along many shorelines. This wave action begins when a wave approaches the shallow waters of the coast and starts to *touch bottom*—that is, when the wave base comes into contact with the sea bed. The resultant friction has two effects: It causes the wave to scour the sea bed, suspending mud and transporting sand toward the shoreline, and it decreases the wave velocity and wavelength while it increases wave height. Eventually, the wave crest collapses or breaks against the shore, and the breakwaters rush landward as turbulent **surf**. This surf pounds and erodes coastal headlands and gently **swashes** up and down the faces of coastal beaches, where it deposits sand on their **foreshores**. Finally, the surf and swash return seaward via **backwash** and **rip currents**, transporting mud in suspension to the deeper waters of the continental shelf.

Waves do not usually approach the coast directly head on, but rather at some angle. This *oblique* approach generates **longshore currents** in the nearshore zone which flow parallel to the shoreline and distribute sediment (called **longshore drift**) down its length. In addition, it causes the approaching wave crest to bend, or **refract**, as one end of the wave touches bottom and slows down before the other end. This concentrates the erosive powers of the waves against coastal headlands, which leads to their erosion and retreat, the deposition of **pocket beaches** between them, and the slow and steady straightening of the shoreline.

TIDES AND TIDAL ACTION

Tides are regular rises and falls of the surfaces of the oceans. They are the result of the combined effects of two astronomical processes: the gravitational pulls of the Moon and Sun, which draw ocean water upward into **tidal bulges**; and the rotation of the Earth, which moves shorelines in and out of these tidal bulges. Rising tide occurs when a shoreline rotates into a tidal bulge, whereas falling tide occurs when a shoreline rotates out of a tidal bulge.

Rising and falling tides generate **flood** and **ebb tidal currents**, which displace water landward and seaward. The velocities of tidal currents are controlled by two factors: the **tidal range** (the vertical distance between the high- and low-tide lines) and the shape of the coast. Generally, a high tidal range produces stronger tidal currents than a low tidal range. However, tidal ranges and tidal velocities can be increased, or **amplified**, in coastal embayments such as **estuaries** and **bays**, where the tidal waters are confined to flow through narrow passages.

Along many shorelines, flood tidal currents are commonly stronger and longer lasting than ebb tidal currents. This condition, which is called **time-velocity asymmetry**, results in the net landward movement of sediment into estuaries and bays and its deposition in **tidal flats** along their shores.

SEA-LEVEL FLUCTUATIONS

Sediment supply, waves, and tides are important controls on the morphology of the shoreline over short periods of geologic time. However, there is another factor which affects shorelines over long periods of geologic time: sea-level fluctuations.

Sea level is rarely constant for long periods of time, but rather fluctuates in response to tectonic and climatic processes. Some of these processes are global and produce **eustatic** sea-level fluctuations along all the shorelines in the world. For example, rifting and the formation of oceanic ridges are commonly accompanied by a eustatic rise in sea level, whereas global cooling and the start of an ice age are usually accompanied by a eustatic fall in sea level. On the other hand, some of these processes are regional and produce **relative** sea-level fluctuations only in their immediate areas. For example, tectonic uplift along a shoreline causes a relative fall in sea level there, whereas local subsidence causes a relative sea-level rise.

The major effect of sea-level fluctuations is to cause landward or seaward shifts in the position of the shoreline. Sea-level fall causes the **emergence** of the coast and the seaward shift, or **regression**, of the shoreline. On the other hand, sea-level rise causes the **submergence** or **transgression** of the coast, the landward retreat of the shoreline, and the creation of estuaries and bays in **drowned river mouths** and glacial fjords.

EXERCISE 14–1	Wave Action, United States Coasts

The strength of wave impact against the shoreline is greatly affected by the slope of the continental shelf, which determines where the waves start to touch bottom and lose velocity to friction with the sea bed. This exercise examines this relationship along three stretches of coast around the United States.

1. Return to the shaded relief map of the United States (Map 10–1). The continental shelf is shaded in a lighter color than the deeper parts of the ocean on this map. The boundary between the shelf and deep ocean (the **shelf-slope break**) lies 200 meters below sea level. Trace the shelf-slope break with a colored pencil or marker.

2. Examine the continental shelves offshore from Savannah, Georgia; Galveston, Texas; and Santa Barbara, California. Measure the width of each shelf (in kilometers) and calculate the overall slope between the shoreline and shelf-slope break. Record your data on the following table.

SHELF	WIDTH	SLOPE	TECTONIC SETTING
Savannah			
Galveston			
Santa Barbara			

3. Briefly characterize the tectonic setting of each shelf. You may want to refer to Figure 9–13 and the Heezen–Tharp world ocean map. What is the relationship between their tectonic settings and slopes?

4. Which of the four shorelines would be most affected by the impact of an incoming storm wave with a wave base of 60 meters? Explain your answer.

Some environmentalists fear that global warming due to the greenhouse effect might cause the melting of the Antarctic and Greenland ice caps, which would raise sea level and drown many coastal cities.

1. Calculate the volume of water (in cubic miles) which would be released by the complete melting of these two ice caps. Use the following dimensions in your calculations:
    a. The Antarctic ice cap covers approximately 5.5 million square miles of land and averages about 7000 feet thick.
    b. The Greenland ice cap covers approximately 700,000 square miles of land and averages about 4500 feet thick.

2. The total surface area of the world's oceans is approximately 150 million square miles. Calculate the approximate rise in sea level (in feet) which would result from the complete melting of the two ice caps.

3. Examine the topographic map of the United States (Map 14–1). Draw the position of the shoreline after this sea-level rise.
    a. Measure the maximum inland extent of invasion by the sea. _____
    b. Which major cities would be flooded, and which inland cities would become coastal resorts?

A constructive shoreline is formed wherever the rate of sediment supply exceeds the capacity of waves and tides to rework and disperse it, so that sediment accumulates along the shoreline in beaches, barrier islands, deltas, spits, and tidal flats. Constructive shorelines are found near the mouths of rivers with large sediment loads, along the margins of oceans and lakes with low to moderate wave energy and tidal ranges, along the inner margins of estuaries and bays, and along emergent coasts.

Three types of constructive coasts can be distinguished on the basis of the relative balance between sediment supply and reworking and dispersal by waves and tides: **fluvial-dominated**, **wave-dominated**, and **tide-dominated shorelines**.

FLUVIAL-DOMINATED SHORELINES

Fluvial-dominated constructive shorelines form around the mouths of large rivers that empty into quiet-water oceans and lakes. They are characterized by the presence of **birdfoot deltas**, which are the extensions of alluvial plains that protrude out into oceans and lakes.

Birdfoot deltas consist of branching networks of **distributary channels**, which are bordered by high levees and separated by vast, low-lying **swamps** and **interdistributary bays**. The distributary channels funnel sediment from the al-luvial valley to the shoreline, where the sand is deposited in **distributary-mouth bars** and the mud is spewed into the open seas. The swamps are highly vegetated flood plains where organic-rich deposits such as **peat** and **lignite** accumulate. The bays are shallow-water environments that are often filled by sand-rich, fan-shaped **crevasse splays**, which enter through breaches or **crevasses** in the channel levees.

WAVE-DOMINATED SHORELINES

Wave-dominated constructive shorelines are characterized by the reworking of shoreline sediment by waves and its deposition in **beaches**, **barrier islands**, **spits**, and **baymouth bars** (Fig. 14–2).

A beach is a wave-washed strip of sand which is attached to the mainland. It is composed of a **foreshore** (the zone of wave action between the low- and high-tide lines) and a **backshore** (the zone of wind action above high tide). The boundary between the foreshore and backshore is usually marked by vegetated accumulations of windblown sand called **foredune ridges**. The seaward growth of a beach produces a distinct ridge-and-swale topography consisting of a series of such foredune ridges separated by low troughs (or **cat-eye ponds**).

Barrier islands are beaches which are separated from the

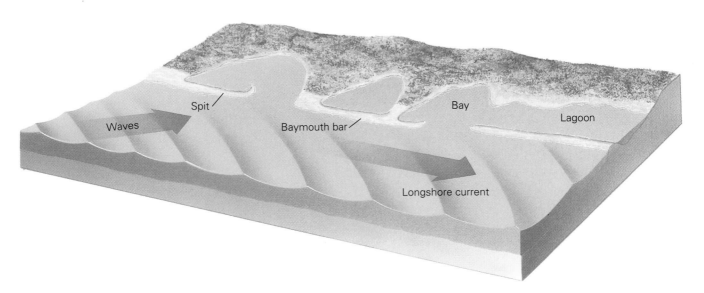

Figure 14–2 The topography of wave-dominated depositional shorelines.

337

mainland by open bodies of water called **lagoons** and **bays**. They occur in long chains that are broken periodically by tidal inlets, which connect the back-barrier lagoons and bays to the open sea. The frequency of tidal inlets varies with the tidal range: They are fairly rare along **microtidal** coasts, where the tidal range is less than 2 meters, but more common along **mesotidal** coasts, where the tidal range is between 2 and 4 meters.

The frequency of tidal inlets has a great effect on deposition in back-barrier lagoons and bays. Microtidal lagoons and bays are typically filled with windblown sediment from their barrier-island side, fluvial sediment from their landward side, and locally produced carbonates and evaporites. On the other hand, mesotidal lagoons and bays contain significant volumes of tidal sediment, which enters through the tidal inlets and is deposited in **tidal flats** and **tidal creeks**.

Spits are sandy deposits of longshore drift which partially stretch across the mouths of bays and other types of coastal embayments. They are generally attached to the mainland on their updrift sides and grow in the downdrift direction as they are supplied with sand by longshore currents. Baymouth bars are spits which have completely crossed and blocked off the mouths of coastal bays.

TIDE-DOMINATED SHORELINES

Tide-dominated constructive shorelines are characterized by the reworking of shoreline sediment by tides and its deposition in tidal flats. Tidal flats are commonly formed along the inner margins of estuaries and open bays along **macrotidal** shorelines, where the tidal range is greater than 4 meters, as well as in the back-barrier estuaries and barred bays of wave-dominated mesotidal shorelines.

Tidal flats are generally broad, low-relief features that are dominated by the deposition of mud by flood tidal currents. They are also dissected by small, sandy tidal creeks which drain the tidal flats during ebb tide. Tidal flats are bordered on their landward sides by supratidal marshes, which are typically highly vegetated and rich in organic sediment.

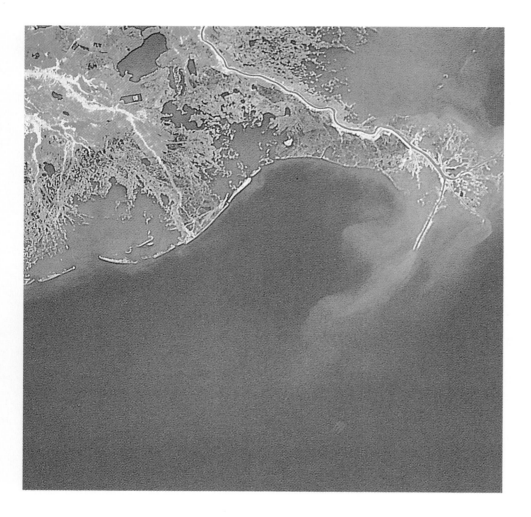

False-color image of the constructive shoreline of eastern Louisiana. The vegetated swamps and levees of the Mississippi Delta are colored red in this image; the muddy discharge of the Mississippi River is colored light blue; and the open marine waters of the Gulf of Mexico are colored dark blue. The eastern half of the image shows the modern Belize or bird-foot lobe of the Mississippi Delta and a plume of muddy river sediment that has washed out into the Gulf of Mexico. The western half of the image shows several barrier islands that were formed by wave-reworking of older abandoned delta lobes.

EXERCISE 14–3   Constructive Shorelines of the United States

The eastern and southern shores of the United States from Cape Cod to Texas are largely constructive, with some exceptions. Examine these shores on the shaded relief map of the United States (Map 10–1), and locate the coastal features which are listed in the following questions. You may want to refer to an atlas or other maps to locate them. Use different colored pencils or markers to indicate the different classes of features on the map. Create a key in a corner of the map to identify the feature represented by each color.

1. Locate and shade with a colored pencil the following deltas. Record the locations of their headwaters in the spaces provided.

DELTA	HEADWATER LOCATION	FLUVIAL OR WAVE DOMINATED
Santee, South Carolina		Wave
Apalachicola, Florida		
Mississippi, Louisiana		Fluvial
Brazos–Colorado, Texas		
Rio Grande, Texas–Mexico		

2. Locate and shade with a colored pencil the following estuaries:
   Narragansett Bay, Rhode Island
   New York Bay, New York–New Jersey
   Delaware Bay, Delaware–New Jersey
   Chesapeake Bay, Maryland–Virginia

3. Use a colored pencil to trace the lengths of the beaches, barrier islands, and spits of the eastern and southern shores. Locate and label the following:
   Cape Cod, Massachusetts
   Cape Hatteras, North Carolina
   Sea Islands, South Carolina–Georgia
   Cape Canaveral, Florida
   Miami Beach, Florida
   Padre Island, Texas

4. Locate and label the following back-barrier lagoons and bays:
   Albemarle Sound
   Mobile Bay
   Galveston Bay
   Matagorda Bay
   Corpus Christi Bay
   Laguna Madre

The most famous example of a fluvial-dominated constructive shoreline is found along the coast of Louisiana around the mouth of the Mississippi River. The Mississippi is the largest river system in North America, draining over 3 million square kilometers of the continent. It deposits slightly less than a million tons of sediment each day into the northern Gulf of Mexico, which has modest waves and a tidal range of less than 1 foot. This situation of high sediment supply and low reworking has prevailed for the last 7000 years, during which time the Mississippi has deposited seven separate delta lobes, advancing the Louisiana shoreline 130 miles into the Gulf of Mexico and creating almost 24,000 square kilometers of new land.

Figure 14-3 is a false color image of the modern birdfoot delta lobe of the Mississippi River. The muddy waters of the river appear orange and red in this image; the waters of the Gulf of Mexico appear dark blue and purple; the vegetated swamps appear light blue and green; and cultivated and cleared land and roads appear light green and white.

1.   Use a colored pencil or marker pen to highlight the channel of the Mississippi River and the four principal distributaries of the birdfoot delta. Label them: They are (from the southwest to the northeast): Southwest Pass (SwP), South Pass (SP), Southeast Pass (SeP) and Pass a Loutre (Pal).

2.   Locate and label these interdistributary bays of the Mississippi delta: West Bay (WB), which lies to the north of Southwest Pass; East Bay (EB), between Southwest and South Pass; and Garden Island Bay (GIB), between South Pass and Southeast Pass.

3.   Garden Island Bay is partly filled with a crevasse splay. Outline and label (with the letters CS) this splay, and use a colored pencil or marker pen to trace its distributary channels. Describe its shape and channel pattern.

4.   Outline and label the other major crevasse splays on the delta.

5.   The Mississippi River and its distributaries are all lined on both sides by narrow levees. Outline and label (with the letter L) the levees. Note that there are several crevasses in the levees of the river. Label them with the letter C.

6.   Locate and label (with the letter S) the swamps of the delta. What kind of deltaic deposits form the foundation upon which these swamps are established? _____

7.   What correlation do you see between the morphology of the delta and the locations of cultivated land and roads?

8.   This image of the birdfoot delta was taken during a period of flooding and high water and sediment discharge. Outline and label with arrows the large **plumes** of mud-rich water that are being discharged from various places on the delta. Describe the shapes of the plumes. What is the overall dispersal pattern of mud in and around the delta: that is, through which avenues is it being discharged, and in which directions?

9.   Examine the false color image of the eastern Louisiana shoreline on page 338. The eastern part of the image shows the modern birdfoot delta; the western part shows the remains of an older abandoned delta of the Mississippi River, the Lafourche. The Lafourche delta is slowly subsiding beneath the waters of the Gulf of Mexico, and its surface is being transgressed and flooded by a relative rise in sea level. Its shoreline is marked by a chain of barrier islands, products of wave reworking of the deltaic deposits by the landward-shifting coastal zone.

   a. Use a colored pencil or marker pen to highlight the distributary channels of the modern and abandoned deltas; the barrier islands of the Lafourche delta; and the edge of the birdfoot delta.
   b. Compare and contrast the shapes of the fluvial-dominated and constructive eastern shoreline with the wave-dominated and destructive western shoreline.

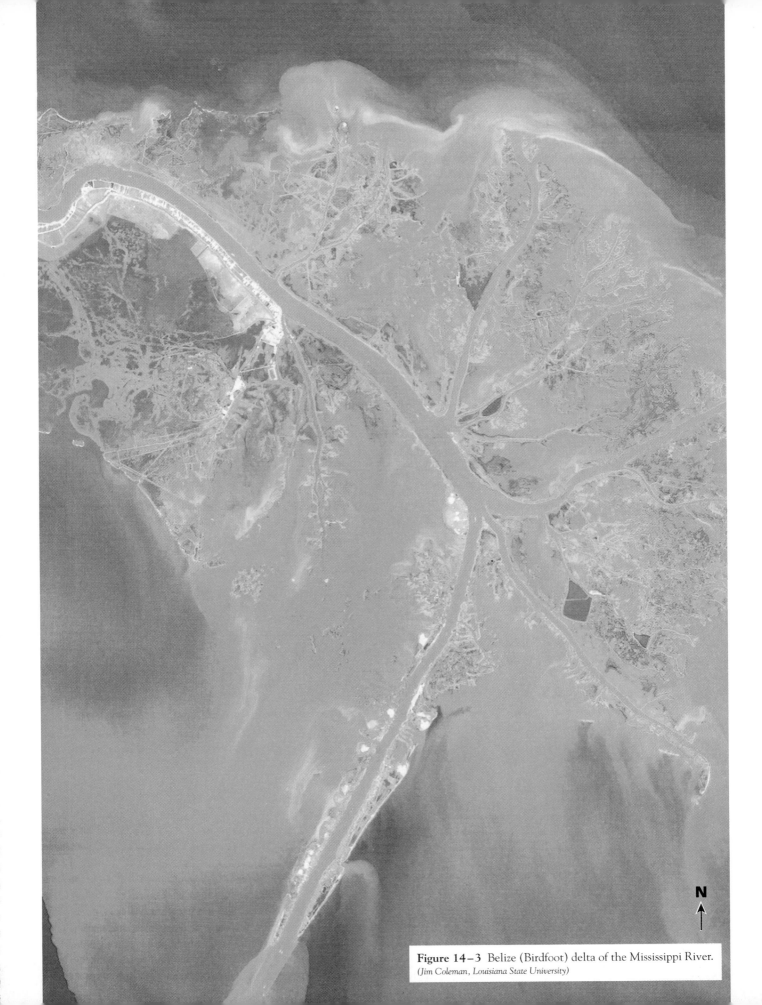

Figure 14–3 Belize (Birdfoot) delta of the Mississippi River.
(Jim Coleman, Louisiana State University)

To the west of the muddy shores of Louisiana are the beaches and barriers of the South Texas coast (Map 14–2). The principal sources of sediment for this coast are a series of small- and moderate-sized rivers that drain the Gulf Coastal Plain and parts of the Great Plains and southern Rocky Mountains. Shoreline conditions here are similar to the Louisiana coast: microtidal, with moderate waves driven by southeasterly winds throughout most of the year.

1. What would be the general orientation of the crests of waves created by the prevailing southeasterly winds of the region? _____

2. Draw a line on the map to represent the wave crests when they reach the barrier islands in the northern, central, and southern parts of the map.

3. Use an arrow to indicate the directions of longshore currents that are generated by wave approaches at these same locations. How do the directions of longshore current change along the length of the coast?

4. What is the likely source of sediment for South Padre Island, the southernmost barrier island on the map? _____

Three large rivers (the Rio Grande, Brazos, and Colorado) deposit sediments directly onto the shores of Gulf of Mexico. There, the sediments are reworked and molded by waves into cuspate deltas with delta fronts lined by sandy beaches and barriers. Some of this sand is also transported by longshore currents down the coast, where it has accumulated in several barrier islands. The smaller rivers of Texas deposit their sediment in coastal bays sheltered from open Gulf water by barrier islands.

5. The Brazos and Colorado rivers have together constructed a single wave-dominated cuspate delta at their mouths. Locate this delta, and trace the paths of the two rivers and the creeks and bayous that are their distributary channels on Map 14–2.

6. Locate and trace the front of the cuspate delta, where the rivers and their distributary channels empty. Describe and illustrate the shapes of the delta front and the delta as a whole. Estimate the distance that the delta has advanced into the Gulf of Mexico.

7. What is the direction of the longshore currents along the seaward edge of the delta? _____ What coastal feature appears to have been formed as a result of these currents? _____

8. Locate the Rio Grande, and trace its distributary channels and delta front. Describe and sketch the general shape of the delta.

9. Compare and contrast the shapes of fluvial-dominated deltas (represented by the Mississippi delta) and the two wave-dominated deltas of the Texas coast.

10. Return to Map 10–1. Trace the outlines of the drainage basins of the Rio Grande, Brazos, Colorado, and Mississippi rivers.
   a. Why does the Mississippi form a fluvial-dominated delta and the remaining rivers form wave-dominated deltas along the shores of the same ocean basin?

   b. The Rio Grande, Brazos, and Colorado drain large parts of the Great Plains of west Texas and eastern New Mexico. How does the climate of this region contribute to the shape of their deltas?

11. The northern Gulf of Mexico continental shelf was exposed by a major fall in sea level during the late Pleistocene and early Holocene. During this low stand, the Rio Grande flowed due east across the shelf to the present shelf margin (approximately the 50 fathom line), where it deposited its sediment load in a wave-dominated cuspate delta. At the same time, the Brazos and Colorado flowed roughly south to the present shelf margin, where they also formed a delta. What evidence do you see in the bathymetry of the Texas shelf for the existence of these ancient shelf margin deltas? Label them on Map 14−2.

12. Examine the shape of the Apalachicola delta on Map 10−1. What kind of delta is it: wave or fluvial dominated? Summarize your answer on the first table in Exercise 14−3.

13. Locate and label these barrier islands and their respective back-barrier lagoons and bays on Map 14−2. You may need to refer to an atlas to locate some of them.
Padre Island and Laguna Madre
Mustang Island and Corpus Christi Bay
Matagorda Island and San Antonio Bay
Matagorda Peninsula and Matagorda Bay

The Texas and South Carolina coasts are both lined by some of the most beautiful wave-washed beaches and barrier islands in the world. However, the Texas coast is set in a microtidal regime in which tides play a minor role in coastal sedimentation, whereas the South Carolina coast is set in a mesotidal regime in which tides are more important.

14. Examine the maps of these two states (Maps 14−2 and 14−3). Use a pencil or marker pen to mark the tidal inlets (including all bays, sounds, and river mouths that open to the sea) along each coast.

15. Calculate the average frequency (number per distance) of tidal inlets along the Texas coast by counting the number of inlets and then dividing by the length of the coast. Do the same for the South Carolina coast.
    a. What is the average frequency of inlets along the Texas coast? _____
    b. What is the average frequency of inlets along the South Carolina coast? _____
    c. How has the difference in the tidal regime affected the morphology of these two coasts?

16. Examine the Santee Delta on Map 14−3. Describe and sketch its shape. What kind of delta is it: wave or fluvial dominated? (Complete the first table in Exercise 14−3 with your answer.)

EXERCISE 14–6 Kiawah Island, South Carolina

Examine the stereophotographs of Kiawah Island, a mesotidal barrier island on the coast of South Carolina (Fig. 14–4).

1. Describe the relief of the barrier island. What features form its topographic highs?

2. Locate and label examples of the following topographic features on the stereophotograph: beach (B), modern foredune ridge (FD), older foredune ridge (OFD), cat-eye (CE) pond, tidal flat (TF), and tidal creeks (TC).

3. Describe the drainage patterns of the smaller tidal creeks in the back-barrier area.

4. Examine the stereophotographs carefully. Trace the crests of some incoming waves. What is the general direction of wave approach to Kiawah Island? ______________ What is the direction of longshore drift on the beach? Indicate your answer with an arrow on the stereophotograph.

Figure 14–4 Kiawah Island, South Carolina. ▶

Cape Cod, Massachusetts, is a large spit which was created by wave reworking of late Pleistocene glacial deposits on the southeastern New England coast. Waves approach this spit from many different directions throughout the seasons, but the strongest waves come out of the northwest and west.

1.   Examine the satellite photograph of Cape Cod (Fig. 14–5). Use a red marker or colored pencil to illustrate the crests of northwesterly waves as they might approach and refract around the spit. Indicate the directions of the long-shore currents which would be generated by the incoming waves along various parts of the spit with red arrows. Do the same for northeasterly waves, using a blue marker to illustrate their crests and currents.

2.   What shoreline features have formed as a result of these longshore currents?

3.   Examine the topographic map of the northern tip of Cape Cod (Map 14–4). Construct a topographic profile along a line which runs from the breakwater in Provincetown Harbor through Oak Head to the northern shore of the spit.

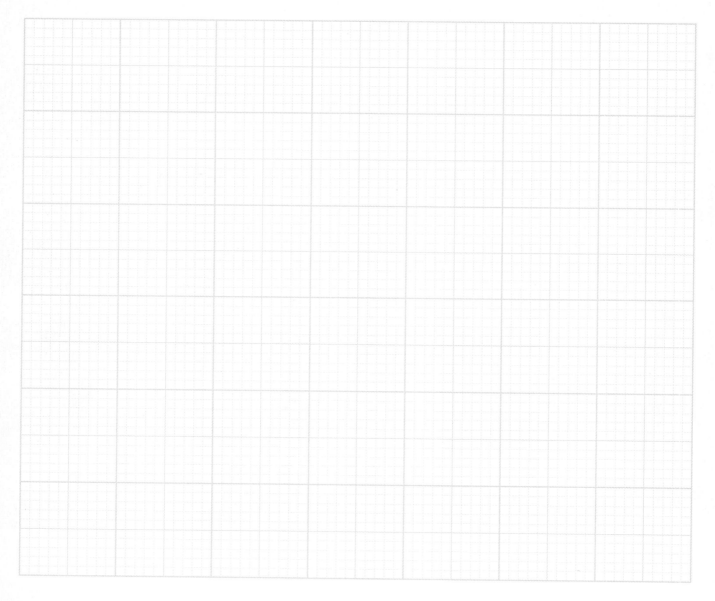

▼ **Figure 14–5** Cape Cod, Massachusetts.

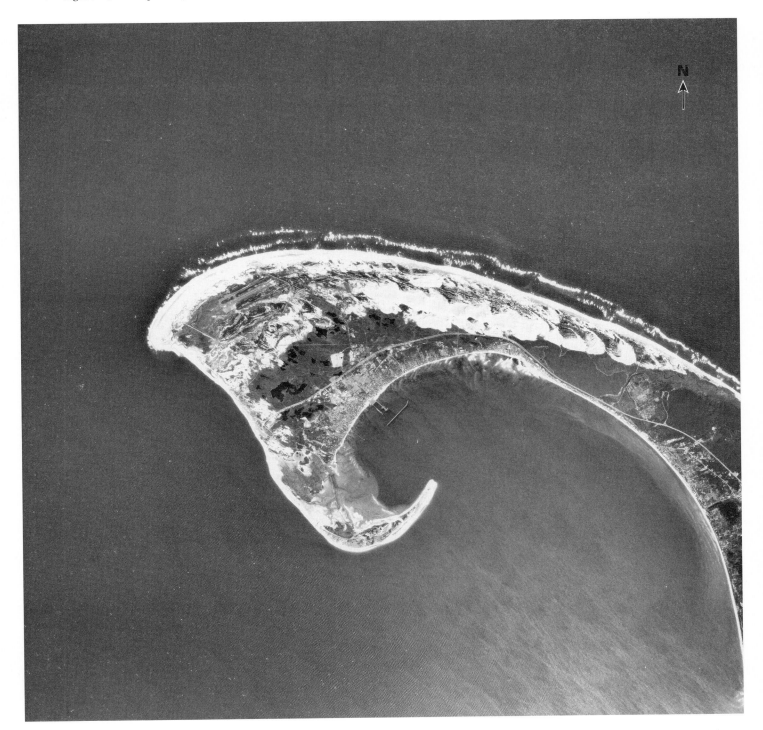

a. What is the origin of the topographic high along the northern shores of the spit? Trace its crest on the topographic map with a solid line.

b. What are the origins of the topographic highs and lows in the interior of the spit? Trace the crests of the topographic highs on the map with dashed lines.

c. What are the relative ages of all the topographic highs, and what do they indicate about the history and growth of the northern tip of Cape Cod?

4. Trace the approach and refraction of the crests of westerly waves around Long Point. Indicate the directions of longshore currents with arrows.

5. Assuming that there is a steady supply of sediment to the longshore currents, what are the probable fates of the tidal flats at Long Point and the southern part of Provincetown Harbor? Explain your answer.

EXERCISE 14–8   Delaware Bay, New Jersey–Delaware

Delaware Bay is a coastal estuary which was formed by the drowning of the mouth of the Delaware River. The estuary is presently being filled with sediments, which are accumulating along its shores in tidal flats.

1. Examine the photograph of the northern shore of the bay (Fig. 14–6). Locate and label examples of tidal flats, tidal creeks, and supratidal marshes, and indicate the position of the high-tide line.

2. Trace and describe the drainage pattern of the tidal creeks.

Figure 14–6 Delaware Bay, Delaware–New Jersey.

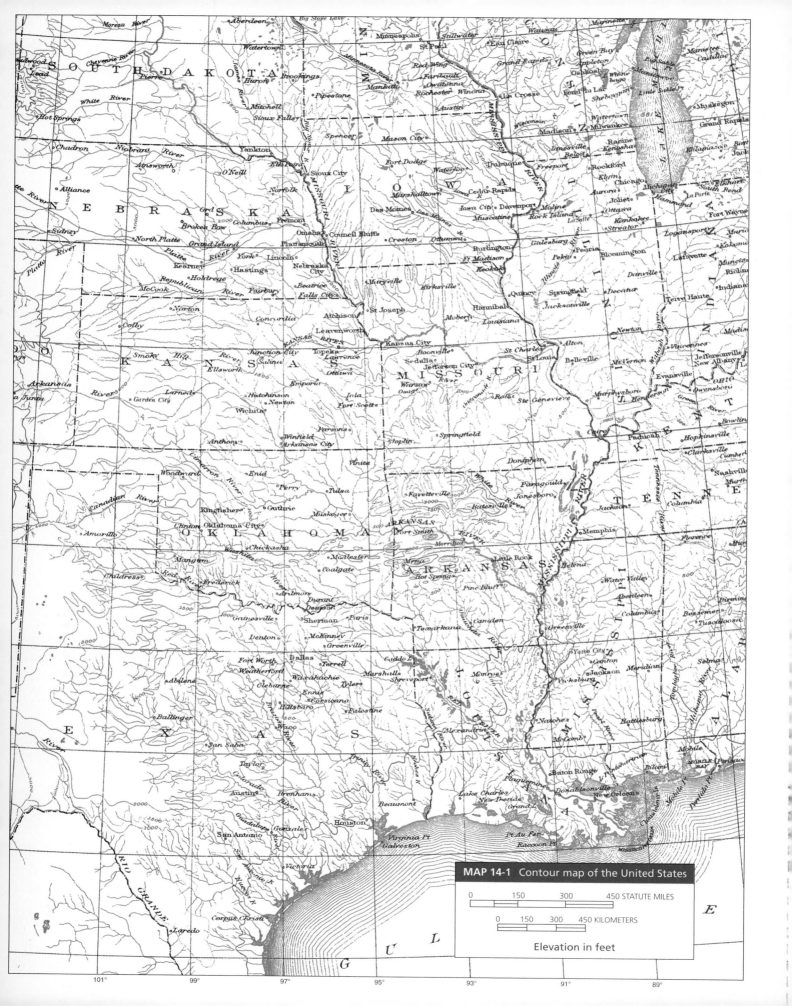

MAP 14-1 Contour map of the United States

0   150   300   450 STATUTE MILES

0   150   300   450 KILOMETERS

Elevation in feet

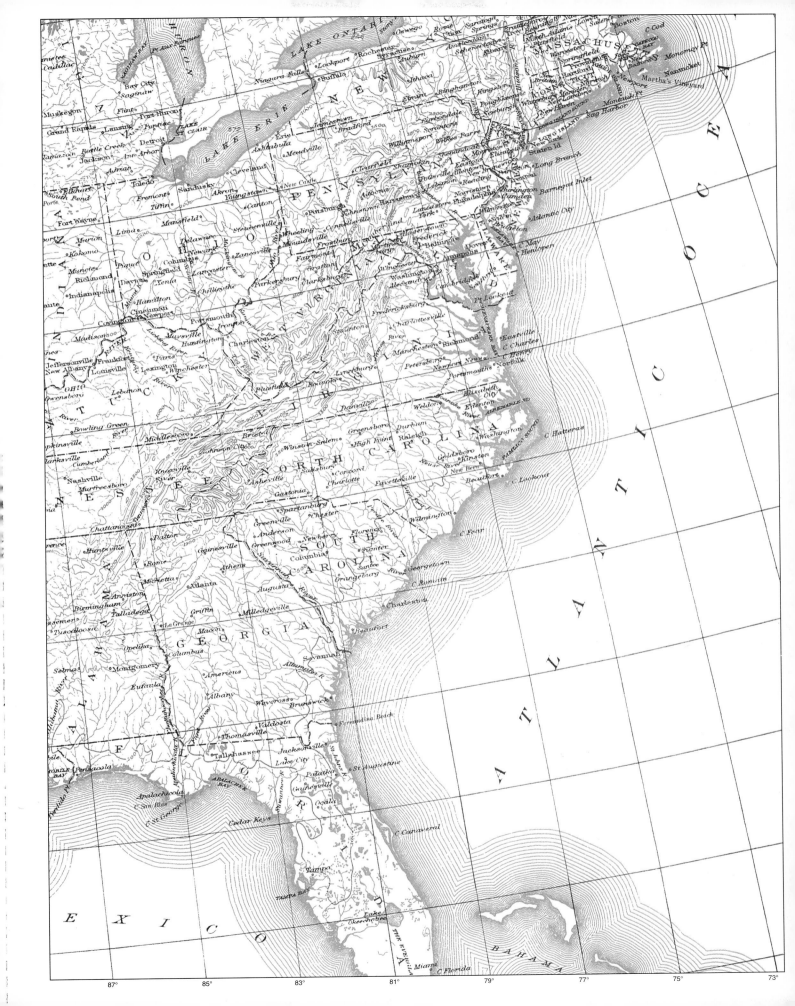

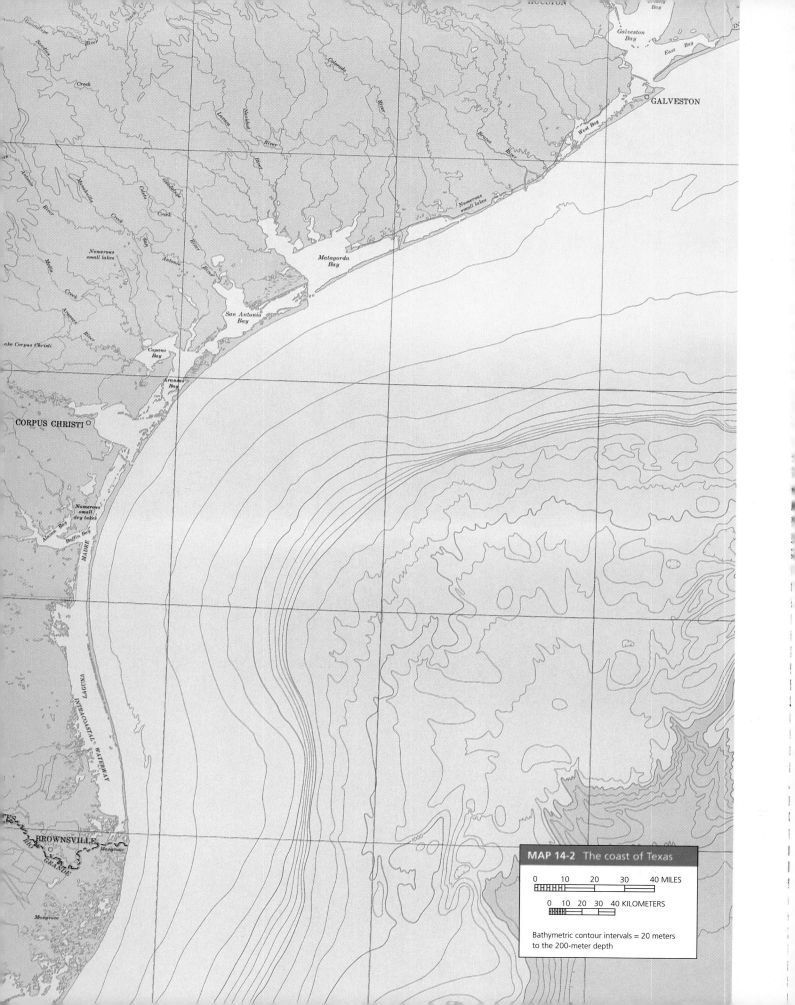

CORPUS CHRISTI ○

GALVESTON

Galveston Bay

East Bay

West Bay

HOUSTON

Colorado River

Matagorda Bay

San Antonio Bay

Copano Bay

Aransas Bay

Numerous small lakes

Numerous small dry lakes

Baffin Bay

MADRE

LAGUNA

INTRACOASTAL WATERWAY

BROWNSVILLE

RIO GRANDE

Mangrove

Mangrove

TEXAS

Lake Corpus Christi

Brazos River

Guadalupe River

Navidad River

Lavaca River

San Antonio River

Aransas River

Medio Creek

Manahuilla Creek

Coleto Creek

Sandies Creek

Guadalupe River

MAP 14-2 The coast of Texas

0   10   20   30   40 MILES

0   10   20   30   40 KILOMETERS

Bathymetric contour intervals = 20 meters
to the 200-meter depth

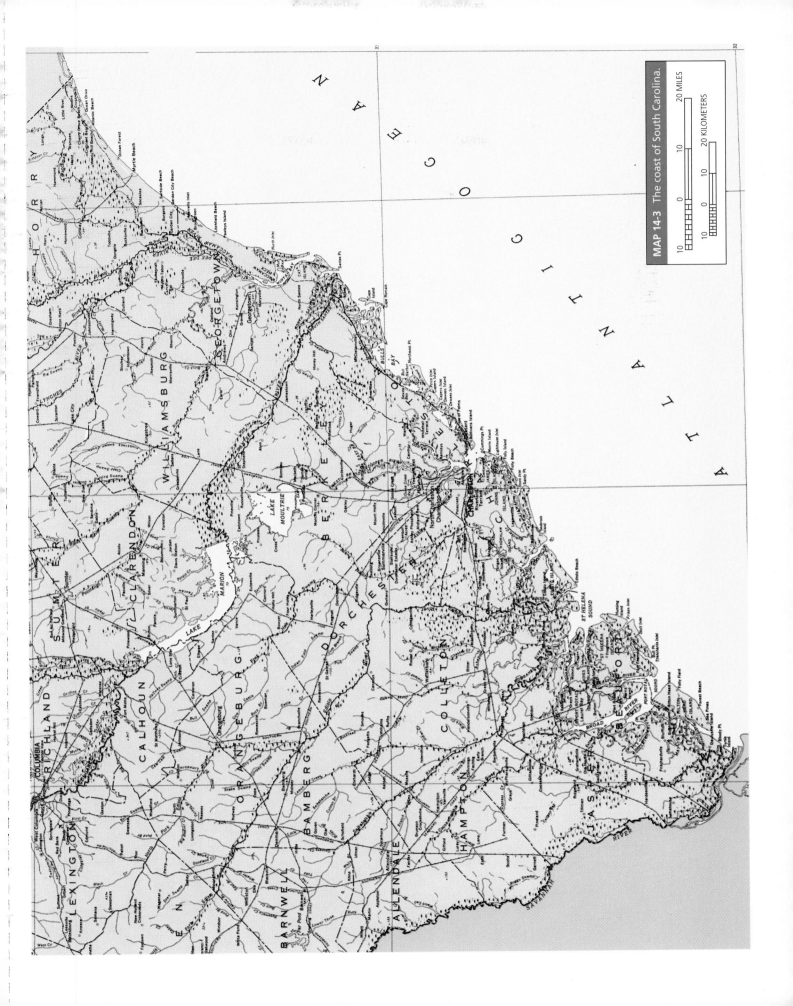

MAP 14-3 The coast of South Carolina.

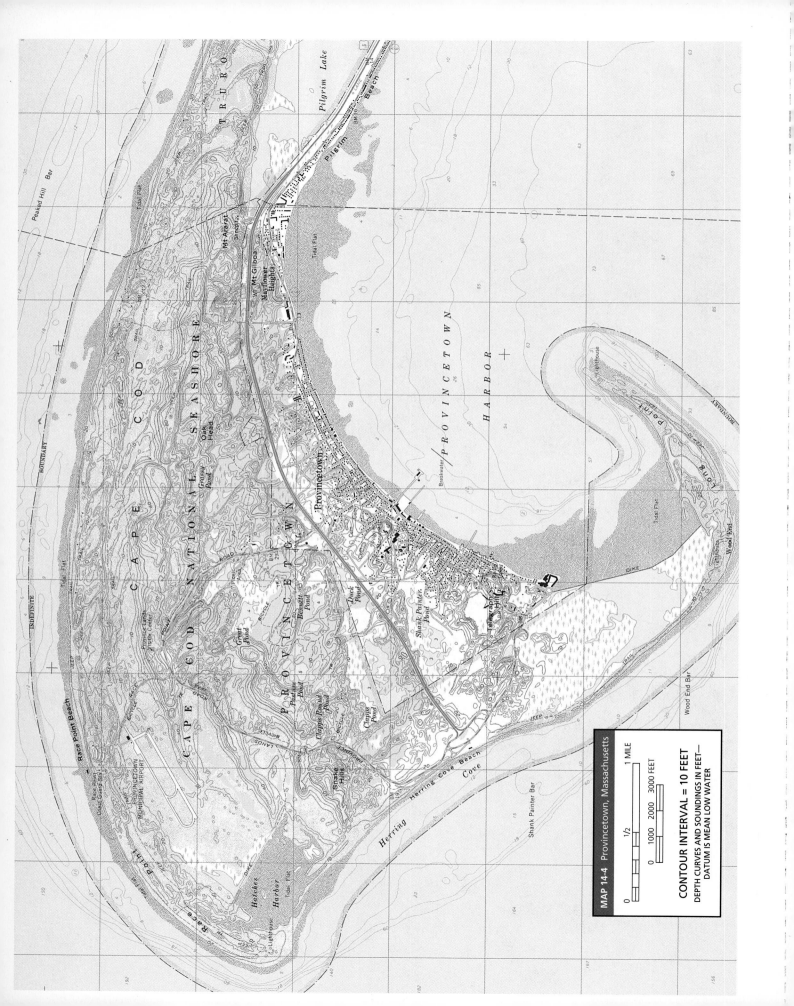

MAP 14-4 Provincetown, Massachusetts

CONTOUR INTERVAL = 10 FEET
DEPTH CURVES AND SOUNDINGS IN FEET—
DATUM IS MEAN LOW WATER

0   1/2   1 MILE

0   1000   2000   3000 FEET

A destructive, or erosive, shoreline is formed wherever marine processes are sufficiently strong to completely rework and erode all the available sediment, sweeping it off into deeper offshore waters and leaving behind a barren and rocky coast. Destructive shorelines are typically found along the margins of oceans with strong waves and high tides, along desert coasts, near the mouths of rivers with low sediment loads, and along submerged coasts. Some of the topographic features which characterize destructive shorelines are shown in Figure 14–7. They include:

- **Wave-cut cliffs**: Steep exposures of bedrock. The faces of these cliffs are sometimes dotted with **sea caves**, which are carved out of the more erodable rocks within the exposures.
- **Wave-cut platforms**: Flat or gently sloping benches at the bases of the wave-cut cliffs. The erosion of these platforms undercuts the cliffs and causes their collapse and retreat.

- **Wave-cut terraces**: Wave-cut platforms that have been elevated above sea level by tectonism.
- **Sea arches** and **sea stacks**: Short-lived remnants of the wave-cut cliffs which are surrounded by the sea. They are generally composed of a more resistant rock type than the remainder of the cliffs.

Sediment deposits are rare along destructive shorelines because they are easily washed away by waves, tides, and longshore currents. Beaches are perhaps the only common depositional feature of erosive shorelines. They are typically found at the feet of wave-cut cliffs, and they are generally steep and narrow. They are composed of very coarse sands and gravels which are eroded from their adjacent wave-cut cliffs. This coarse-grained sediment accumulates during the summer months, but it is usually washed away by strong winter storm waves. Other depositional features seen along destructive shorelines are spits and baymouth bars.

Figure 14–7 The topography of an erosive shoreline.

EXERCISE 14-9   Destructive Shorelines of the United States

The Pacific coast of the United States and the shores of the Gulf of Maine are both generally destructive. Examine these coasts on the shaded relief map (Map 10–1).

1. Use a colored pencil or marker to highlight these shorelines on the map. (Add this color to the key on Map 10–1.) In addition, locate and label the mouths of the Sacramento and Columbia Rivers.
    a. What kind of coastal landforms are the river mouths?
    b. Describe the shapes of the two shorelines and the factors which contribute to their shapes.

2. Note the two small bays (Willapa Bay and Gray's Harbor) on the southern shore of Washington state. The mouths of the bays are partly blocked by sandy coastal deposits.
    a. What kind of landforms are they?
    b. What does the morphology of these landforms indicate about conditions along this stretch of the coast? about the probable source of the sand?

3. Point Reyes is the seaward tip of an outcrop of resistant rocks which juts out into the Pacific Ocean a few miles north of San Francisco. Examine the topographic map of this coastal headland (Map 14–5).
    a. Locate and label examples of the following landforms: wave-cut cliffs, wave-cut platforms, drowned river mouths, sea stacks, spits, and baymouth bars.
    b. What is the prevailing wind direction in this area? _____
    c. Assume that waves approach from approximately the same direction as the prevailing wind. Illustrate the approach and refraction of incoming waves around Point Reyes and in Drake's Bay. Indicate with arrows the directions of longshore drift along the Pacific coast and the shores of Drake's Bay.
    d. Examine the shores of Drake's Bay. What coastal landforms are present which support your model for wave refraction and longshore drift in the bay?

    e. Examine the directions of wave approach around Point Reyes. How has wave refraction contributed to its erosion?

4. Examine the topographic map of an emergent destructive shoreline along the southern California coast (Map 14–6).

    a. Construct a topographic profile along line A-A′.

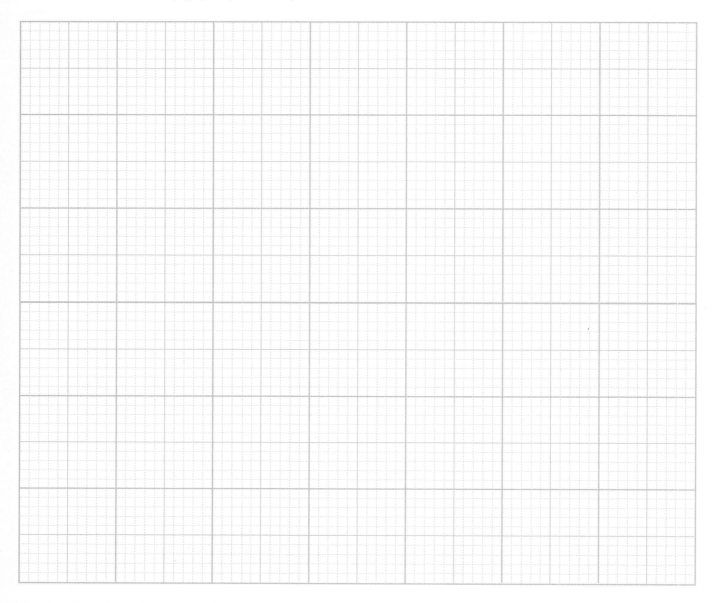

    b. What is the origin of the distinct breaks in slope along this profile?

5. Examine the stereophotograph of the California coast near Covelo, California (Fig. 14–8). Note the small river which empties onto this shoreline. Locate and label the steep wave-cut cliffs (C) and narrow beaches (B) north of the river's mouth, and the sea stacks (S) south of it.

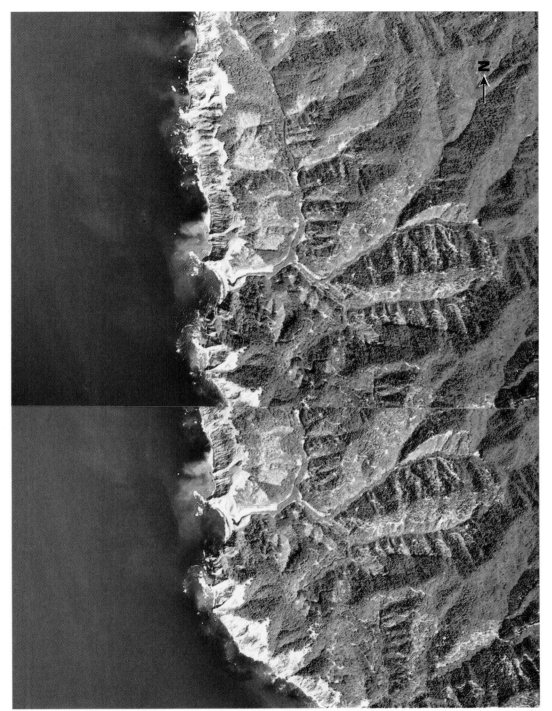

Figure 14–8 Covelo, California.

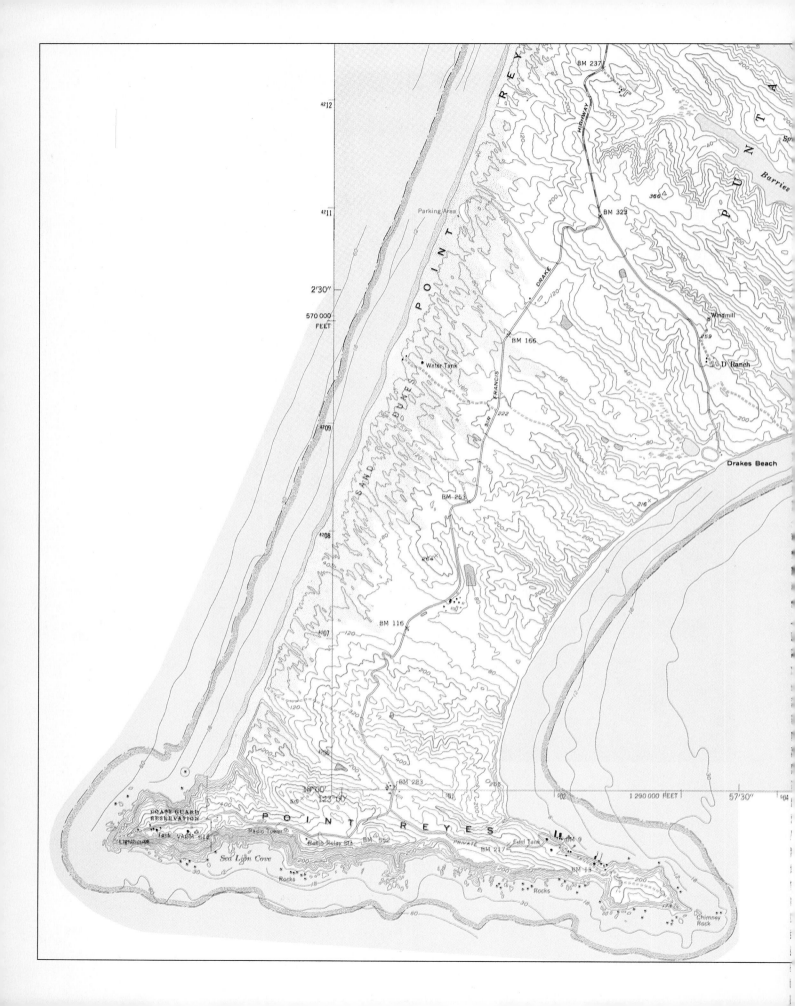

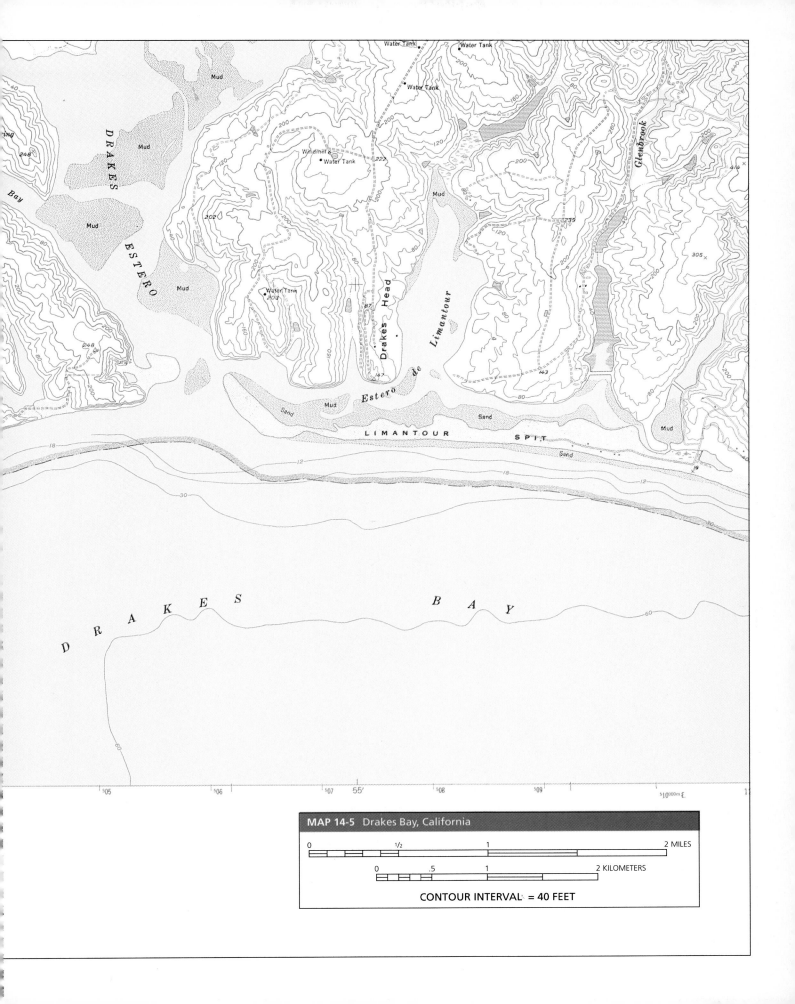

MAP 14-5 Drakes Bay, California

0   1/2   1   2 MILES

0   .5   1   2 KILOMETERS

CONTOUR INTERVAL = 40 FEET

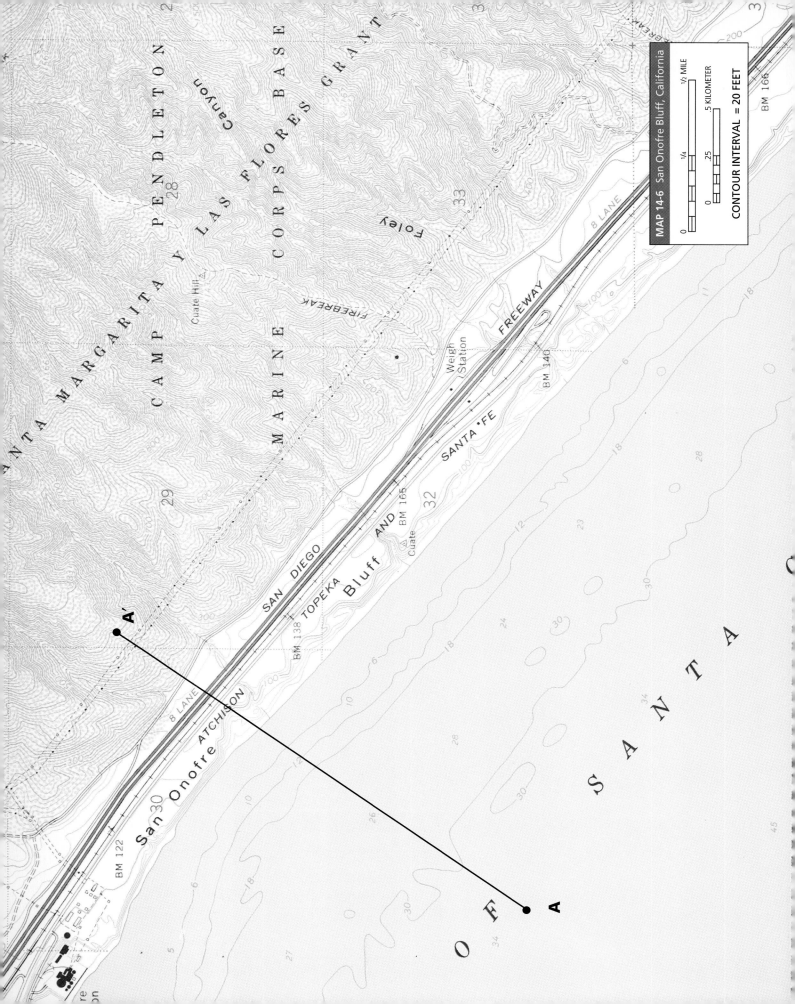

MAP 14-6 San Onofre Bluff, California

CONTOUR INTERVAL = 20 FEET